AF323854

NONADIABATIC TRANSITION

Concepts, Basic Theories and Applications

NONADIABATIC TRANSITION

Concepts, Basic Theories and Applications

by

Hiroki Nakamura

Institute for Molecular Science and

The Graduate University for Advanced Studies,
Okazaki, Japan

Published by

World Scientific Publishing Co. Pte. Ltd.

P O Box 128, Farrer Road, Singapore 912805

USA office: Suite 1B, 1060 Main Street, River Edge, NJ 07661

UK office: 57 Shelton Street, Covent Garden, London WC2H 9HE

British Library Cataloguing-in-Publication Data
A catalogue record for this book is available from the British Library.

NONADIABATIC TRANSITIONS: CONCEPTS, BASIC THEORIES AND APPLICATIONS

ISBN 981-02-4719-2

Printed in Singapore by World Scientific Printers (S) Pte Ltd

Preface

"Nonadiabatic transition" is a very multi-disciplinary concept and phenomenon which constitutes a fundamental mechanism of state and phase changes in various dynamical processes in physics, chemistry, and biology. This book has been written on the opportunity that the complete solutions of the basic problem have been formulated for the first time since the pioneering works done by Landau, Zener, and Stueckelberg in 1932. It not only contains this new theory, but also surveys the history and theoretical works in the related subjects without going into much details of mathematics. Both time-independent and time-dependent phenomena are discussed. Since the newly completed theory is useful for various applications, the final recommended formulas are summarized in Appendix in directly usable forms. Discussions are also devoted to intriguing phenomena of complete reflection and bound states in the continuum, and further to possible applications of the theory such as molecular switching and control of molecular processes by external fields.

This book assumes the background knowledge of the level of graduate students and is basically intended as a standard reference for practical uses in various research fields of physics and chemistry.

The writing of this book was suggested and recommended by Professor Kazuo Takayanagi and Professor Phil G. Burke. My thanks are also due to my many collaborators with whose help a lot of works written in this book have been accomplished.

Acknowledgment is also due to the following for permission of reproducing various copyright materials: American Institute of Physics, American Physical Society, Institute of Physics, John-Wiley & Sons Inc., Marcel Dekker Inc., Annual Reviews, Gordon & Breach Science Publisher, Physical Society of Japan,

American Chemical Society, Elsevier Science Publisher, and Royal Society of Chemistry.

Finally, I would like to thank my wife, Suwako, for her continual support in my life.

Okazaki, Japan
February 2001

Hiroki Nakamura

Contents

Chapter 6   Basic Two-State Theory for Time-Dependent Processes   **121**

Chapter 7   Two-State Problems   **145**

Chapter 1

Introduction: What is "Nonadiabatic Transition"?

"Adiabaticity" or "adiabatic state" is a very basic well-known concept in natural sciences. This concept implies that there are two sets of variables which describe the system of interest and the system can be well characterized by the eigenstates defined at each fixed value of one set of variables which change slowly compared to the other set. This slowly varying set of variables are called "adiabatic parameters" and the eigenstates are called "adiabatic states." Good adiabaticity means that a system stays mostly on the same adiabatic state, as the adiabatic parameter changes slowly. So, if we can find such a good adiabatic parameter, it would be very useful and helpful to describe and understand static properties and dynamic behaviour of the system. However, if the good adiabaticity holds all the time, then no state change happens at all and nothing exciting occurs. This world would have been dead. Namely, we expect that in some regions of the adiabatic parameter the adiabaticity breaks down and transitions among the adiabatic states are induced somehow. This transition, i.e. a transition among adiabatic states, is called "nonadiabatic transition." The adiabaticity breaks down when the adiabatic parameter changes quickly, because the rapidly changing set of variables cannot fully follow the change of the adiabatic parameter and the state of the system changes accordingly. In this world, this kind of transition is occurring everywhere, not only in natural phenomena but also in social phenomena. Whatever rather abrupt changes of states occur, they can be understood as nonadiabatic transitions. So, in a sense, we may say that nonadiabatic transitions represent one of the very basic mechanisms of the mutability of the universe.

One of the most well-known examples of the adiabatic approximation and nonadiabatic transitions is the Born–Oppenheimer approximation to define molecular electronic states and transitions among them. Since the mass of an electron is so light compared to that of nuclei and the electron moves so quickly, the internuclear coordinates, collectively denoted as R, can be considered as a very good adiabatic parameter. The adiabatic states, i.e. Born–Oppenheimer states, are functions of R and generally describe molecule quite well. In some regions of R, however, two or more Born–Oppenheimer states happen to come close together. At theses positions, a small amount of electronic energy change is good enough to induce a transition between the adiabatic electronic states and the transition can be actually achieved rather easily by gaining that energy from the nuclear motion. This energy transfer between the electronic degrees of freedom and those of nuclei is nothing but a nonadiabatic transition. It is easily conjectured that this transition occurs more effectively when the nuclei move fast. That is to say, nonadiabatic transitions occur effectively at high energies of nuclear motion, as can be easily understood from the definitions of "adiabaticity" and "adiabatic parameter." Various molecular spectroscopic processes, molecular collisions and chemical reactions can all be described by this kind of idea. The basic equations are, of course, the time-independent Schrödinger equations. The "fast" or "slow" change of the adiabatic parameter in this case means "high" or "low" energy of the nuclear motion, as is easily guessed. If the adiabatic parameter R is an explicit function of time, i.e. if, for instance, a time-dependent external field is applied to the system, then the time-dependent field parameter is considered to be the adiabatic parameter R and the problem is described by the time-dependent Schrödinger equations. The notions of "adiabaticity" and "nonadiabatic transition" can, of course, be applied to these time-dependent problems. The concept of "fast" and "slow" holds as it is. As mentioned above, nonadiabatic transitions are induced by the dependence of the adiabatic eigenstates on the adiabatic parameters. In the case of molecular electronic transitions, the Born–Oppenheimer electronic wave function parametrically depends on the nuclear coordinate R and the derivative of that wave function with respect to the nuclear coordinate causes the nonadiabatic transition. In the case of time-dependent process, time-derivative of the adiabatic state wave function plays that role. Recent remarkable progress of laser technologies makes control of molecular processes realizable and further endorses the importance of various kinds of time-dependent nonadiabatic transitions. By applying a strong laser field, molecular energy levels or potential curves can be shifted up and down by the amount corresponding to the photon energy. Thus potential curve crossings can be artificially created and nonadiabatic transitions can be induced there. The diabatic coupling there is proportional to the transition dipole moment between the relevant two states and the square root of the laser intensity. This idea is not just restricted to laser, but is applicable also to magnetic and electric fields. The high technological manipulation of such external fields is expected to open up a new field of science and nonadiabatic transitions again play a crucial role there.

Because of the fundamentality and significance of nonadiabatic transitions in various branches of natural sciences, there is a long history of theoretical investigation of nonadiabatic transitions. The pioneering works done by Landau, Zener, and Stueckelberg date back to 1932 [1–3]. Since then a lot of researches have been carried out by many investigators. This book tries not only to present the historical survey of the research briefly and explain the basic concepts, but also to introduce the recently completed new theories of nonadiabatic transitions. Various chapters are arranged in such a way that those readers who are interested in nonadiabatic transitions but are not very familiar with the basic mathematics can skip the details of mathematical descriptions and utilize the new theories directly.

This book is organized as follows. In the next chapter multi-disciplinary and fundamentality of the concept of nonadiabatic transition are explained by taking various examples in a wide range of fields of physics, chemistry, biology, and economics. A brief historical survey is given in Chap. 3. Essential ideas and basic formulas are explained and summarized for the Landau–Zener–Stueckelberg type curve-crossing problems, the Rosen–Zener type non-curve-crossing problems, the Nikitin's exponential potential model, and a unified treatment of rotaional or Coriolis coupling problems. The background mathematics is explained in Chap. 4. The most basic semiclassical theory, i.e. the WKB (Wentzel–Kramers–Brilloin) theory, and the Stokes phenomenon associated with asymptotic behaviour of the WKB solutions of differential equations are presented. Those who are not interested in the mathematics can skip this chapter. Chapter 5 presents recent new results of basic two-state time-independent problems (Zhu–Nakamura theory). First, the quantum mechanically exact solutions of the linear curve crossing problems are provided. Then, their generalizations to general two-state curve-crossing problems are presented together with numerical demonstrations. Recent developments about the other cases, i.e. the non-curve-crossing case and the exponential potential model are further discussed. Some new mathematical developments associated with the exact solutions are also provided there. The final recommended formulas with empirical corrections for the Landau–Zener–Stueckelberg (LZS) curve crossing problems (Zhu–Nakamura theory) are presented in Appendix A. These empirical corrections which are not given in Chap. 5 are introduced in order to cover thoroughly the whole ranges of energy and coupling strength, although the original formulas given in Chap. 5 are accurate enough except in some small regions of parameters. Appendix A is arranged in a self-contained way as far as the basic formulas are concerned, and the formulas given there can be directly applied to various problems and must be quite useful for those people who are eager about immediate applications. Chapter 6 provides the similar basic theories for time-dependent processes. Exact solutions can be obtained for the quadratic potential model based on those of the time-independent linear potential model. Furthermore, as in the time-independent case, accurate semi-classical formulas can be derived for general curve-crossing problems. Even the case of diabatically avoided crossing can be accurately treated. Other exactly soluble models are also discussed here. The time-dependent version of the final recommended Zhu–Nakamura theory is presented in Appendix B. They can be directly utilized for various applications. Chapter 7 discusses various two-state problems in atomic and molecular processes such as inelastic scattering, elastic scattering with resonances and predissociation, and perturbed bound states, and demonstrate how to utilize the formulas presented in Chaps. 5, 6 and Appendix A. The basic theories of two-state nonadiabatic transitions provide us with local transition matrices in curve-crossing regions which describe the distribution of probability amplitudes due to the nonadiabatic transition. This is quite useful, since the application of the theories is not only restricted to a particular type of two-state problem such as inelastic scattering or bound state problem, but can also be made to any problems involving the same type of curve-crossing. Even applications to multi-channel problems become possible. For this the diagrammatic techniques are very useful. The diagram representing the nonadiabatic transition matrix can be plugged into a whole framework of the problem, whatever the system is. This diagrammatic technique is explained in this chapter. Finally, the time-dependent periodic crossing problems are also discussed. Effects of dissipation and fluctuation in condensed matter are discussed in Chap. 8. Within the framework of the original simple Landau–Zener formula, effects of fluctuation of diagonal (energy levels) and off-diagonal (adiabatic coupling) elements and of energy dissipation are analyzed. Utilizations of the accurate one-dimensional two-state theories presented in Chaps. 5 and 6 in multi-channel and multi-dimensional problems are explained in Chaps. 9 and 10 together with some other exactly solvable models. These are definitely important issues from the view point of applications to realistic practical systems. Since exactly solvable multi-channel problems are naturally very much limited and intrinsically multi-dimensional analytical theory is almost impossible to be developed, it is very important to formulate practically useful theories based on the achievements of the one-dimensional two-state theories. Usefulness of the accurate two-state theories can be demonstrated for multi-channel problems. Applications to multi-dimensional systems have been just started, presenting a very important subject in future. Intriguing phenomenon of complete reflection occurs at some discrete energies when the two adiabatic potential curves cross with opposite signs of slopes. This is quite a unique phenomenon due to quantum mechanical interference effect and provides us with new interesting possibilities. One is a possibility of bound states in the continuum, because a positive energy wave can be trapped in between the two completely reflecting curve-crossing units. This is discussed in Chap. 11. Another possibility is a new idea of molecular switching in a periodic array of curve-crossing potential units. Complete transmission, which is always possible in a periodic potential system, can be switched off and on by somehow clicking the system and creating the complete reflection condition. This is demonstrated in Chap. 12. Finally, in Chap. 13 an interesting subject of controlling nonadiabatic dynamic processes by time-dependent external fields is discussed. Especially, the recent remarkable progress of laser technologies enables us to create effective curve-crossings and to control the nonadiabatic transitions there, thus to control various molecular processes. This is explained in more detail in Chap. 13. Chapter 14 provides short concluding remarks.

Chapter 2

Multi-Disciplinarity

As was mentioned in Introduction, the concept of nonadiabatic transition is very much multi-disciplinary and plays essential roles in state/phase changes in various fields of physics, chemistry, and biology. Not only in natural sciences, but also even in social sciences the concept must be useful to analyze various phenomena. In this chapter some practical examples in physics, chemistry, and biology are presented to emphasize the significance of the concept and to help the reader's deep understanding. Possible applications in economics will also be touched upon briefly.

2.1. Physics

The most typical and well known example is electronic transitions in atomic and molecular collisions [4–14, 17]. Since the electron mass is so light compared to nuclear masses, the electronic motion is first solved at a fixed inter-nuclear distance and the electronic potential energy curves are defined as a function of the inter-nuclear distance R. This R plays a role of adiabatic parameter and its motion induces a transition between different electronic states when the electronic potential energy curves come close together. If we can assume that the two states $V_1(R)$ and $V_2(R)$ as a function of R cross somewhere at R_X and are coupled by $V(R)$, then the corresponding adiabatic states are given by

$$E_{1,2}(R) = \frac{1}{2}\{V_1(R) + V_2(R) \pm [(V_1(R) - V_2(R))^2 + 4V(R)^2]^{1/2}\}. \qquad (2.1)$$

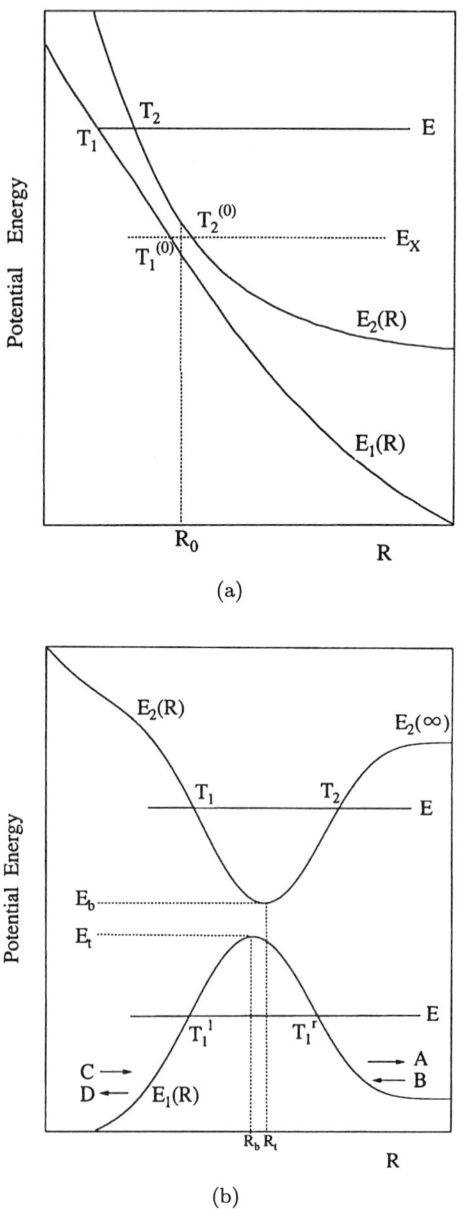

Fig. 2.1.   (a) Landau–Zener type curve crossing and (b) Nonadiabatic tunneling type curve crossing. (Taken from Ref. [183] with permission.)

Unless the coupling $V(R)$ becomes zero at R_X accidentally, the adiabatic states $E_1(R)$ and $E_2(R)$ come close together but never cross on the real axis (see Figs. 2.1). This is called avoided crossing.

The transition between the two states $E_1(R)$ and $E_2(R)$ occur most effectively at this avoided crossing, because the necessary energy transfer between two different degrees of freedom, i.e. between the electronic and nuclear degrees of freedom, is minimum there. In general, the smaller is the necessary energy transfer between different kinds of degrees of freedom, the more probably the dynamic process, the transition between the two states, occurs. The states $V_1(R)$ and $V_2(R)$ are called "diabatic states", and the coupling $V(R)$ is called "diabatic coupling." The representation in theses states is called "diabatic-state representation." On the other hand, the representation in $E_1(R)$ and $E_2(R)$ is called "adiabatic-state representation." These states are coupled by the nuclear kinetic energy operator, namely through the dependence of the adiabatic electronic eigenfunction on the nuclear coordinate R. This is called nonadiabatic coupling, the explicit form of which will be given later (see Eq. (3.6)). If the symmetries of the two states $V_1(R)$ and $V_2(R)$ are different, then the diabatic coupling $V(R)$ should be zero and the two states can cross on the real axis. This occurs, for instance, in the case of states of different electronic symmetries such as Σ and Π states of diatomic molecules. The transitions among them are caused by the nuclear rotational motion, i.e. by the Coriolis coupling or the coupling between electronic and nuclear rotational angular momenta. In any case, we may say that if some processes accompanying transitions among electronic states occur effectively, there must exist avoided crossings of potential energy curves somewhere. Hereafter the following two types of curve crossing are clearly distinguished: the Landau–Zener (LZ) type in which two diabatic potentials cross with the same sign of slopes, and the nonadiabatic tunneling (NT) type in which the two diabatic potentials cross with opposite signs of slopes and a potential barrier is created.

In nuclear collisions and reactions, nuclear molecular orbitals can be defined and transitions among them can be analyzed in terms of nonadiabatic transitions [19, 20]. Adiabaticity is worse compared to atomic and molecular systems, because the mass disparity among nucleons is much smaller than that between electron and nucleus. The similar pictures can, however, hold as those in atomic and molecular processes. Many dynamic processes on solid surfaces are also induced effectively by such nonadiabatic transitions. Examples are neutralization of an ion by a collison with surface and molecular desorption

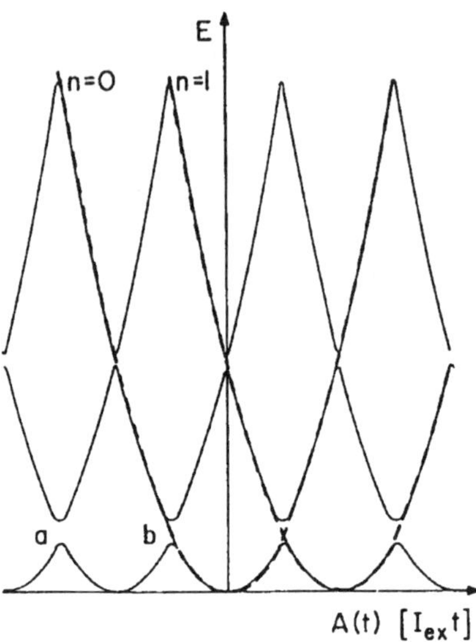

Fig. 2.2. Energy levels of a normal tunnel junction as a function of the vector potential $A(t)$. (Taken from Ref. [29] with permission.)

from a solid surface [21]. So called radiationless transitions in condensed matter such as the quenching of F-colour center and the self-trapping of exciton are other good examples of nonadiabatic transitions in solid state physics [22]. In these examples the abscissa is always a certain spatial coordinate, but in many other examples this is time. When we apply a certain time-dependent external field, nonadiabatic transitions are induced by the change of the field with respect to time. This is called time-dependent nonadiabatic transition. The adiabatic states are defined as eigenstates of the system at each fixed time. Examples are (i) transitions among Zeeman or Stark states in an external magnetic or electric field [23], (ii) transitions induced by laser [24–26], (iii) tunneling junction and Josephson junction in an external magnetic or electric field [27–29]. Figure 2.2 shows an energy level diagram of a tunnel junction as a function of the time-dependent vector potential [29]. Nonadiabatic transitions are induced at avoided crossings by a change of the magnetic field.

Thanks to the recent rapid technological developments of laser, magnetic and electric fields, it has become feasible to control state transitions and dynamic processes by manipulating the external fields and using nonadiabatic

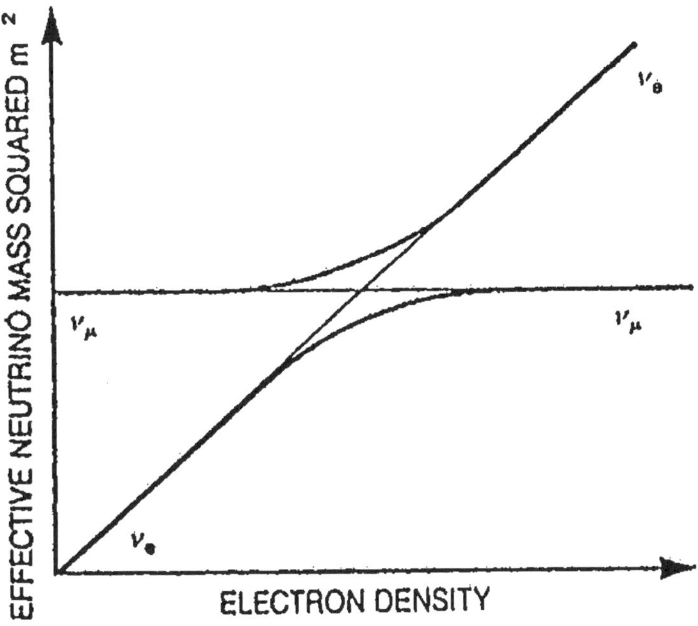

Fig. 2.3. Resonant neutrino conversion in the Sun. ν_e — electron neutrino, ν_μ — muon neutrino. (Taken from Ref. [33] with permission.)

transitions efficiently. This branch of science will become very important in the 21st century, since various dynamic processes might be controlled as we wish. Nonadiabatic transitions may also present one of the crucial ingredients to cause so called chaotic behaviour and are even related to soliton-like structure [30–32]. In all the cases mentioned above the ordinate still represents a certain potential energy. However, the ordinate also is not necessarily an ordinary potential energy. It can be anything, in principle. The most exotic example is the neutrino conversion in the Sun [33, 34] (see Fig. 2.3).

In this case the ordinate is the neutrino mass squared or the flavour of neutrino and the abscissa is electron density. The neutrino mass represents different kinds of neutrino and the conversion between them is induced by the time-variation of the enviromental electron density. This nonadiabatic transition between different kinds of neutrinos is deterministic to judge the existence of finite masses of neutrinos.

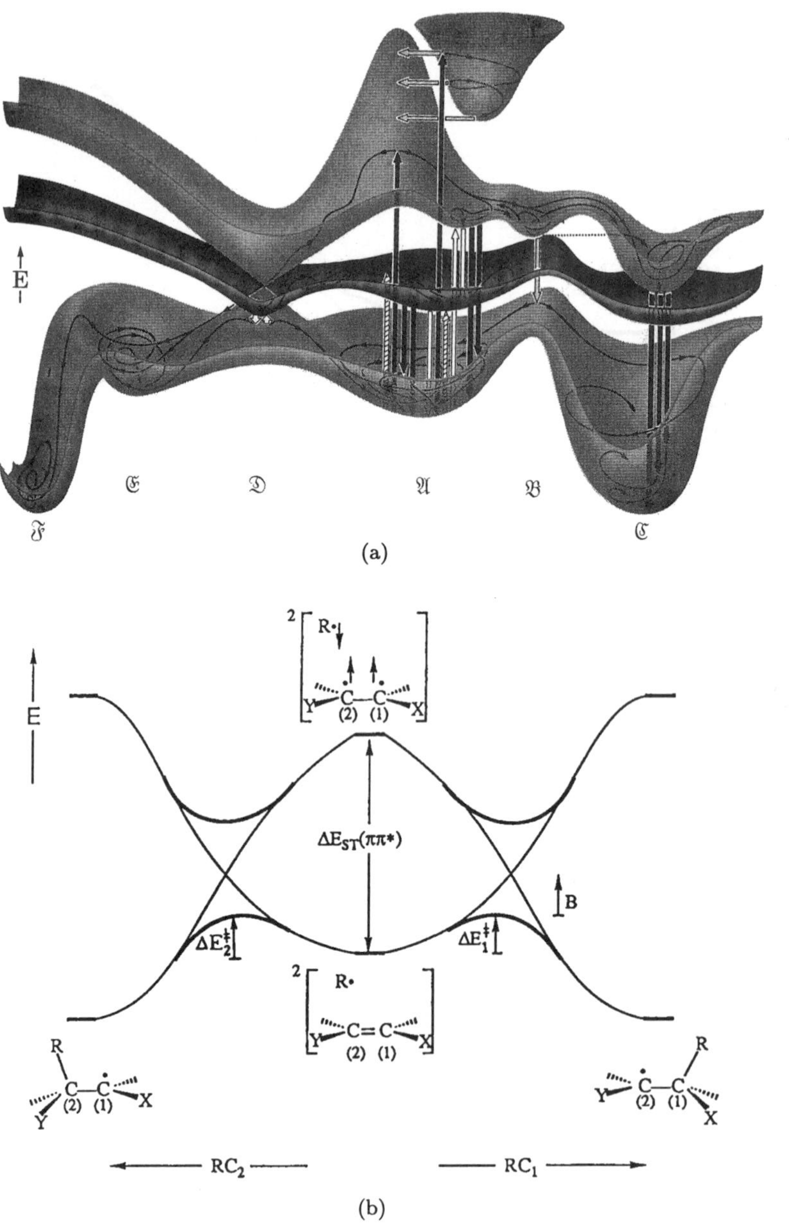

Fig. 2.4.   (a) A schematic potential energy surface representation of a photochemical process. (Taken from Ref. [35] with permission.) (b) Avoided crossing diagrams for the two regiochemical pathways of radical addition to olefins. (Taken from Ref. [36] with permission.)

2.2. Chemistry

Chemical reactions and various spectroscopic processes of molecules are mostly induced by nonadiabatic transitions due to potential energy curve or surface crossings [5, 7, 9, 10, 15]. Without potential-energy-surface (PES) crossings (or avoided crossings) these dynamic processes cannot occur effectively. Even complicated reactions such as photochemical reactions and organic reactions must proceed, if they occur efficiently, via many steps of these nonadiabatic transitions (see Figs. 2.4) [35, 36].

Actually, organic chemical reactions are tried to be classified in terms of potential curve crossing schemes [36]. In big molecules, there might be even unknown stable structures which are connected through PES crossings with ground states. These structures might be utilized as new functions of molecules such as molecular photo-elements. Another good example is the recent development of femto-second dynamics of molecules [37]. This clearly reveals the importance of nonadiabatic transitions due to potential curve (or PES) crossing. Control of chemical reactions by lasers, being now one of the hot topics in chemistry, also endorses the importance of nonadiabatic transitions. By applying a strong laser field, we can create molecules dressed with photons, i.e. dressed states, and shift up and down the molecular energy levels or potential curves by the amount of correspondig photon energies [24–26]. This means that we can induce PES crossings among dressed states and control the nonadiabatic transitions there. The dressed states are schematically shown in Figs. 2.5(a) and 2.5(b). Figure 2.5(a) shows potential energy curves as a function of coordinate. The curves represent the dressed states. The number n represents the photon number absorbed or emitted. Figure 2.5(b), on the other hand, depicts variations of energy levels as a function of laser frequency ω. This time, the slopes of curves represent the number of photons absorbed (positive slope) or emitted (negative slope).

Here we present an interesting example to demonstrate the significance of nonadiabatic transition. This is a photodissociation of bromoacethylchloride at the photon energy which is only a little bit higher than the dissociation enegy [38].

Figure 2.6(a) shows a schematic potential diagram along the dissociation coordinate. In spite of the fact that the potential barrier for the C-Cl bond fission is higher than that for the C-Br fission and the photon enegy is only a little bit higher than the former barrier, the C-Cl fission was observed to occur

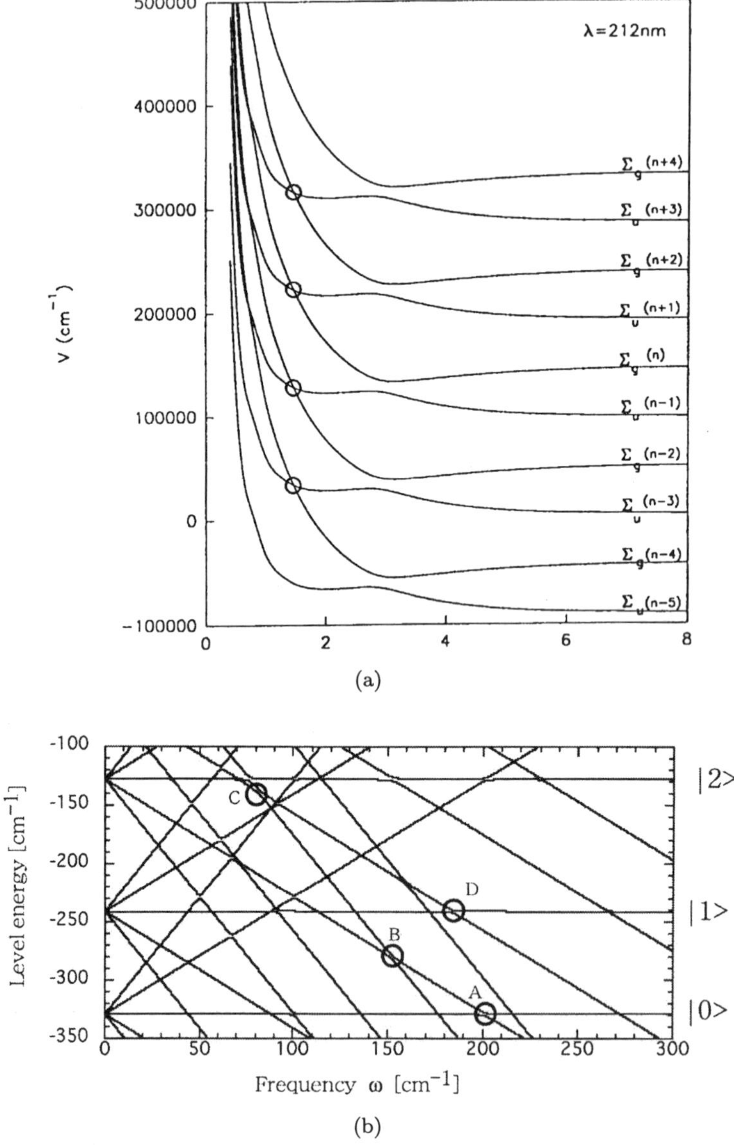

Fig. 2.5. (a) Adiabatic potential versus R (a.u.) in the dressed state picture of $^2\Sigma_g^+$ and $^2\Sigma_u^+$ of H_2^+. The field is $\lambda = 532$ nm and I(intensity) $= 10^{14}$ W/cm^2. (Taken from Ref. [178] with permission.) (b) Dressed-state diagram of vibrational levels as a function of laser frequency ω (see Fig. 13.8). (Taken from Ref. [222] with permission.)

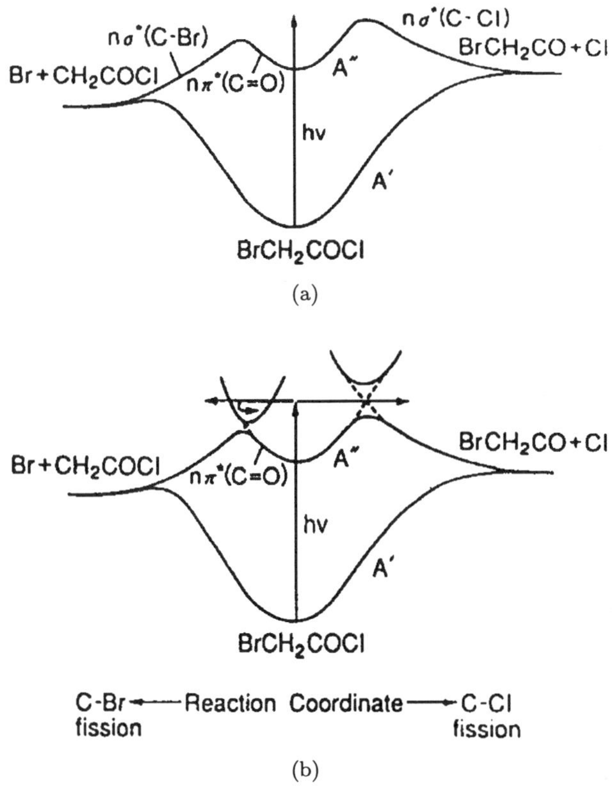

Fig. 2.6. (a) Schematic potential diagram for C-Cl and C-Br fission by the 248 nm photodissociation of $Br(CH_2)_n COCl$. (Taken from Ref. [38] with permission.) (b) The same as Fig. 2.6(a) except that the upper adiabatic states are properly taken into account. (Taken from Ref. [38] with permission.)

more efficiently. This contradicts the well known fact that the over-barrier transmission becomes simply more effective as the photon energy increases. This experimental fact can be interpreted in terms of the effects of nonadiabatic coupling which creates the potential barriers. The potential barriers are actually created by crossings of the two diabatic potential curves (surfaces) with slopes of different signs, as shown in Fig. 2.6(b). The energy separation between the two adiabatic potentials at the top of the barrier is equal to two times of the constant diabatic coupling between the original crossing diabatic states. This means that the diabatic coupling in the dissociation of C-Br bond is smaller than that in the C-Cl bond dissociation, and thus the former dissociation occurs less efficiently than the latter. This can be understood rather easily from the following fact. If there were no diabatic coupling between the two crossing diabatic potentials of different signs of slopes, no transmission would be possible. This warns us that we have to be very careful about the nature and origin of potential barriers, whenever we encounter them, namely how they are created and whether the upper state is located nearby or not. When the barrier is created by two crossing states, the nonadiabtic coupling between the two adiabatic states should be properly taken into account. We cannot neglect the effects of the upper adiabatic potential, unless that is located far above or we are interested in the energy region far below the potential barrier. Namely, the effects of the upper adiabatic potential cannot be neglected even at energies lower than the barrier top of the lower adiabatic potential. The transmission probability is always smaller than the tunneling probability in the case of single adiabatic potential barrier with the upper one neglected. This fact is not properly recognized. This kind of transmission phenomenon is called "nonadiabatic tunneling", i.e. tunneling (or transmission) affected by nonadiabatic coupling.

2.3. Biology

Electron and proton transfers are now well known to play important roles to promote and keep necessary functions in various biological systems [39–41]. For instance, photosynthetic reactions tranform light energy to chemical energy by the actions of proton and electron transfers which are caused most effectively through potential energy surface crossings [40].

Many other biological proceses are also supposed to be induced by nonadiabatic transitions. Even the origin of life might have been affected by nonadiabatic transitions. This is very natural, because in biological systems various kinds of chemical reactions play crucial roles and many chemical reactions proceed effectively only through nonadiabatic transitions.

2.4. Economics

As explained in the previous subsections, "nonadiabatic transitions" appear as important basic mechanisms in various fields of natural sciences. The concept, however, is more general and is not just restricted to natural sciences. Many phenomena in social sciences, even many incidents occuring in daily social life, can be viewed as nonadiabatic transitions. An abrupt change of state, whatever it is, can be considered to be a nonadiabatic transition. Of course, many of them may not be expressed easily in terms of mathematics, say in terms of differential equations, for instance; however, the phenomenon itself may be interpreted by the concept of nonadiabatic transition. For instance, an interesting example is found in a Japanese traditional comic story, "God of Death" [42, 43]. More serious examples may be found in economics. A possible example is an interaction between spot market and future market or an interaction between monetary supply and money market [44]. This is not a traditional idea in economics, of course, but could open a new way of formulating some fields of economics.

Chapter 3

Historical Survey of Theoretical Studies

The theoretical studies of nonadiabatic transitions between potential energy curves date back to 1932, when Landau [1], Zener [2], and Stueckelberg [3] published pioneering papers independently. Very interestingly, in the same year Rosen and Zener [16] published another pioneering paper on the non-curve-crossing problem, which is now known as Rosen–Zener model. Landau, Zener, and Stueckelberg discussed potential curve-crossing problems and formulated the theory now known as the Landau–Zener formula and the Landau–Zener–Stueckelberg (LZS) theory. Since these pioneering works, a large number of authors contributed to the subjects in order to try to improve the pioneers' works. It is almost impossible to list up all these original papers. Here are cited only review articles and books [4–15]. This simply reflects the fact that the importance of nonadiabatic transition is well recognized among scientists. In this section, a brief historical survey of the theoretical studies is provided.

3.1. Landau–Zener–Stueckelberg Theory

Landau discussed the potential energy curve crossing problem by using the complex contour integral method [1, 45]. In general, the transition probability p in the first order perturbation theory is given by

$$p \simeq \left| \int \chi_1(R) f(R) \chi_2(R) dR \right|^2 , \tag{3.1}$$

where $f(R)$ represents the coupling (can be an operator) between the two states and $\chi_1(\chi_2)$ is the nuclear wave function in channel 1(2). Since the nuclear wave functions oscillate rapidly on the real R-axis because of the heavy mass and the ordinary WKB functions are not applicable in the vicinity of turning points, it is more convenient to move into the complex R-plane. Using the primitive WKB functions for $\chi_j(j = 1, 2) \sim \exp[i \int^R k_j(R)dR]$, and picking up only the main term in the upper half-plane, he finally obtained the expression

$$p \simeq p_{\mathrm{LS}} \equiv \exp\left\{ -2\,\mathrm{Im} \int_{\mathrm{Re}(R_*)}^{R_*} [k_1(R) - k_2(R)]dR \right\} \qquad (3.2)$$

where subscript LS stands for Landau–Stueckelberg and

$$k_j(R) = \left\{ \frac{2\mu}{\hbar^2}[E - E_j(R)] \right\}^{1/2}, \qquad (3.3)$$

where μ is the mass of the system and $E_2(R) > E_1(R)$. Im and Re in Eq. (3.2) indicate to take the imaginary and real parts, respectively. R_* represents the complex crossing point of the adiabatic potentials: $E_1(R_*) = E_2(R_*)$. If we further assume that the adiabatic potentials $E_j(R)$ derived from the linear diabatic potentials and the constant coupling between them and that the relative nuclear motion is described by a straight line trajectory with the constant velocity v, then we can finally obtain the famous Landau–Zener formula,

$$p_{\mathrm{LZ}}^{(0)} = \exp\left[-\frac{2\pi A^2}{\hbar v|\Delta F|} \right], \qquad (3.4)$$

where A is the diabatic coupling and $\Delta F = F_1 - F_2$ is the difference of the slopes of the diabatic potentials. The above mentioned method, generally called Landau method, is very instructive and useful, because it is quite general. It should be noted, however, that the original coupling $f(R)$ in Eq. (3.1) has disappeared and the pre-exponential factor is taken to be unity in Eqs. (3.2) and (3.4). It is a big mystery how Landau did this, but the formula (3.4) is actually correct within the time-dependent Landau–Zener model (linear potential, constant coupling and constant velocity), as was proved by Zener [2]. Landau himself probably did not care about the pre-exponential factor, because the exponent is the most decisive factor in any case. But, it is interesting to note that if we employ the adiabatic-state representation (otherwise, we cannot obtain the exponents of Eqs. (3.2) and (3.4)), the coupling $f(R)$ should be the nonadiabatic coupling and is actually given by

$$f(R) \propto T_{12}^{\mathrm{rad}} \frac{\partial}{\partial R} \quad \text{(derivative operator!)}, \tag{3.5}$$

where

$$T_{12}^{\mathrm{rad}} = \left\langle \psi_1^{(a)} \left| \frac{\partial}{\partial R} \right| \psi_2^{(a)} \right\rangle$$

$$= \left\langle \psi_1^{(a)} \left| \frac{\partial H_{\mathrm{el}}}{\partial R} \right| \psi_2^{(a)} \right\rangle \bigg/ [E_1(R) - E_2(R)]^2 . \tag{3.6}$$

Here $\psi_j^{(a)} (j = 1, 2)$ are the electronic wave functions of the adiabatic states. Since $E_1(R) - E_2(R)$ has a complex zero R_* of order one-half (see Eq. (2.1)), T_{12}^{rad} has a pole of order unity there. If we evaluate the contour integral with the residue of the pole taken into account, we cannot get unity as a pre-exponential factor, naturally, although the exponent is the same as that in Eq. (3.2) even with the operator given by Eq. (3.5). So the expressions (3.2) and (3.4) are clearly beyond the first order perturbation theory, and the pre-exponential factor ($=$ unity) is the result of a kind of renormalization based on the above analytical properties. Namely, the formula (3.2) cannot be obtained simply by the ordinary perturbation theory. This is because all the higher order terms contain the same exponential factor and we cannot truncate the perturbation series at any finite order. This is a general feature of the nonadiabatic transition in the adiabatic state representation.

Zener employed the time-dependent Schrödinger equation in the diabatic representation [2],

$$i\hbar \frac{d}{dt} \begin{pmatrix} c_1(t) \\ c_2(t) \end{pmatrix}$$

$$= \begin{pmatrix} 0, & V(t) \exp\left[\dfrac{i}{\hbar} \displaystyle\int^t (V_1 - V_2)dt'\right] \\ V(t) \exp\left[-\dfrac{i}{\hbar} \displaystyle\int^t (V_1 - V_2)dt'\right], & 0 \end{pmatrix}$$

$$\times \begin{pmatrix} c_1(t) \\ c_2(t) \end{pmatrix}, \tag{3.7}$$

where the total wave function is expanded as

$$\Psi = c_1(t)\exp\left[-\frac{i}{\hbar}\int^t V_1(t')dt'\right]\varphi_1^{(d)}$$

$$+ c_2(t)\exp\left[-\frac{i}{\hbar}\int^t V_2(t')dt'\right]\varphi_2^{(d)}. \tag{3.8}$$

As can be seen in Eq. (3.7), the coefficients $c_1(t)$ and $c_2(t)$ do not depend on the diabatic potentials separately, but on the difference $V_1(t) - V_2(t)$, which is assumed to be a linear function of time,

$$V_1(t) - V_2(t) = |\Delta F|vt. \tag{3.9}$$

If we further assume that the diabatic coupling $V(t)$ is constant, then Eq. (3.7) can be solved exactly in terms of the Weber function. Then final transition probability is exactly equal to Eq. (3.4). It should be noted that the linearity in time t is very much different from the linearity in coordinate R and the effects of turning points are completely neglected in the former approximation. Since time is uni-directional, the linear potential in time means that the energy changes forever to $\pm$ infinity as time goes and there is no turning point. In order to express the effect of turning point as in the time-independent linear potential model, at least quadratic dependence on time is required. In the Landau's treatment, the time-dependent linear approximation corresponds to the assumption of the common classical straightline trajectory with constant velocity. Namely, the coordinate R is assumed to increase linealy as s function of time, $R = vt$. This is why the Landau–Zener formula Eq. (3.4) presents the exact result for the time-dependent linear model, but in the time-independent linear potential model this is valid only at collision energies much higher than the crossing point.

Stueckelberg did the most elaborate analysis among the three [3]. He applied the approximate complex WKB analysis to the fourth order differential equation obtained from the original second order coupled Schrödinger equations. In the complex R-plane he took into account the so called Stokes phenomenon (which will be explained in detail in Sec. 4.2) associated with the asymptotic solutions in an approximate way, and finally derived not only the Landau–Zener transition probability but also the total inelastic transition probability P_{12} as

$$P_{12} \simeq 4p_{\mathrm{LS}}(1 - p_{\mathrm{LS}})\sin^2\tau, \tag{3.10}$$

where p_{LS} is the same as Eq. (3.2),

$$\tau = \int_{T_1}^{R_X} k_1(R)dR - \int_{T_2}^{R_X} k_2(R)dR, \tag{3.11}$$

and T_j is the turning point on the adiabatic potential $E_j(R)$ $(j = 1, 2)$. Equation (3.10) contains the effects of turning points and can be interpreted nicely, as described below, in terms of localized nonadiabatic transition at $R_X = \mathrm{Re}(R_*)$ and adiabatic wave propagation along the adiabatic potentials. Since there are two possible (classical) paths for the $1 \to 2$ inelastic transition (see Fig. 3.1), the corresponding scattering matrix elements S_{21} may be expressed as

$$\begin{aligned}
S_{21} &= \sqrt{p_{LS}(1 - p_{LS})}e^{i(\eta_1 + \tau_1)}e^{i(\eta_2 - \tau_2)} - \sqrt{p_{LS}(1 - p_{LS})}e^{i(\eta_1 - \tau_1)}e^{i(\eta_2 + \tau_2)} \\
&= 2ie^{i(\eta_1 + \eta_2)}\sqrt{p_{LS}(1 - p_{LS})}\sin\tau
\end{aligned} \tag{3.12}$$

with

$$\tau = \tau_1 - \tau_2, \tag{3.13}$$

where the first (second) term in the first equation of Eq. (3.12) corresponds to the path I (II) in Fig. 3.1, η_j represents the elastic scattering phase shift

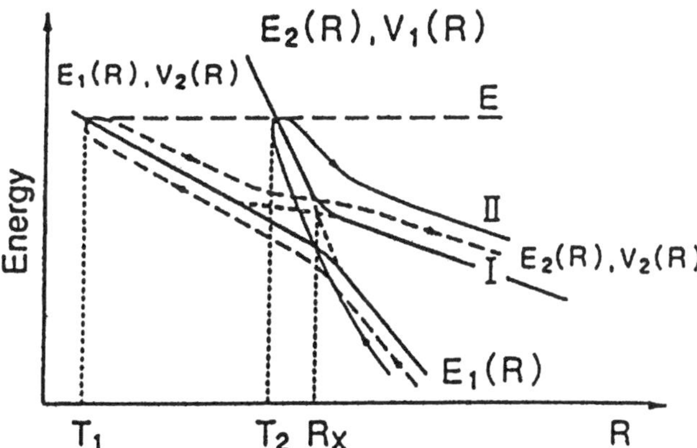

Fig. 3.1. Schematic potential energy curves in the case of Landau-Zener type crossing. $V_j(R)(j = 1, 2)$ are the diabatic potentials and the corresponding adiabatic potentials are $E_j(R)(j = 1, 2)$ shown by solid lines with $E_2(R) > E_1(R)$. I and II indicate the possible paths at energy E for the $1 \to 2$ inelastic transition.

by the potential $E_j(R)$ $(j = 1, 2)$, and τ_j is the phase integral from T_j to R_X along $E_j(R)$ $(j = 1, 2)$. Thus $P_{12} = |S_{21}|^2$ is equal to Eq. (3.10). This idea of decomposing the whole process into sequential events of nonadiabatic transitions and adiabatic wave propagations is very physical and constitutes the basis of semiclassical theory [7, 8].

Figure 3.2 clearly demonstrates the insufficiency of the simple Landau–Zener formula Eq. (3.4). This figure depicts the results for the linear potential

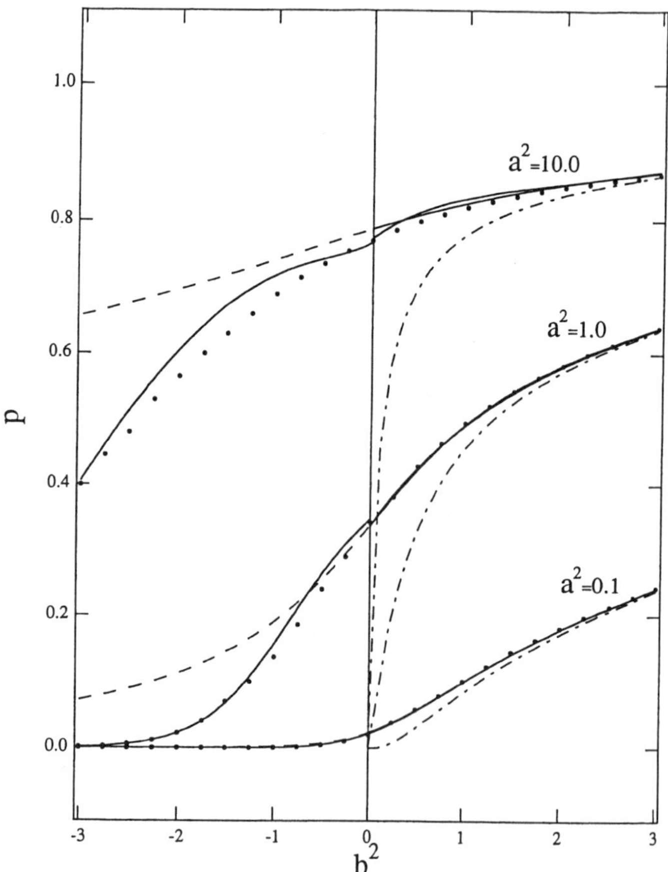

Fig. 3.2. Nonadiabatic transition probability p against b^2 of Eq. (3.14) in the linear potential model. $\cdots$: exact numerical result, ———: new analytical formula given in Chap. 5. - - -: extension of the solid line. (Taken from Ref. [180] with permission.)

model (in R), in which, as will be shown in Chap. 5, the quantum mechanically exact analytical solutions have been obtained.

Here the following two basic parameters, which describe the linear model completely, are introduced:

$$a^2 = \frac{\hbar^2}{2\mu} \frac{F(F_1 - F_2)}{8A^3}, \qquad b^2 = (E - E_X)\frac{F_1 - F_2}{2FA} \tag{3.14}$$

with

$$F = \sqrt{F_1|F_2|}, \tag{3.15}$$

where A is the constant diabatic coupling, E_X is the energy at crossing point, and $F_1 > 0$ and $F_1 > F_2$ are assumed without loss of generality. These parameters a^2 and b^2 effectively represent the coupling strength and the collision energy. Large (small) a^2 corresponds to weak (strong) diabatic coupling regime, and $a^2 \simeq 1$ corresponds to intermediate coupling strength which usually appears in many applications as an important case. It should be noted that a^2 is always positive but b^2 can be negative at $E < E_X$. Figure 3.2 clearly tells that Eq. (3.4) is usable only at high energies. As is clearly exemplified in this figure, the LZS theory contains many defects as described below, although the qualitative physical idea is all right. Despite the fact that a lot of efforts by many authors have been made in order to improve the theory since the pioneering works done by Landau, Zener, and Stuckelberg, many defects have been left unsolved. The defects may be summarized as follows: (a) The Landau–Zener formula does not work at energies near and lower than the crossing point. (b) No good formula exists for the transmission when the two diabatic potentials cross with opposite signs of slopes. (c) The available accurate formulas, which are valid only at energies higher than the crossing point, contain inconvenient complex contour integrals and are not very useful for experimentalists. (d) The Landau–Zener formula requires knowledge of diabatic potentials, which cannot be uniquely obtained from adiabatic potentials. (e) The accurate phases to define the scattering matrices are not available for all cases. As for the efforts paid by many investigators since Landau, Zener and Stueckelberg, readers can refer to many review articles and books [4–9, 11–14].

All the defects listed above are completely solved recently by Zhu and Nakamura (Zhu–Nakamura theory), which are summarized in Chap. 5.

The best formulas up to ~ 1991 before the complete solutions by Zhu and Nakamura may be summarized as follows: The 2×2 scattering matrix corresponding to Fig. 3.1 is given by

$$S = P_{\infty X} O_X P_{XTX} I_X P_{X\infty}, \tag{3.16}$$

where $P_{A \ldots B}$ is a diagonal matrix, representing the adiabatic wave propagation from B to A, and $I_X(O_X)$ is a non-diagonal matrix, expressing the nonadiabatic transition at the avoided crossing point R_X in the incoming (outgoing) segment. These matrices are explicitly given as

$$[P_{\infty X}]_{nm} = [P_{X\infty}]_{nm}$$

$$= \delta_{nm} \exp\left\{ i \int_{R_X}^{\infty} [k_n(R) - k_n(\infty)]dR - ik_n(\infty)R_X \right\}, \tag{3.17}$$

$$[P_{XTX}]_{nm} = \delta_{nm} \exp\left[2i \int_{T_n}^{R_X} k_n(R)dR + \frac{i\pi}{2} \right], \tag{3.18}$$

$$I_X = \begin{pmatrix} \sqrt{1 - p_{\mathrm{LS}}}\, e^{i\phi_S} & -\sqrt{p_{\mathrm{LS}}}\, e^{i\sigma_0^{\mathrm{LS}}} \\ \sqrt{p_{\mathrm{LS}}}\, e^{-i\sigma_0^{\mathrm{LS}}} & \sqrt{1 - p_{\mathrm{LS}}}\, e^{-i\phi_S} \end{pmatrix} \tag{3.19}$$

and

$$O_X = \tilde{I}_X \quad (\text{transpose of } I_X), \tag{3.20}$$

where

$$p_{\mathrm{LS}} = e^{-2\delta}, \tag{3.21}$$

$$\sigma_0^{\mathrm{LS}} + i\delta = \int_{R_X}^{R*} [k_1(R) - k_2(R)]dR, \tag{3.22}$$

and

$$\phi_S(\delta) = \frac{\delta}{\pi} \ln\left(\frac{\delta}{\pi}\right) - \frac{\delta}{\pi} - \frac{\pi}{4} - \arg\Gamma\left(i\frac{\delta}{\pi}\right). \tag{3.23}$$

This phase $\phi_S(\delta)$ is called Stokes phase. This Stokes phase correction ϕ_S is equal to $\pi/4(0)$ in the limit of zero (infinitely strong) diabatic coupling, i.e., in the limit of $\delta = 0$ ($\delta = \infty$). Thus the total inelastic transition probability $P_{12} = |S_{21}|^2$ is given by

$$P_{12} \simeq 4p_{\mathrm{LS}}(1 - p_{\mathrm{LS}})\sin^2(\sigma + \phi_{\mathrm{S}}) \tag{3.24}$$

with

$$\sigma = \sigma_0^{\mathrm{LS}} + \tau. \tag{3.25}$$

The elastic scattering phase shift η_n is given by

$$e^{2i\eta_n} = (P_{\infty X}P_{XTX}P_{X\infty})_{nn}. \tag{3.26}$$

With the phase corrections σ_0^{LS} and ϕ_{S}, Eq. (3.24) works quite well, but still cannot be free from the following defects: (1) δ requires the analytic continuation of $E_n(R)$ into the complex R-plane and the complex contour integral Eq. (3.22) which is not easy for experimentalists to use. (2) The formula works only at energies higher than E_X, although Eq. (3.21) works better than the original Landau–Zener formula Eq. (3.4). At energies slightly lower than E_X some empirical modification was introduced to δ, but no analytical formula has actually been developed to cover the whole range of energy lower than E_X.

In the nonadiabatic tunneling (NT) case (Fig. 2.1(b)) in which two diabatic potentials $V_1(R)$ and $V_2(R)$ cross with different signs of slopes, the semiclassical idea of sequential events of nonadiabatic transition and adiabatic wave propagation can also be usefully utilized at energies higher than the bottom of the upper adiabatic potential. The S-matrix in this case is expressed as

$$S_{nm} = S_{nm}^{\mathrm{R}}e^{i(\eta_n^{\mathrm{NT}} + \eta_m^{\mathrm{NT}})} \tag{3.27}$$

with

$$\eta_n^{\mathrm{NT}} = \lim_{R \to \pm\infty}\left[\pm\int_{R_X}^{R}k_1(R')dR' \mp k_1(R)R + \frac{\pi}{4}\right] \quad \text{for } n = \begin{pmatrix}1\\2\end{pmatrix}, \tag{3.28}$$

where S^{R} is called reduced scattering matrix and is given by

$$S_{11}^{\mathrm{R}} = \frac{p_{\mathrm{LS}}}{1 + (1 - p_{\mathrm{LS}})e^{2i\psi_{\mathrm{LS}}}}e^{2i\gamma_2(T_2^{\mathrm{L}},R_X) - 2i\sigma_0^{\mathrm{LS}}}, \tag{3.29}$$

$$S_{22}^{\mathrm{R}} = \frac{p_{\mathrm{LS}}}{1 + (1 - p_{\mathrm{LS}})e^{2i\psi_{\mathrm{LS}}}}e^{2i\gamma_2(R_X,T_2^{\mathrm{R}}) + 2i\sigma_0^{\mathrm{LS}}}, \tag{3.30}$$

and

$$S_{12}^{R} = S_{21}^{R} = \frac{2(1 - p_{LS})^{1/2}}{1 + (1 - p_{LS})e^{2i\psi_{LS}}} \cos\psi_{LS} e^{i\gamma_2(T_2^{L}, T_2^{R})} \,, \tag{3.31}$$

where

$$\psi_{LS} = -\phi_S(\delta) + \gamma_2(T_2^{L}, T_2^{R}) \tag{3.32}$$

and

$$\gamma_n(a, b) = \int_a^b k_n(R)dR \,. \tag{3.33}$$

The physical processes in this NT-case are quite different from those in the LZ-case: The diagonal (off-diagonal) elements of S-matrix represent reflection (transmission) with the channel 1 (2) designating the wave on the right (left) side of the potential barrier. The denominator in Eqs. (3.29)–(3.31) manifests the trapping by the upper adiabatic potential. Another very interesting feature is "complete reflection ($S_{12} = S_{21} = 0$)" which occurs at certain discrete energies where $\cos\psi_{LS} = 0$ is satisfied. This will be discussed in more detail in Chaps. 11–13. It should be noted again that Eqs. (3.29)–(3.31) are valid only at energies higher than the bottom (E_b) of the upper adiabatic potential and that no accurate formulas have been available at energies lower than that.

3.2.   Rosen–Zener–Demkov Theory

There is another kind of radially induced non-adiabatic transition, called Rosen–Zener–Demkov type [4–9, 11–13, 16].

In contrast to the curve crossing case discussed above, the two diabatic potentials have very weak R-dependence (actually, their difference is assumed to be constant in the basic model) and the diabatic coupling has a strong ($\sim$ exponential) R-dependence (see Fig. 3.3).

The nonadiabatic transition is not so effective compared to the curve crossing case unless the energy defect is quite small; but this case also presents an important transition, especially when the two potentials are in near resonance asymptotically.

The name of Rosen–Zener came from their original work on the time-dependent theory of the double Stern–Gerlach experiment [16]. This problem of Stern–Gerlach experiment is totally different from our present subject,

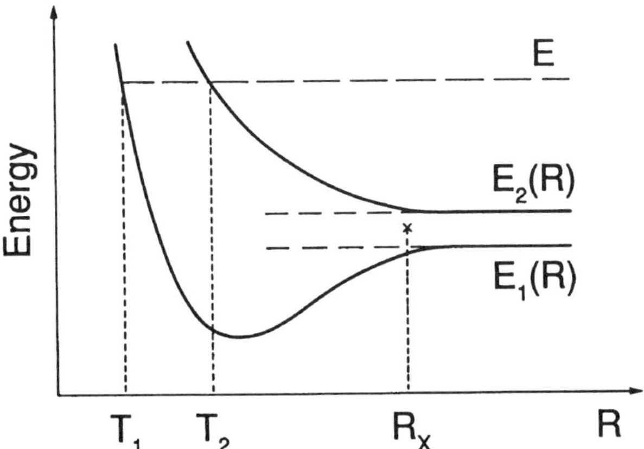

Fig. 3.3. Schematic potential energy curves of the Rosen–Zener–Demkov type. (Taken from Ref. [8] with permission.)

but the basic mathematics corresponds to that of nonadiabatic transition: the corresponding potential difference and diabatic coupling are constant and sechyperbolic function of time, respectively. They solved the time-dependent Schrödinger equation exactly. Later in 1963, Demkov [14, 17] discussed the near resonant charge transfer process again in the time-dependent formalism and obtained essentially the same formula as that of Rosen and Zener. He assumed the constant energy difference and the exponential function for the coupling. The overall inelastic transition probability $P_{12} = |S_{21}|^2$ is given by

$$P_{12} \simeq \operatorname{sech}^2 \left(\frac{\pi \Delta}{2\hbar \beta v} \right) \sin^2 \left(\frac{2V_0}{\pi \alpha v} \right), \tag{3.34}$$

where Δ is the difference (constant) of the two diabatic potentials, β and V_0 are the exponent and the preexponential factor of the diabatic coupling. Although the adiabatic potentials have no conspicuous avoided crossing, the nonadiabatic transition occurs quite locally at $R_X = \operatorname{Re}(R_*)$, where the adiabatic potentials start to diverge (see Fig. 3.3). R_* is the complex crossing point closest to the real axis. In this case there are infinite number of complex crossing points which are distributed with equal distance in parallel with the imaginary axis; but it suffices to take into account the only one closest to the real axis. As the non-crossing of potential curves implies, there is no switching of the character of electronic state and the nonadiabatic transition probability does not approach to unity but to one-half at high energy limit. The nonadiabatic transition probability p_{RZ} by one passage of R_X is obtained from Eq. (3.34) as

$$p_{\mathrm{RZ}}^{(0)} = \left[1 + \exp\left(\frac{\pi\Delta}{\hbar\beta v}\right)\right]^{-1}. \tag{3.35}$$

The same semiclassical idea of decomposing the whole process into sequential events of nonadiabatic transition and adiabatic wave propagation can be applied together with the complex phase integral for nonadiabatic transition. Thus the scattering matrix can be expressed in the same way as Eqs. (3.16)–(3.20). The difference from the Landau–Zener case appears in the matrices I_X and $O_X(=\tilde{I}_X)$ as

$$I_X = \begin{pmatrix} \sqrt{1-p_{\mathrm{RZ}}}\,e^{i\phi} & -\sqrt{p_{\mathrm{RZ}}}\,e^{-i\psi} \\ \sqrt{p_{\mathrm{RZ}}}\,e^{i\psi} & \sqrt{1-p_{\mathrm{RZ}}}\,e^{-i\phi} \end{pmatrix} \tag{3.36}$$

with

$$p_{\mathrm{RZ}} = [1 + e^{2\delta}]^{-1}, \tag{3.37}$$

$$\phi = \gamma(\delta) - \gamma(2\delta), \tag{3.38}$$

$$\psi = \phi - \sigma_0^{\mathrm{LS}}, \tag{3.39}$$

$$\gamma(X) = X\ln X - X - \arg\Gamma(iX), \tag{3.40}$$

where σ_0^{LS} and δ are defined in the same way as before by Eq. (3.22). In the original Rosen–Zener model, σ_0^{LS} is given by (see Sec. 5.4, especially Eq. (5.261))

$$\sigma_0^{\mathrm{LS}} = \frac{\pi\Delta}{\hbar\beta v}\left[2\sqrt{2} + \ln\frac{\sqrt{2}-1}{\sqrt{2}+1}\right]. \tag{3.41}$$

It should be noted that the expressions of the I-matrix for the Rosen–Zener model given in the following references are in error: Eqs. (3.34), (3.22) and (3.36) in [7–9], respectively. The total inelastic transition probability is expressed as

$$P_{12} \simeq 4p_{\mathrm{RZ}}(1-p_{\mathrm{RZ}})\sin^2\sigma = \mathrm{sech}^2\,\delta\,\sin^2\sigma, \tag{3.42}$$

where σ is defined by Eq. (3.25). It should be noted that the Stokes phase ϕ_{S} does not appear in σ [11].

3.3. Nikitin's Exponential Model

In the case of the original Landau–Zener model the diabatic potentials depend strongly on R (or time) and the diabatic coupling is constant. On the other hand, in the Rosen–Zener model the potential difference is constant and the diabatic coupling strongly depends on R (or time), i.e. exponentially. What happens, if both of them depend on R (or time) strongly? As such an example, we can think of the exponential model in which both diabatic potentials and coupling are exponential functions. This actually presents an interesting generalization of the Landau–Zener and the Rosen–Zener models.

Nikitin [4] considered such a model for the first time. The model potentials he employed are the following in diabatic representation:

$$V_1 = V_0(x) - \frac{1}{2}\delta\epsilon[1 - \cos 2\theta_0 e^{-x}]$$

$$V_2 = V_0(x) + \frac{1}{2}\delta\epsilon[1 - \cos 2\theta_0 e^{-x}] \tag{3.43}$$

$$v(x) = \frac{1}{2}\delta \sin 2\theta_0 e^{-x} ,$$

where

$$x \equiv \alpha(R - R_{\mathrm{p}}) , \tag{3.44}$$

and $V_0(x)$ is a certain function whose functionality is not important, since he considered only the time-dependent version which requires only the difference of the two diabatic potentials. The two diabatic potentials cross at $x_X = \ln(\cos 2\theta_0)$ when $0 \le \theta_0 < \pi/4$ and do not cross when $\pi/4 \le \theta_0 \le \pi/2$. The complex crossing points of the adiabatic potentials are given by

$$x_{\mathrm{c}} = i(\pm 2\theta_0 \pm 2n\pi) \qquad (n = 0, \pm 1, \pm 2, \ldots) . \tag{3.45}$$

As in the Rosen–Zener case there are infinite number of complex crossing points.

He simplified the problem by introducing the linear trajectory approximation $R = R_{\mathrm{p}} + v_{\mathrm{p}}t$, where v_{p} is the constant velocity and t is time. Introducing the new variables

$$\tau = \alpha v_{\mathrm{p}}t = x, \qquad \xi = \frac{\delta\epsilon}{\hbar\alpha v_{\mathrm{p}}} , \tag{3.46}$$

we can write down the time-dependent coupled differential equations (see Eqs. (3.7))

$$i\hbar\frac{dc_1}{d\tau} = \sqrt{\xi_\mathrm{p}(\xi - \xi_\mathrm{p})}e^{-\tau}\exp\left[i\int_0^\tau\{\xi - (\xi - 2\xi_\mathrm{p})e^{-\tau}\}d\tau\right]c_2,$$

$$i\hbar\frac{dc_2}{d\tau} = \sqrt{\xi_\mathrm{p}(\xi - \xi_\mathrm{p})}e^{-\tau}\exp\left[-i\int_0^\tau\{\xi - (\xi - 2\xi_\mathrm{p})e^{-\tau}\}d\tau\right]c_1. \tag{3.47}$$

These coupled equations can be solved in terms of confluent hypergeometric functions [4]. Only the final results for the nonadiabatic transition probability p_N for one-passage of transition region and the corresponding dynamical phases are given here. The corresponding nonadiabatic transition matrix (see Eq. (3.19)) is given by

$$I_X = \begin{pmatrix} \sqrt{1 - p_\mathrm{N}}\,e^{i\phi} & -\sqrt{p_\mathrm{N}}\,e^{-i\psi} \\ \sqrt{p_\mathrm{N}}\,e^{i\psi} & \sqrt{1 - p_\mathrm{N}}\,e^{-i\phi} \end{pmatrix}, \tag{3.48}$$

where

$$p_\mathrm{N} = \frac{e^{-\pi\xi_\mathrm{P}}\sinh\pi(\xi - \xi_\mathrm{p})}{\sinh\pi\xi}, \tag{3.49}$$

$$\phi = \gamma(\xi_\mathrm{p}) - \gamma(\xi), \tag{3.50}$$

$$\psi = \gamma(\xi - \xi_\mathrm{p}) - \gamma(\xi) - 2\left[\sqrt{\xi\xi_\mathrm{p}} + \frac{\xi - \xi_\mathrm{p}}{2}\ln\frac{\sqrt{\xi} - \sqrt{\xi_\mathrm{p}}}{\sqrt{\xi} + \sqrt{\xi_\mathrm{p}}}\right], \tag{3.51}$$

$$\gamma(\delta) = \pi/4 + \delta\ln\delta - \delta - \arg\Gamma(1 + i\delta) = \delta\ln\delta - \delta - \arg\Gamma(i\delta). \tag{3.52}$$

As mentioned in the beginning, the nice feature of this exponential model is that the nonadiabatic transition probabilities in the Landau–Zener and the Rosen–Zener models can be derived as certain limiting cases from this model. When $\theta_0 \ll 1$, $\xi_\mathrm{p} \to \xi\theta_0^2$ and

$$p_\mathrm{N} \longrightarrow e^{-\pi\xi\theta_0^2}, \tag{3.53}$$

which is equal to the Landau–Zener transition probability with $A = \delta\epsilon \cdot \theta_0$ and $|\Delta F| = \delta\epsilon \cdot \alpha$ (see Eq. (3.4)). On the other hand, when $\theta_0 = \pi/4$, $\xi_{\mathrm{p}} \to \xi$ and

$$p_{\mathrm{N}} \longrightarrow \frac{e^{-\pi\xi}}{(1 + e^{-\pi\xi})}, \tag{3.54}$$

which is equal to the nonadiabatic transition probability in the Rosen–Zener model with $\Delta = \delta\epsilon$ (see Eq. (3.35)). Generalizations of the Nikitin model will be discussed in Sec. 5.4.

3.4. Nonadiabatic Transition Due to Coriolis Coupling and Dynamical State Representation

There is another kind of nonadiabatic transition in molecules, which is induced by Coriolis interaction (or rotational coupling) [4, 5, 7, 8, 11]. This has a very different character from the radially induced nonadiabatic transitions discussed so far. The transition is not localized at the crossing point even in the Born–Oppenheimer representation, because the Coriolis coupling is proportional to R^{-2} and is most dominant at the turning point, not at the crossing point (see Eq. (3.57), below). The theories developed for the curve crossing and the non-crossing cases discussed above cannot be applied directly to this case. However, if we introduce a new representation called "dynamical state representation" in which Coriolis coupling is diagonalized, then we can make them applicable [7–9].

This is explained in this section by considering a diatomic molecule as an example. Let us start with the Hamiltonian $\tilde{H}$ of a diatomic molecule, which is given in the body fixed coordinate system as (the mass polarization and relativistic effects are disregarded)

$$\tilde{H} = -\frac{\hbar^2}{2\mu}\frac{1}{R^2}\left(\frac{\partial}{\partial R}R^2\frac{\partial}{\partial R}\right) + \frac{\hbar^2}{2\mu R^2}(\mathcal{J} - \mathcal{L})^2 + H_{\mathrm{el}}$$

$$= -\frac{\hbar^2}{2\mu}\frac{1}{R^2}\left(\frac{\partial}{\partial R}R^2\frac{\partial}{\partial R}\right) + H_{\mathrm{rot}} + H_{\mathrm{cor}} + H' + H_{\mathrm{el}}, \tag{3.55}$$

where

$$H_{\mathrm{rot}} = -\frac{\hbar^2}{2\mu R^2}(\mathcal{J}^2 - 2\Lambda^2), \tag{3.56}$$

$$H_{\text{cor}} = -\frac{\hbar^2}{2\mu R^2}\left(\mathcal{L}_+\mathcal{U}_+ + \mathcal{L}_-\mathcal{U}_-\right), \tag{3.57}$$

and

$$H' = \frac{\hbar^2}{2\mu R^2}\mathcal{L}^2. \tag{3.58}$$

Here, $\mathcal{J}(\mathcal{L})$ is the total (electronic) angular momentum operator, R is the internuclear distance as before, μ is the reduced mass, and Λ is the component of $\mathcal{L}$ along the molecular axis. The ladder operators $\mathcal{L}_\pm$ and $\mathcal{U}_\pm$ are explicitly given as follows:

$$\mathcal{L}_\pm = \mathcal{L}_\xi \pm i\mathcal{L}_\eta \tag{3.59}$$

and

$$\mathcal{U}_\pm = \mp\frac{\partial}{\partial\Theta} + \frac{i}{\sin\Theta}\frac{\partial}{\partial\Phi} + L_\zeta\cot\Theta, \tag{3.60}$$

where $\mathcal{L}_\xi$, $\mathcal{L}_\eta$ and $\mathcal{L}_\zeta$ are the components of $\mathcal{L}$ in the body-fixed coordinate system with the ζ axis along the molecular axis, and (Θ, Φ) are the ordinary angle variables to define the molecular axis orientation in the space-fixed coodinate system. It should be noted that the Schrödinger equation $\tilde{H}\tilde{\Psi} = E\tilde{\Psi}$ and the Hamiltonian $\tilde{H}$ can be transformed to

$$H\Psi = E\Psi \tag{3.61}$$

and

$$H = R\tilde{H}R^{-1} = -\frac{\hbar^2}{2\mu}\frac{\partial^2}{\partial R^2} + H_{\text{rot}} + H_{\text{cor}} + H' + H_{\text{el}} \tag{3.62}$$

with

$$\Psi = R\tilde{\Psi}. \tag{3.63}$$

The ordinary Born–Oppenheimer adiabatic states are defined by the eigenvalue problem of H_{el}:

$$H_{\text{el}}\psi_n^{(a)}(\mathbf{r}:R\,|\,\Lambda) = E_n(R:\Lambda)\psi_n^a(\mathbf{r}:R\,|\,\Lambda). \tag{3.64}$$

Nonadiabatic transitions among these states are induced by either the first term of Eq. (3.62) or H_{cor}. The states of the same Λ are coupled by the first term of Eq. (3.62) (see also Eqs. (3.5) and (3.6)). Transitions between the states of different electronic symmetries (different $|\Lambda|$) are induced by the Coriolis coupling H_{cor} and have quite different properties from the radially induced transitions. In order to look into this in more detail, let us introduce the electronic-rotational basis functions defined as

$$\Phi_n^{J\pm}(\mathbf{r}) = \frac{1}{\sqrt{2}}\{\psi_n^{(a)}(\mathbf{r}:R\,|\,\Lambda^+) \pm \psi_n^{(a)}(\mathbf{r}:R\,|\,\Lambda^-)\}Y(\hat{R}:J\Lambda) \quad \text{for } \Lambda \neq 0 \quad (3.65)$$

and

$$\Phi_n^{J\pm}(\Sigma) = \psi_n^{(a)}(\mathbf{r}:R\,|\,\Sigma^\pm)Y(\hat{R}:J\Sigma) \quad \text{for } \Lambda = 0, \quad (3.66)$$

where $\Lambda^\pm = \pm|\Lambda|$ and $Y(\hat{R}:J\Lambda)$ is the eigenfunction of H_{rot}:

$$H_{\mathrm{rot}}Y(\hat{R}:J\Lambda) = [J(J+1) - 2\Lambda^2]Y(\hat{R}:J\Lambda). \quad (3.67)$$

These functions $\Phi_n^{J\pm}$ are the eigenfunctions of $H_{\mathrm{el}} + H_{\mathrm{rot}}$,

$$(H_{\mathrm{el}} + H_{\mathrm{rot}})\Phi_n^{J\pm}(\Lambda) = \left\{E_n^\Lambda(R) + \frac{\hbar^2}{2\mu R^2}[J(J+1) - 2\Lambda^2]\right\}\Phi_n^{J\pm}(\Lambda). \quad (3.68)$$

There is no coupling between the two manifolds $\{\Phi_n^{J+}(\Lambda)\}$ and $\{\Phi_n^{J-}(\Lambda)\}$. Now, the Coriolis coupling (or nonadiabatic rotational) matrix element within each manifold is given by

$$T_{1,2}^{\mathrm{rot}}(\Lambda_1, \Lambda_2) = \langle\Phi_1^{J\pm}(\Lambda_1)|H_{\mathrm{cor}}|\Phi_2^{J\pm}(\Lambda_2)\rangle$$

$$= -\frac{\hbar^2}{2\mu R^2}\{\lambda_-(J,\Lambda_2^+)\delta(\Lambda_1^+, \Lambda_2^+ - 1)\langle\psi_1^{(a)}(\Lambda_1^+)|\mathcal{L}_-|\psi_2^{(a)}(\Lambda_2^+)\rangle$$

$$+ \lambda_+(J,\Lambda_2^+)\delta(\Lambda_1^+, \Lambda_2^+ + 1)\langle\psi_1^{(a)}(\Lambda_1^+)|\mathcal{L}_+|\psi_2^{(a)}(\Lambda_2^+)\rangle\},$$

$$\text{for } \Lambda_1, \Lambda_2 \neq 0, \quad (3.69)$$

where

$$\lambda_\pm(J,\Lambda) = \sqrt{(J+\Lambda)(J\pm\Lambda+1)}. \quad (3.70)$$

If one of the states is a Σ-state, then we have

$$T_{1,2}^{\mathrm{rot}}(\Sigma, \Pi) \equiv \langle\Phi_1^J(\Sigma)|H_{\mathrm{cor}}|\Phi_2^{J\pm}(\Pi)\rangle$$

$$= \begin{cases} -\dfrac{\hbar^2}{\sqrt{2}\mu R^2}\sqrt{J(J+1)}\langle\psi_1^{(a)}(\Sigma)|L_-|\psi_2^{(a)}(\Pi^+)\rangle & \text{for } + \\ 0 & \text{for } - . \end{cases} \quad (3.71)$$

The Coriolis coupling is usually not very strong, unless the nuclear kinetic energy is very high; but it plays an important role, because it couples the states which cannot be coupled by the radial coupling T^{rad} (Eq. (3.6)). In spectroscopic problems, this coupling is called "heterogeneous perturbation" in contrast with the "homogeneous perturbation" for the radial coupling case [46]. As can be easily conjectured from Eqs. (3.6) and (3.69) (or (3.71)), the two kinds of nonadiabatic transitions have quite different properties from each other. In the radially induced case the coupling has a pole of order unity at the complex crossing point R_*, where $\Delta E(R_*) = E_1(R_*) - E_2(R_*)$ has a zero of order one-half (see Eqs. (2.1) and (3.6)). This analytical property underlies the curve-crossing problem and is the reason why the pre-exponential factor of the Landau–Zener transition probability is exactly unity. The Coriolis coupling, on the other hand, has a pole of order two at $R = 0$ (see Eq. (3.57)) and the corresponding potential curves $E_n^{\Lambda}(R)$ can have a real curve crossing at finite R or at $R = 0$ (united-atom degeneracy).[a] This suggests that the semiclassical theories developed for the radial coupling problem cannot be directly applied to the Coriolis coupling problem. However, this problem can be treated by introducing the new representation, called "dynamical state representation" [7, 8, 11]. The dynamical states (DS) are defined as the eigenstates of the total Hamiltonian of Eq. (3.55) with the first term excluded, i.e.,

$$H_{\text{dyn}}\Psi_n^{J\pm}(\mathbf{r}, \hat{R} : R) \equiv (H_{\text{el}} + H_{\text{rot}} + H_{\text{cor}} + H')\Psi_n^{J\pm}(\mathbf{r}, \hat{R} : R)$$

$$= W_n^{J\pm}(R)\Psi_n^{J\pm}(\mathbf{r}, \hat{R} : R), \tag{3.72}$$

where $\Psi_n^{J\pm}$ may be expanded in terms of electronic-rotational basis functions as

$$\Psi_n^{J\pm}(\mathbf{r}, \hat{R} : R) = \sum_{\Lambda} C_n^{J\Lambda\pm} \Phi_n^{J\pm}(\mathbf{r}, \hat{R} : R \,|\, \Lambda). \tag{3.73}$$

Transitions among the dynamical states are exclusively induced by the radical coupling given by Eq. (3.6) with $\psi^{(a)}$'s replaced by the relevant DS's $\Psi_n^{J\pm}$. In this representation Λ is not a good quantum number anymore, and the Neuman–Wigner's non-crossing rule applies to all the dynamical states. The analytical properties of the coupling and energy difference ΔW in this

---

[a]When the relevant two electronic states $E_n^{\Lambda}(R)$ correlate to the same atomic orbital in the united atom limit $R \to 0$, $\Delta E(R) \propto R^2$ and the electronic matrix element of $\mathcal{L}_+$ (or $\mathcal{L}_-$) in Eq. (3.57) is not equal to zero at $R = 0$.

Table 3.1. Analytical properties of the various nonadiabatic coupling schemes.

Coupling scheme	Potential energy difference $\Delta E \propto$	Coupling $T \propto$
Adiabatic-state representation		
Radial (R_*:complex) rotational	$(R - R_*)^{1/2}$	$(R - R_*)^{-1}$
(a) Degeneracy at $R = 0$	R^2	
(b) Crossing at finite $R = R_x$	$R - R_X$	R^{-2}
(c) No crossing	Constant	
Dynamical-state representation		
Any transition (R_*:complex)	$(R - R_*)^{1/2}$	$(R - R_*)^{-1}$

representation are the same as those of the original radial coupling problems; and thus the semiclassical theories developed for the latter can now be applied in a unified way to any transitions in the new representation. This idea can be, in principle, generalized to more complicated systems by using the hyperspherical coordinate system [7, 8]. The analytical properties of the various nonadiabatic transitions are summarized in Table 3.1. Figures 3.4 demonstrate the localization of rotationally induced transition in the DS-representation and the effectiveness of the semiclassical theory in this representation [8]. This is the case of the two-state ($1\pi_u$ and $2\sigma_u$ states) problem in the $Ne^+ + Ne$ collision. The corresponding potential curves are shown in Fig. 3.5. Further numerical examples are found in Refs. [7] and [8].

It is interesting to note that the radial coupling T^{rad} loses its personality in the complex R-plane, especially at $R \sim R_*$. From Eqs. (2.1) and (3.6) we can easily show

$$T_{12}^{\mathrm{rad}} = \left\langle \psi_1^{(a)} \left| \frac{\partial}{\partial R} \right| \psi_2^{(a)} \right\rangle \simeq \frac{\partial \theta}{\partial R} \simeq \frac{i}{4} \frac{1}{R - R_*}, \tag{3.74}$$

where

$$\theta = \frac{1}{2} \tan^{-1} \left(\frac{2V(R)}{V_2(R) - V_1(R)} \right). \tag{3.75}$$

Equation (3.74) holds irrespective of the forms $V_1(R)$, $V_2(R)$ and $V(R)$. Furthermore, the radial coupling should satisfy the following condition even on

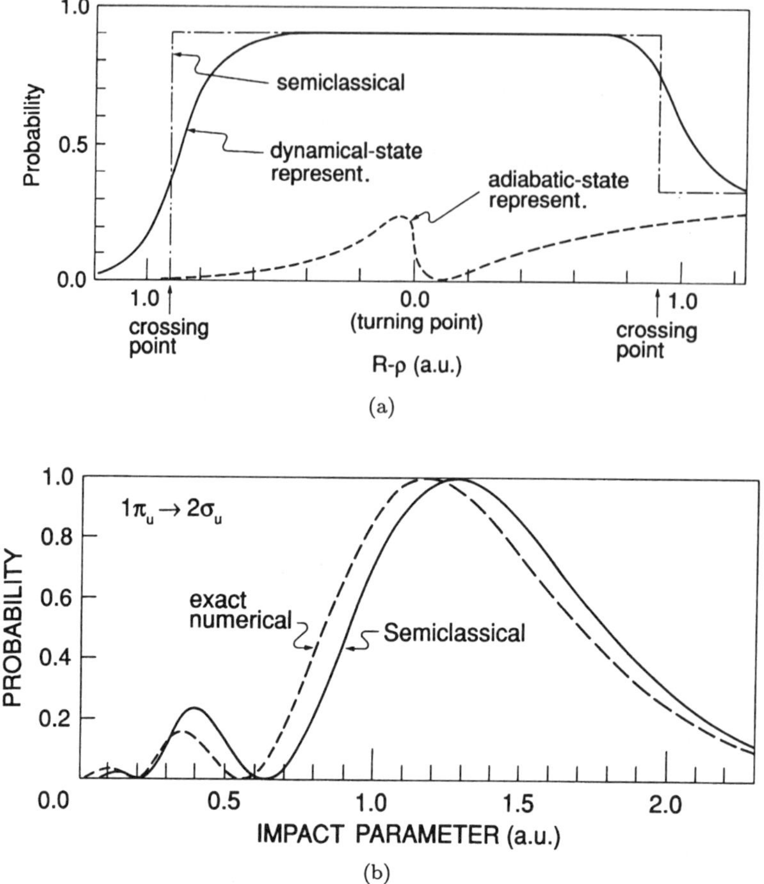

Fig. 3.4.   (a) Localization of the rotationally induced transition $1\pi_u \to 2\sigma_u$ in the Ne^+-Ne system in the dynamical-state representation. The electronic energy diagram is shown in Fig. 3.5 and (b) Impact parameter dependence of the transition probability for $1\pi_u \to 2\sigma_u$ at v (velocity) $= 0.9$ a.u. in the Ne^+-Ne system (see Fig. 3.5). (Taken from Ref. [8] with permission.)

the real axis in the case of curve crossing:

$$\int_{-\infty}^{\infty} T_{12}^{\rm rad} dR = \pi/2 \,. \tag{3.76}$$

As a consequence of these properties, direct information on $T^{\rm rad}$ is not required in the semiclassical analytical theory, as was demonstrated already in

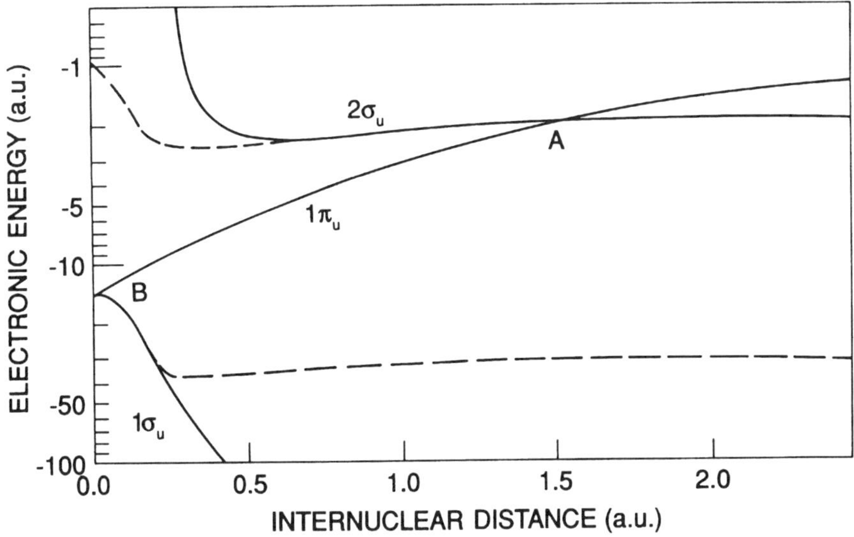

Fig. 3.5. Electronic energy diagram of the $1\sigma_u$, $1\pi_u$, and $2\sigma_u$ states of the Ne$^+$–Ne system. - - -: variable screening model; ———: model potential used. (Taken from Ref. [8] with permission.)

the previous section. That information is replaced by the analytical continuation of the adiabatic potentials into complex R-plane (see Eq. (3.22)). In order to carry out the quantum mechanical numerical calculations, however, we always stay on the real R-axis and need the explicit information of the nonadiabatic couplings. Even in the diabatic representation, which is often employed because of its convenience, the nonadiabatic couplings are necessary to obtain the diabatic couplings [18], unless the diabatic potential matrix is known from the beginning.

Chapter 4

Background Mathematics

4.1. Wentzel–Kramers–Brillouin Semiclassical Theory

If we could know such wavefunctions in the whole range of coordinate space
that satisfy necessary physical boundary conditions, then we could solve all the
corresponding physical problems completely. This is not usually the case, however, and we definitely need approximate analytical wavefunctions in order to
formulate basic physical problems. Such approximate analytical wave functions
are provided by the Wentzel–Kramers–Brillouin approximation in the case of
potential problem. Suppose a particle of mass μ moving in a one-dimensional
potential $V(x)$. The wavefunction of this particle is given by

$$\psi_{\mathrm{I}} = c_+ \phi_+^{\mathrm{I}} + c_- \phi_-^{\mathrm{I}} \tag{4.1}$$

in a region where the classical motion is allowed or by

$$\psi_{\mathrm{II}} = c_+' \phi_+^{\mathrm{II}} + c_-' \phi_-^{\mathrm{II}} \tag{4.2}$$

in a region where the classical motion is not allowed. Here the functions $\phi_\pm^{\mathrm{I,II}}$
are given by

$$\phi_\pm^{\mathrm{I}} = \frac{\exp\left[\pm i \int_a^x k(x)dx\right]}{\{k(x)\}^{1/2}}, \tag{4.3}$$

$$\phi_\pm^{\mathrm{II}} = \frac{\exp\left[\pm \int_a^x |k(x)|dx\right]}{\{|k(x)|\}^{1/2}}, \tag{4.4}$$

where $k(x)$ is the local wavenumber in the potential $V(x)$ under the total energy E (see Eq. (3.3)), a is a turning point where $E = V(a)$, and c_+, c'_+, etc. are the coefficients to be determined by appropriate boundary conditions. These solutions are valid at x far away from the turning point a, because $k(x)$ becomes zero there. This divergence problem can be solved by using the so called uniform approximation [5, 12, 13]. In order to obtain physical quantities such as scattering matrix or tunneling probability, we have to know the relation between the two sets of coefficients $c_\pm$ and $c'_\pm$. It is well known that the following connections should be satisfied in the case that the turning point is a zero of order unity of $E - V(x)$:

$$\frac{c}{2} \exp\left[-\left|\int_a^x k(R)dR\right|\right] \Leftrightarrow c \sin\left[\int_a^x k(R)dR + \frac{\pi}{4}\right] \tag{4.5}$$

and

$$-ic \exp\left[\left|\int_a^x k(R)dR\right|\right] \Leftrightarrow c \exp\left[i \int_a^x k(R)dR - \frac{\pi}{4}i\right]. \tag{4.6}$$

The above simple example instructs us the following general fact. Independent analytical solutions can be provided by the WKB approximation in asymptotic regions, but any linear combination of them holds only in a certain restricted region of the complex configuration space. The coefficients of the linear combination cannot be the same beyond that region. In order to obtain the unique solution in the whole space that satisfies the proper physical boundary condition, we have to connect them beyond the boundaries of these regions. These boundaries in the asymptotic region are called Stokes lines emanating from the turning points which are complex in general. Thus, the connections of WKB solutions beyond the Stokes lines present a very basic problem in the WKB type semiclassical theory. In the case of simple potential scattering with a single turning point, the connection formula, Eqs. (4.5) and (4.6), are applied and the scattering phase shift can be obtained as

$$\delta = \lim_{x \to \infty} \left[\int_a^x k(x)dx - k(x)x\right] + \frac{\pi}{4}. \tag{4.7}$$

Here, the classically forbidden region is assumed to be located in $x < a$. This connection formula tells the physically natural fact that the standing wave in the classically allowed region connects to the exponentially decaying wave in the classically inaccessible region (see Eq. (4.5)) and the additional phase

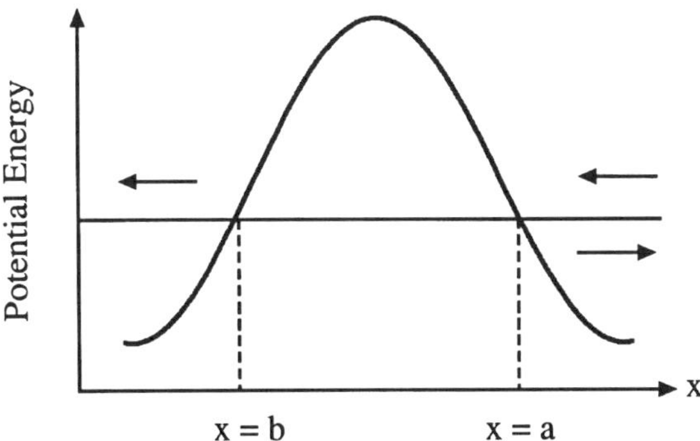

Fig. 4.1.   Potential barrier penetration.

$\pi/4$ is created due to the turning point. In the case of tunneling, another similar connection is necessary at the boundary $x = b$ (see Fig. 4.1), where the outgoing wave in the classically allowed region is connected to the exponentially rising wave in the classically inaccessible region, namely the connection formula Eq. (4.6) is used.

With use of these two connection formulas, the well-known expression for tunneling probability can be derived,

$$P_{\text{tunnel}} = \exp\left(-2\int_b^a |k(x)|dx\right). \tag{4.8}$$

The next naive question is how the connection formulas, Eqs. (4.5) and (4.6), can be derived mathematically. And also, what happens when the energy is close to the potential barrier top which cannot be approximated by a linear curve but is quadratic? These questions are related to the so called Stokes phenomenon of asymptotic solutions of ordinary differential equations of the second order. This will be explained in the next subsection. Another important concept is the comparison equation method. We can use the connection formulas, Eqs. (4.5) and (4.6), even if the potential is not exactly linear. They can be used when the turning points are well separated from each other and are zeros of order unity. Namely, once we know exact solutions or complete knowledge of Stokes phenomenon about a certain basic problem, say linear potential problem, then we can use the corresponding solutions for other general cases which have the same analytical structure. Suppose we know the exact solution of the following differential equation,

$$\left[\frac{d^2}{d\xi^2} + j^2(\xi)\right]\phi(\xi) = 0. \tag{4.9}$$

With use of this solution, $\phi(\xi)$, we try to solve a more general equation which has the same analytical structure,

$$\left[\frac{d^2}{dx^2} + k^2(x)\right]\Psi(x) = 0, \tag{4.10}$$

in the form

$$\Psi_{\text{app}}(x) = A(x)\phi(\xi(x)). \tag{4.11}$$

In order for the function Ψ_{app} to be a good approximation to $\Psi(x)$, the WKB solutions of Eqs. (4.9) and (4.10),

$$\Psi_{\text{WKB}}(x) = k(x)^{-1/2}\exp\left(i\int_{x_0}^{x} k(x')dx'\right), \tag{4.12}$$

$$\psi_{\text{WKB}}(\xi) = j(\xi)^{-1/2}\exp\left(i\int_{\xi_0}^{\xi} j(\xi')d\xi'\right), \tag{4.13}$$

should coincide. In other words, the following relations should be satisfied:

$$\int_{x_0}^{x} k(x')dx' = \int_{\xi_0}^{\xi} j(\xi')d\xi' \tag{4.14}$$

and

$$A(x) = \left[\frac{j(\xi)}{k(x)}\right]^{1/2}. \tag{4.15}$$

From these equations we can show that $\Psi_{\text{app}}(x)$ satisfies the equation,

$$\left[\frac{d^2}{dx^2} + k(x)^2 + \gamma(x)\right]\Psi_{\text{app}}(x) = 0, \tag{4.16}$$

where

$$\gamma(x) = -\left(\frac{d\xi}{dx}\right)^{1/2}\frac{d^2}{dx^2}\left(\frac{d\xi}{dx}\right)^{-1/2} = -\left(\frac{k(x)}{j(\xi)}\right)^{1/2}\frac{d^2}{dx^2}\left(\frac{k(x)}{j(\xi)}\right)^{-1/2}. \tag{4.17}$$

Thus $\Psi_{\mathrm{app}}(x)$ can be a good approximation, when the following condition is satisfied:

$$\gamma(x) \ll k(x)^2 \,. \tag{4.18}$$

The important question about this comparison equation method is what "the same analytical structure" means. If we try to obtain the solution for the equation which has a zero at x_0, namely $k(x_0) = 0$, from the exact solution for $j(\xi) = $ constant, $k(x)/j(\xi)$ becomes zero at x_0 and $\gamma(x)$ diverges there. This indicates that we can use the comparison equation method, only when the order and number of zeros are the same. Namely, if the conditions

$$k(x_i) = j(\xi_i) = 0 \qquad (i = 1, 2, \ldots) \tag{4.19}$$

and

$$\int_{x_i}^{x_{i+1}} k(x)dx = \int_{\xi_i}^{\xi_{i+1}} j(\xi)d\xi \tag{4.20}$$

are satisfied, then we can construct the good approximate solution $\Psi_{\mathrm{app}}(x)$ from the solution $\phi(\xi)$ of the standard equation (4.9). Then the next natural question is how to solve the standard equation when the number and order of zeros are given.

4.2. Stokes Phenomenon

The very basic mathematics, i.e. Stokes phenomenon, which underlies the semiclassical theory, is briefly explained in this section by taking the Airy function as an example. The Stokes constant and connection matrix in the case of Weber equation are also provided, since the Weber function is useful in many applications.

Let us first consider the Airy's differential equation,

$$\frac{d^2 w(z)}{dz^2} - h^2 z w(z) = 0 \,, \tag{4.21}$$

where h is a certain large positive parameter. The WKB type of asymptotic solutions in the complex z-plane are given by

$$(\cdot, z) = z^{-1/4} \exp\left(h \int^{z} z^{1/2} dz \right) = z^{-1/4} \exp\left(\frac{2}{3} h z^{3/2} \right) \tag{4.22}$$

and

$$(z, \cdot) = z^{-1/4} \exp\left(-h \int^z z^{1/2} dz\right) = z^{-1/4} \exp\left(-\frac{2}{3} h z^{3/2}\right). \qquad (4.23)$$

These are called standard WKB solutions. A general solution in the asymptotic region is, of course, given by a linear combination of these two solutions,

$$w(z) \simeq A(\cdot, z) + B(z, \cdot), \qquad (4.24)$$

where A and B are arbitrary constants. Can this be a single-valued function in the whole asymptotic region of complex-z plane with the same coefficients A and B? Answer is no. This is obvious, since Eq. (4.21) is nothing but the Schrödinger equation with the linear potential $h^2 x$ at zero total energy. Equation (4.24) represents a running wave at $x = \mathrm{Re}(z) < 0$, but includes the unphysical exponentially growing wave in the classically forbidden region $x = \mathrm{Re}(z) > 0$. This suggests that the coefficients should be changed from region to region. What does this mean? The functions (4.22) and (4.23) are multi-valued functions with branch point at $z = 0$. This branch point comes from the zero of the coefficient of Eq. (4.21). This is called transition point of order unity. There emanate two kinds of lines from $z = 0$: One is defined by $\mathrm{Im}(z^{3/2}) = 0$ and is called Stokes line, and the other $(\mathrm{Re}(z^{3/2}) = 0)$ is called anti-Stokes line. Figure 4.2 shows these lines (dashed line for Stokes and solid line for anti-Stokes) together with a branch cut (wave line).

The two solutions (4.22) and (4.23) are approximate ones; and one of them is exponentially large (dominant), while the other is exponentially small (subdominant). Thus the coefficient of the subdominant solution is affected by the error of the dominant solution. This effect can be taken into account by assigning a certain constant T (Stokes constant) to each Stokes line, acrossing which the coefficient A of the subdominant solution is changed to $A + BT$ with B the coefficient of the dominant solution. Stokes constant is assigned to Stokes line, because the dominant (subdominant) solution becomes most dominant (subdominant) there. Thus the asymptotic solutions have to be changed from sector to sector in the complex z-plane. This is called Stokes phenomenon [47, 48]. Acrossing the anti-Stokes line, the dominancy changes, namely dominant (subdominant) solution becomes subdominant (dominant). When we cross the branch cut in counter clockwise, the solution $(\cdot, z)[(z, \cdot)]$ changes to $-i(z, \cdot)[-i(\cdot, z)]$. According to these rules, the solution given by Eq. (4.24) in

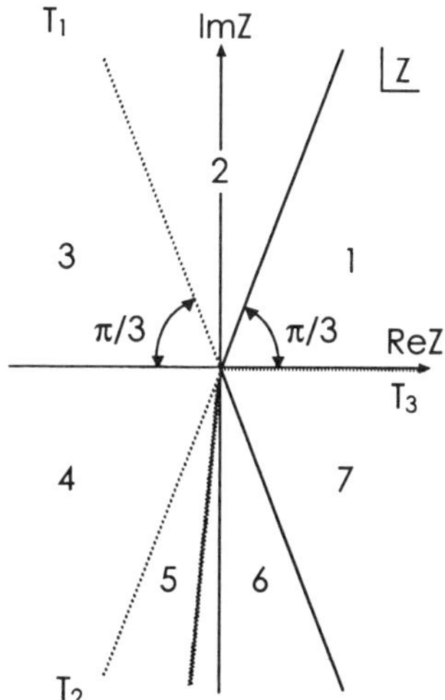

Fig. 4.2.   Stokes and anti-Stokes lines in the case of Airy function. - - -: Stokes line, ———: anti-Stokes line. (Taken from Ref. [9] with permission.)

region 1 of Fig. 4.2 changes as follows:

$$\text{region } 1: \qquad A(\cdot, z) + B(z, \cdot)_s$$

$$\text{region } 2: \qquad A(\cdot, z) + B(z, \cdot)_d$$

$$\text{region } 3: \qquad (A + BT_1)(\cdot, z)_s + B(z, \cdot)_d$$

$$\text{region } 4: \qquad (A + BT_1)(\cdot, z)_d + B(z, \cdot)_s$$

$$\text{region } 5: \quad (A + BT_1)(\cdot, z)_d + [B + T_2(A + BT_1)](z, \cdot)_s \qquad (4.25)$$

$$\text{region } 6: -i(A + BT_1)(z, \cdot)_d - i[B + T_2(A + BT_1)](\cdot, z)_s$$

$$\text{region } 7: -i(A + BT_1)(z, \cdot)_s - i[B + T_2(A + BT_1)](\cdot, z)_d$$

$$\text{region } 1: \quad -i\{(A + BT_1) + T_3[B + T_2(A + BT_1)]\}(z, \cdot)_s$$
$$-i[B + T_2(A + BT_1)](\cdot, z)_d \,.$$

Since the last equation of Eq. (4.25) should coincide with the first one of Eq. (4.25) for arbritary A and B, we can easily obtain

$$T_1 = T_2 = T_3 = i\,. \qquad (4.26)$$

Now, we have the single-valued general solution valid in the whole asymptotic region. For instance, we can obtain the following connection formula for the physical solution on the real axis:

$$(\cdot,x) - i(x,\cdot) \quad (x<0) \longleftrightarrow -i(x,\cdot) \quad (x>0) \,. \tag{4.27}$$

The ordinary Airy function $A_i(z)$ corresponds to this solution with $A = 0$. Equation (4.27) represents the famous connection formula of the WKB solutions acrossing the turning point (see Eq. (4.5)). As can now be easily understood, once we know all Stokes constants, the connections among asymptonic solutions are known and thus the physical quantities such as scattering matrix can be derived. However, the Airy function is exceptionally simple and Stokes constants are generally not known except for some special cases [49]. When the coefficient of differential equation is the nth order polynomial, the $(n+2)$ Stokes lines run radially in the asymptotic region. There are thus $(n+2)$ unknown Stokes constants, but only three independent conditions are obtained from the single-valuedness as demonstrated in the case of Airy function.

Let us next consider the Weber equation,

$$\frac{d^2w(z)}{dz^2} + h^2(z^2 - \epsilon^2)w(z) = 0 \,, \tag{4.28}$$

which presents a very basic model in quantum mechanics. There are two first order transition points $z = \pm\epsilon$ and four Stokes lines in the asymptotic region as is shown in Fig. 4.3. Here, expressions of the Stokes constant and

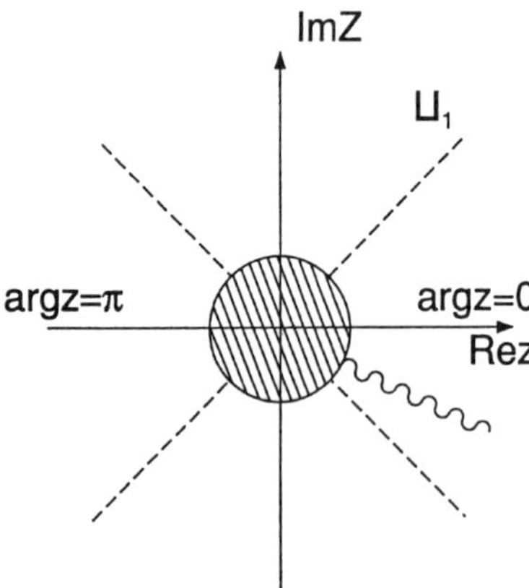

Fig. 4.3.   Stokes and anti-Stokes lines in the asymptotic region in the case of Weber function. - - -: Stokes line, ———: anti-Stokes line, ∿∿: branch cut. (Taken from Ref. [9] with permission.)

the connection matrices are provided without the details of derivation. More details can be found in [47, 49, 50].

Asymptotic solutions of Eq. (4.28) on $\arg z = 0$ and π are expressed in terms of the ordinary WKB functions as

$$w(z) \xrightarrow[z \to +\infty]{} Aq^{-1/4}(z) \exp\left[i \int_0^z q^{1/2}(z)dz\right]$$
$$+ Bq^{-1/4}(z) \exp\left[-i \int_0^z q^{1/2}(z)dz\right] \tag{4.29}$$

and

$$w(z) \xrightarrow[z \to -\infty]{} Cq^{-1/4}(z) \exp\left[i \int_0^z q^{1/2}(z)dz\right]$$
$$+ Dq^{-1/4}(z) \exp\left[-i \int_0^z q^{1/2}(z)dz\right], \tag{4.30}$$

where

$$q(z) = h^2(z^2 - \epsilon^2). \tag{4.31}$$

The standard WKB solutions are

$$(\,\cdot\,,z) = z^{-1/2}e^{iP_W(z)} \tag{4.32}$$

and

$$(z,\,\cdot\,) = z^{-1/2}e^{-iP_W(z)} \tag{4.33}$$

with

$$P_W(z) = \frac{1}{2}h(z^2 - \epsilon \ln z). \tag{4.34}$$

The asymptotic solutions Eqs. (4.29) and (4.30) are rewritten as

$$\phi_W(z) \xrightarrow[z \to +\infty]{} A'(\,\cdot\,,z) + B'(z,\,\cdot\,) \tag{4.35}$$

and

$$\phi_W(z) \xrightarrow[z \to -\infty]{} C'(\,\cdot\,,z) + D'(z,\,\cdot\,). \tag{4.36}$$

The primed coefficients are connected by

$$\begin{pmatrix} C' \\ D' \end{pmatrix} = \begin{pmatrix} 1 & U \\ -(1 + e^{2\pi\beta})/U & e^{-2\pi\beta} \end{pmatrix} \begin{pmatrix} A' \\ B' \end{pmatrix}, \tag{4.37}$$

where U is the Stokes constant on the Stokes line $\arg z = \pi/4$ which is given by

$$U = i \frac{\sqrt{2\pi}}{\Gamma[(1/2) - i\beta]} e^{-\pi\beta/2} e^{-i\beta \ln(2h)} \tag{4.38}$$

and

$$\beta = \frac{1}{2} h\epsilon^2 . \tag{4.39}$$

Since the phase parts in Eqs. (4.29)–(4.30) and (4.35)–(4.36) are related as

$$\pm i \int_0^z q^{1/2}(z)dz \xrightarrow[z \to \pm\infty]{} \pm i P_W(z) + \delta_\pm^W , \tag{4.40}$$

where

$$\delta_+^W = ih \left(-\frac{1}{4}\epsilon^2 - \frac{1}{2}\ln 2 + \frac{1}{2}\epsilon^2 \ln \sqrt{e^{\pi i}\epsilon^2} \right) \tag{4.41}$$

and

$$\delta_-^W = -\delta_+^W + \frac{\pi}{2}h\epsilon^2 , \tag{4.42}$$

we can finally obtain the connection matrix between the coefficients in Eqs. (4.29) and (4.30) as

$$F(\beta) = \begin{pmatrix} \dfrac{\sqrt{2\pi}\exp(\pi/2\beta - i\beta + i\beta\ln(e^{\pi i}\beta))}{\Gamma(1/2 + i\beta)} & ie^{\pi\beta} \\[2em] -ie^{\pi\beta} & \dfrac{\sqrt{2\pi}\exp[\pi/2\beta + i\beta - i\beta\ln(e^{\pi i}\beta)]}{\Gamma(1/2 - i\beta)} \end{pmatrix}, \tag{4.43}$$

with

$$\begin{pmatrix} C \\ D \end{pmatrix} = F(\beta) \begin{pmatrix} A \\ B \end{pmatrix}. \tag{4.44}$$

Next, let us consider Eq. (4.28) with h replaced by $ih'(h' > 0)$ and the connection matrix along the Stokes lines $\arg z = \pm\pi/2$. The phase integrals in Eqs. (4.29) and (4.30) are replaced by

$$\pm i \int_0^z q^{1/2}(z)dz \longrightarrow \mp \int_0^z q^{1/2}(z)dz\,, \tag{4.45}$$

where $q(z)$ on the right side is defined also by Eq. (4.31) with h replaced by h'. Using the same procedure as above, one can obtain the connection matrix $H(\beta)$ as

$$H(\beta) = \begin{pmatrix} \dfrac{\sqrt{2\pi}\cos(\pi\beta)}{2\Gamma(1/2 - \beta)}e^{\beta - \beta\ln\beta} & -\sin(\pi\beta) \\[2em] \sin(\pi\beta) & \dfrac{\sqrt{2\pi}}{\Gamma(1/2 + \alpha)}e^{-\beta + \beta\ln\beta} \end{pmatrix} \tag{4.46}$$

with

$$\begin{pmatrix} C \\ D \end{pmatrix} = H(\beta)\begin{pmatrix} A \\ B \end{pmatrix}, \tag{4.47}$$

where β is given by Eq. (4.39) with h replaced by h'.

Chapter 5

Basic Two-State Theory for Time-Independent Processes

5.1.   Exact Solutions of the Linear Curve Crossing Problems

The most fundamental quantum mechanical model of curve crossing is the linear potential model (in coordinate R), in which the two diabatic crossing potentials $V_1(R)$ and $V_2(R)$ are linear functions of R and the diabatic coupling $V(R)$ is constant ($\equiv A$). The basic coupled Schrödinger equations are

$$-\frac{\hbar^2}{2\mu}\frac{d^2\varphi_1^{(d)}}{dR^2} + [V_1(R) - (E - E_X)]\varphi_1^{(d)} = A\varphi_2^{(d)} \tag{5.1}$$

and

$$-\frac{\hbar^2}{2\mu}\frac{d^2\varphi_2^{(d)}}{dR^2} + [V_2(R) - (E - E_X)]\varphi_2^{(d)} = A\varphi_1^{(d)}, \tag{5.2}$$

where

$$V_j(R) = -F_j(R - R_X) \qquad (j = 1, 2). \tag{5.3}$$

Without loss of generality, it is assumed that $F_1 > 0$ and $F_1 > F_2$ as before.

5.1.1.   *Landau–Zener type*

The Landau–Zener (LZ) case (see Fig. 2.1(a)), in which the two diabatic potentials cross with the same sign of slopes, corresponds to $F_1 F_2 > 0$.

Transformations

$$\varphi_j^{(d)} = \frac{1}{2\pi} \int_{-\infty}^{\infty} u_j(k) e^{ikR} dR, \qquad (j = 1, 2), \tag{5.4}$$

$$u_j(k) = \frac{2}{|f_j|} A_j(k) \exp\left[\frac{i}{f_j}\left(\epsilon k - \frac{k^3}{3}\right)\right], \qquad (j = 1, 2), \tag{5.5}$$

and

$$B(\xi) = A_1(k) \exp\left[\frac{i}{2}\left(\frac{a^2 \xi^3}{3} - b^2 \xi\right)\right] \tag{5.6}$$

lead to

$$\frac{d^2 B(\xi)}{d\xi^2} + q(\xi) B(\xi) = 0 \tag{5.7}$$

with

$$q(\xi) = \frac{1}{4} - ia^2\xi + \frac{1}{4}(a^2\xi^2 - b^2)^2, \tag{5.8}$$

where

$$\epsilon = \frac{2\mu}{\hbar^2}(E - E_X), \qquad f_j = \frac{2\mu}{\hbar^2} F_j, \qquad \alpha = \frac{2\mu A}{\hbar^2}, \qquad \xi = \frac{2\alpha k}{f}, \tag{5.9}$$

$$a^2 = \frac{f(f_1 - f_2)}{8\alpha^3}, \qquad b^2 = \frac{\epsilon(f_1 - f_2)}{2\alpha f}, \tag{5.10}$$

and

$$f = (f_1|f_2|)^{1/2}. \tag{5.11}$$

The quantities $a^2(\geq 0)$ and $b^2(-\infty < b^2 < \infty)$ are the same basic parameters as before (see Eq. (3.14)), representing effectively the coupling strength and the collision energy, respectively. The most basic physical quantity is the reduced scattering matrix S^{R} defined from the ordinary scattering matrix S as (see Eq. (3.27))

$$S_{mn} = S_{mn}^{\mathrm{R}} e^{i(\eta_m + \eta_n)}, \tag{5.12}$$

where η_m and η_n are the elastic scattering phase shifts which diverge in the present linear potential model. From the solution of Eq. (5.7) the functions $A_j(k)$ $(j = 1, 2)$ are obtained from Eq. (5.6) and

$$A_2(k) = 2i \exp\left[i\left(\frac{a^2}{3}\xi^3 - b^2\xi\right)\right] \frac{dA_1}{d\xi} . \tag{5.13}$$

Then the reduced scattering matrix is defined by

$$\begin{pmatrix} A_1(\infty) \\ A_2(\infty) \end{pmatrix} = S_{\mathrm{LZ}}^{\mathrm{R}} \begin{pmatrix} A_1(-\infty) \\ A_2(-\infty) \end{pmatrix} . \tag{5.14}$$

Thus the key point is to obtain the connections of the asymptotic solutions of Eq. (5.7) at $\xi \to \pm\infty$. The underlying mathematics to carry this out analytically is the Stokes phenomenon of asymptotic solutions of differential equations, the essence of which was explained briefly in Chap. 4.2 by taking the simple examples of Airy and Weber differential equations. This is a very important mathematics for the general semiclassical theory and thus for various physical phenomena. Recently, Zhu and Nakamura carefully analyzed the Stokes phenomenon of Eq. (5.7) and succeeded in deriving not only the quantum mechanically exact analytical, but also new semiclassical expressions of $S_{\mathrm{LZ}}^{\mathrm{R}}$. The quantum mechanically *exact* solutions of the Landau–Zener (LZ) type are given as follows [51–54]:

$$S_{\mathrm{LZ}}^{\mathrm{R}} = \begin{pmatrix} 1 + U_1 U_2 & -U_2 \\ -U_2 & 1 - U_1^* U_2 \end{pmatrix} \tag{5.15}$$

with

$$U_2 = \frac{2i\,\mathrm{Im}\,U_1}{1 + |U_1|^2} . \tag{5.16}$$

The overall transition probability P_{12} and the nonadibatic transition probability p for one passage of the crossing point are given by

$$P_{12} = |(S_{\mathrm{LZ}}^{\mathrm{R}})_{21}|^2 = 4p(1 - p)\sin^2(\arg U_1) \tag{5.17}$$

and

$$p = \frac{1}{1 + |U_1|^2} , \tag{5.18}$$

where $U_1(a^2, b^2)$ is a complex quantity (Stokes constant) and is given in the form of infinite series [51]. In addition to the unitarity, the reduced scattering matrix satisfies the following properties: $(S_{\mathrm{LZ}}^{\mathrm{R}})_{11} = (S_{\mathrm{LZ}}^{\mathrm{R}})_{22}$ and $(S_{\mathrm{LZ}}^{\mathrm{R}})_{12} = (S_{\mathrm{LZ}}^{\mathrm{R}})_{21} =$ pure imaginary. These properties can be proved from the symmetries of the differential equations satisfied by $A_j(k)$'s. Equation (5.17) is a familiar expression for inelastic scattering probability in the semiclassical theory, but it should be noted that this is *exact*.

Below, the mathematical procedure is briefly outlined for deriving the exact expressions of the Stokes constants U_1 and U_2.

Let us first consider the general second order differential equation with quartic polynomial coefficient function [50, 51],

$$\frac{d^2 w(z)}{dz^2} + q(z)w(z) = 0 , \tag{5.19}$$

where

$$q(z) = a_4 z^4 + a_2 z^2 + a_1 z + a_0 , \qquad (a_4 > 0) . \tag{5.20}$$

The standard WKB solutions are given by (see Eqs. (4.22) and (4.23))

$$(\cdot, z) = z^{-1} e^{ip(z)} \tag{5.21}$$

and

$$(z, \cdot) = z^{-1} e^{-ip(z)} \tag{5.22}$$

with

$$p(z) = \frac{i}{3}\sqrt{a_4}\, z^3 + \frac{i}{2\sqrt{a_4}} a_2 z + \frac{i}{2\sqrt{a_4}} a_1 \ln(z) . \tag{5.23}$$

For the later convenience, the following variable transformations are introduced according to Hinton [55]:

$$\zeta = \frac{2}{3}(-a_4)^{1/2} z^3 \quad \text{and} \quad w(z) = \frac{u(\zeta)}{z} . \tag{5.24}$$

Then Eq. (5.19) becomes

$$\frac{d^2 u(\zeta)}{d\zeta^2} + Q(\zeta)u(\zeta) = 0 \tag{5.25}$$

with

$$Q(\zeta) = -\frac{1}{4} + Q_0\zeta^{-2/3} + Q_1\zeta^{-1} + Q_2\zeta^{-4/3} + Q_4\zeta^{-2} . \qquad (5.26)$$

There are six Stokes lines in the asymptotic region of z as is shown in Fig. 5.1. In ζ-plane they coincide with the real axis, since $\arg\zeta = 3\arg z + \pi/2$.

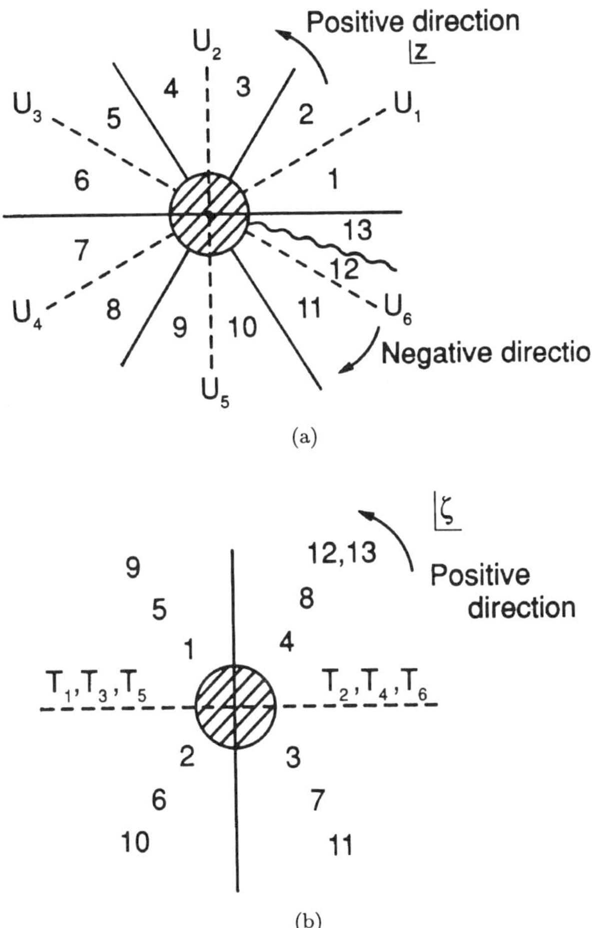

Fig. 5.1. Stokes lines and anti-Stokes lines in the case of Eqs. (5.19) and (5.20). $\zeta = 3\arg z + \pi/2$. - - -: Stokes line, ———: anti-Stokes line, $\sim\!\sim$: branch cut. (Taken from Ref. [50] with permission.)

From the transformation Eq. (5.24), we can obtain the simple relations between the Stokes constant U_j in the z-plane and the Stokes constant T_j in the ζ-plane for $j = 1 \sim 6$. Using the symmetry of the differential equation (5.25) with Eq. (5.26), we can derive the inter-relations among T_j's as

$$T_{j+1} = \bar{T}_j e^{-2\pi i Q_1}, \qquad (j = 1, 3, 5) \tag{5.27}$$

and

$$T_{j+1} = \bar{T}_j e^{2\pi i Q_1}, \qquad (j = 2, 4), \tag{5.28}$$

where "-" means the following transformation:

$$\zeta \to \zeta e^{-\pi i}, \qquad Q_0 \to \omega^2 Q_0, \qquad Q_1 \to \omega^3 Q_1 \quad \text{and} \quad Q_2 \to \omega^4 Q_2 \tag{5.29}$$

with $\omega = e^{-\pi i/3}$. Finally, by transforming Eq. (5.25) into coupled integral equations and utilizing the asymptotic expansion of incomplete Gamma function, the analytical expression of T_1 is obtained in the form of a convergent infinite series. This enables us to derive analytical expressions of all the Stokes constants T_j's.

The mathematical procedures for deriving the expressions of T_j's can be found in Ref. [50] and are not given here. Here, only the final expression of T_1 is provided. The Stokes constant U_1 necessary to obtain the reduced scattering matrix given by Eq. (5.15) is related to T_1 as follows:

$$U_1 = U_1(a_4, Q_0, Q_1, Q_2) = T_1(Q_0, Q_1, Q_2)(2i\sqrt{a_4}/3)^{-2Q_1}, \tag{5.30}$$

where the quantities $Q_0, Q_1,$ and Q_2 (see Eq. (5.26)) are defined by

$$Q_0 = -a_2(12a_4)^{-2/3} e^{i\pi/3}, \tag{5.31}$$

$$Q_1 = -a_1(36a_4)^{-1/2} e^{i\pi/2}, \tag{5.32}$$

$$Q_2 = -a_0(18^2 a_4)^{-1/3} e^{2i\pi/3}. \tag{5.33}$$

The coefficient T_1 is given in the form of a convergent infinite series,

$$T_1 = -2i\pi e^{2i\pi Q_1} \sum_{s=0}^{\infty} \sum_{n=0}^{\infty} \frac{(6Q_0)^n}{n!} \frac{\Delta_s e^{i\pi(n-s-3)3}}{\Gamma(-2Q_1 - (n-s-3)/3)}, \tag{5.34}$$

where

$$\Delta_s = \sum_{n+m=s} B^{(1)}_{n+1}\beta^{(1)}_m, \qquad s \geq 0, \tag{5.35}$$

$$\beta^{(1)}_n = -\frac{3}{n} \sum_{p+q=n-1} B^{(2)}_{p+1}\alpha^{(1)}_{q+1}, \qquad n \geq 1, \tag{5.36}$$

$$\alpha^{(1)}_{n+3} = - \sum_{p+q=n} \Delta_p W_{pq}, \qquad n \geq 0, \tag{5.37}$$

$$W_{pq} = \sum_{m=[q/3]}^{[p/2]} \Theta(3m-q) \sum_{n=3m-q}^{m} \frac{m!}{n!(m-n)!} \frac{\Gamma(2Q_1 - p/3)(6Q_0)^{3m-q}}{\Gamma[2Q_1 - p/3 - (m-n)]}, \tag{5.38}$$

and

$$C_{nr} = \begin{cases} \delta_{n0} & \text{for } r = 0, \\ \sum_{s=0}^{r}(-1)^{r-s}\dfrac{\Gamma(s/3+1)}{s!(r-s)!\Gamma(s/3-n+1)} & \text{for } r = 1 \end{cases} \tag{5.39}$$

with $\beta^{(1)}_0 = 1$ and $\alpha^{(1)}_1 = \alpha^{(1)}_2 = 0$. The coefficients $B^{(1)}_n$ and $B^{(2)}_n$ in Eqs. (5.35) and (5.36) are defined by

$$B^{(1)}_n = \sum_{k=0}^{n-1} B_{n-k}T_k(d_1, d_2), \qquad n \geq 1, \tag{5.40}$$

$$B^{(2)}_n = \sum_{k=0}^{n-1} B_{n-k}T_k(-d_1, -d_2), \qquad n \geq 1, \tag{5.41}$$

where

$$T_k(d_1, d_2) = \sum_{n=[k/2]}^{k} \Theta(2n-k)\frac{(6d_1)^{2n-k}(3d_2)^{k-n}}{(2n-k)!(k-n)!}, \tag{5.42}$$

$$d_1 = -Q_0^2 - Q_2, \tag{5.43}$$

$$d_2 = -2Q_0Q_1. \tag{5.44}$$

The sequence B_n in Eqs. (5.40) and (5.41) is given by

$$B_n = v_n + \sigma_n, \qquad n \geq 1, \tag{5.45}$$

where v_n is obtained from the recurrence relations,

$$v_1^2 - v_1 + P_1 = 0, \tag{5.46}$$

$$v_2 = P_2/(4/3 - 2v_1), \tag{5.47}$$

$$v_n = \frac{\left(\displaystyle\sum_{m=2}^{n-1} v_m v_{n+1-m} + P_n\right)}{[(n+2)/3 - 2v_1]}, \qquad n \geq 3 \tag{5.48}$$

with

$$
\begin{aligned}
P_1 &= 2/9 + Q_1^2 + 2Q_0(Q_0^2 + Q_2), \\
P_2 &= 4Q_0^2 Q_1 + 2Q_1(Q_0^2 + Q_2), \\
P_3 &= 4Q_0 Q_1^2 + (Q_0^2 + Q_2)^2, \\
P_4 &= 4Q_0 Q_1(Q_0^2 + Q_2), \\
P_5 &= 4Q_0^2 Q_1^2, \\
P_n &= 0, \qquad n \geq 6.
\end{aligned}
\tag{5.49}
$$

The constants σ_n are also given by the recurrence relations,

$$
\begin{aligned}
\sigma_1 &= \sigma_2 = 0, \\
\sigma_3 &= 2Q_0/3, \\
\sigma_4 &= Q_1, \\
\sigma_5 &= 4(Q_0^2 + Q_2)/3 + 2Q_0\sigma_3, \\
\sigma_6 &= 10Q_0 Q_1/3 + 2Q_0\sigma_4 + 2Q_1\sigma_3, \\
\sigma_7 &= 2Q_0\sigma_5 + 2Q_1\sigma_4 + 2(Q_0^2 + Q_2)\sigma_3, \\
\sigma_n &= 2Q_0\sigma_{n-2} + 2Q_1\sigma_{n-3} + 2(Q_0^2 + Q_2)\sigma_{n-4} + 4Q_0 Q_1\sigma_{n-5}, \qquad n \geq 8.
\end{aligned}
\tag{5.50}
$$

The various notations appearing in Eqs. (5.38)–(5.42) have the following meanings: $\Gamma(x)$ is the Gamma function, $[x]$ means the largest integer not larger than x, and $\Theta(x)$ is the step function, i.e., $\Theta(x) = 0$ when $x < 0$ and $\Theta(x) = 1$ when

$x > 0$. Although there are two roots for v_1, as is easily seen from Eq. (5.46), we can select the one with smaller real part or any one of the two, if the two roots have the same real part. In the above summations and recurrence relations, we can easily see that all of them are given in the form of finite series except Eq. (5.34). It should also be noted that a direct recurrence relation for Δ_s can be obtained from Eqs. (5.36)–(5.38) as follows:

$$\Delta_s = B_{s+1}^{(1)} - V_{(s-2)2}\Delta_0 W_{00} - \Delta_0 \sum_{n=1}^{s-3} V_{(s-1)(n+2)} W_{0n}$$

$$- \sum_{k=1}^{s-3} \left\{ \sum_{n=k}^{s-3} V_{(s-1)(n+2)} W_{k(n-k)} \right\} \Delta_k \,, \tag{5.51}$$

where

$$V_{pq} = \begin{cases} -3 \sum_{m=q}^{p} \dfrac{B_{p+1-m}^{(1)} B_{m+1-q}^{(2)}}{m+1} & \text{for } p \geq q \geq 0 \\ 0 & \text{otherwise.} \end{cases} \tag{5.52}$$

Now, we can go back to our linear curve crossing problem. What we have to do is to replace Q_0, Q_1, and Q_2 in Eqs. (5.31)–(5.33) by the following expressions (see Eq. (5.8)):

$$Q_0 = b^2 (9a^2)^{-1/3} e^{i\pi/3}/2 \,, \tag{5.53}$$

$$Q_1 = -1/3 \,, \tag{5.54}$$

$$Q_2 = -(1 + b^4)(9a^2)^{-2/3} e^{2i\pi/3}/4 \,. \tag{5.55}$$

The parameters a^2 and b^2 are defined by Eq. (5.10) and Eq. (5.15) with U_1 given by Eqs. (5.30) and (5.34) provides the exact analytical expression of the reduced scattering matrix which can, of course, cover the whole ranges of the two parameters a^2 and b^2, i.e. the whole ranges of energy and coupling strength. One big drawback is, however, that the analytical expression of U_1 is quite cumbersome and not very transparent with respect to the dependencies on a^2 and b^2. Numerical computations confirm that this expression surely gives the exact results, and the infinite series of Eq. (5.34) converges reasonably fast in the region $|b^2/a^2| \leq 1$. In order to give a rough idea about the behaviour of U_1, $|U_1|$ and $\arg U_1$ are shown in Figs. 5.2 as a function of b^2 for some values of a^2.

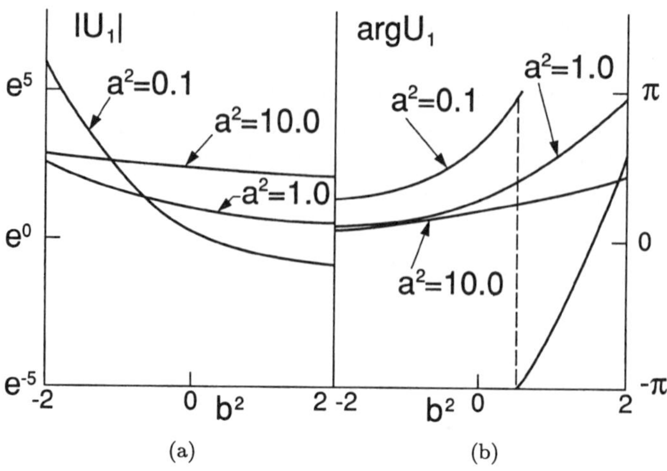

Fig. 5.2. (a) Modulus and (b) argument of the Stokes constant U_1 in the Landau–Zener case. (Taken from Ref. [51] with permission.)

5.1.2. *Nonadiabatic tunneling type*

In the nonadiabatic tunneling (NT) case, the two diabatic potentials cross with different signs of slopes and thus $F_1 F_2 < 0$ (see Fig. 2.1(b)). This linear NT-case presents a rather unique model and even the numerical solution requires a special consideration as is explained in Ref. [56]. With the transformation

$$B(\xi) = A_1(k) \exp\left[-\frac{i}{2}\left(\frac{a^2\xi^3}{3} - b^2\xi\right)\right] \qquad (5.56)$$

together with Eqs. (5.4) and (5.5), we obtain Eq. (5.7) with

$$q(\xi) = -\frac{1}{4} + ia^2\xi + \frac{1}{4}(a^2\xi^2 - b^2)^2, \qquad (5.57)$$

where the parameters $\epsilon, f_j, \alpha, \xi, a^2, b^2$, and f are the same as before (see Eqs. (5.9)–(5.11). In the NT-case $b^2 = 1(-1)$ corresponds to the bottom (top) of the upper (lower) adiabatic potential (see Fig. 2.1(b)).

The reduced scattering matrix $S_{\rm NT}^{\rm R}$ can be defined in the same way as before (see Eq. (5.12)). The physical meaning of the matrix elements is quite different from the LZ-type, though. The off-diagonal (diagonal) elements represent transmission (reflection). From the solutions of Eq. (5.7) the function

$A_j(k)(j = 1, 2)$ are obtained by Eq. (5.56) and

$$A_2(k) = 2i \exp\left[-i\left(\frac{a^2}{3}\xi^3 - b^2\xi\right)\right]\frac{dA_1}{d\xi}.$$ (5.58)

The reduced scattreing matrix is defined by

$$\begin{pmatrix} A_1(\infty) \\ A_2(-\infty) \end{pmatrix} = S_{\mathrm{NT}}^{\mathrm{R}}\begin{pmatrix} A_1(-\infty) \\ A_2(\infty) \end{pmatrix}.$$ (5.59)

It is interesting to note that the apparent small differences in $q(\xi)$ of Eqs. (5.8) and (5.57) between the LZ- and NT-cases make a big difference not only mathematically but also physically. The physical differences are rather obvious, as was pointed out already in Chap. 2, and the mathematical differences will be shown later.

As in the LZ-case, Zhu and Nakamura successfully derived not only quantum mechanically *exact* but also new semiclassical compact expressions of $S_{\mathrm{NT}}^{\mathrm{R}}$ [51, 53, 54]. The exact solutions are given as follows:

$$S_{\mathrm{NT}}^{\mathrm{R}} = \frac{1}{1 + U_1 U_2}\begin{pmatrix} 1 & U_2 \\ U_2 & 1 \end{pmatrix}$$ (5.60)

with

$$U_2 = \frac{2i\,\mathrm{Im}(U_1)}{|U_1|^2 - 1}.$$ (5.61)

The overall transmission probability P_{12} and the nonadiabatic transition probability p for one passage of the crossing point are given by

$$P_{12} = |(S_{\mathrm{NT}}^{\mathrm{R}})_{21}|^2 = \frac{4(\mathrm{Im}\,U_1)^2}{(|U_1|^2 - 1)^2 + 4(\mathrm{Im}\,U_1)^2}$$ (5.62)

$$= \frac{2\sin^2(\arg U_1)}{2\sin^2(\arg U_1) + p^2/[2(1-p)]}, \qquad \text{for } b^2 \geq 1$$ (5.63)

and

$$p = 1 - |U_1|^2, \qquad \text{for } b^2 \geq 1,$$ (5.64)

where $U_1(a^2, b^2)$ is also a complex quantity (Stokes constant) given by an infinite series [51]. It should be noted that p has the physical meaning of nonadiabatic transition for one passage of crossing point only at energies higher than the bottom of the upper adiabatic potential. In the same way as in the LZ-type, the reduced scattering matrix can be proved to satisfy the symmetry relations: $(S_{NT}^R)_{11} = (S_{NT}^R)_{22}$ and $(S_{NT}^R)_{12} = (S_{NT}^R)_{21}$. It should be noted that Eq. (5.63) has also the similar form as that in the semiclassical theory (see Eq. (3.31)); but this is *quantum mechanically exact*. It should also be noted that the same notations U_1 and p as in the LZ-type are used, but they are naturally different. It is further noted that Eq. (5.62) is valid at any energy, but Eq. (5.63) is valid only at $b^2 \geq 1$. It is very interesting to note that complete reflection ($P_{12} = 0$) occurs when

$$\arg U_1 = l\pi, \qquad (l = 0, 1, \ldots). \tag{5.65}$$

is satisfied. This suggests an interesting theoretical possibility of molecular switching and control of molecular processes [57–59]. Although the infinite series for U_1 are quite complicated, they have provided us with a unique opportunity to investigate and improve the validity of various semiclassical solutions.

The exact analytical expression of the Stokes constant U_1 is given by Eqs. (5.30)–(5.55) in the same way as in the LZ-type. Only difference is that we have to insert the follwoing expressions for Q_0, Q_1, and Q_2 into these

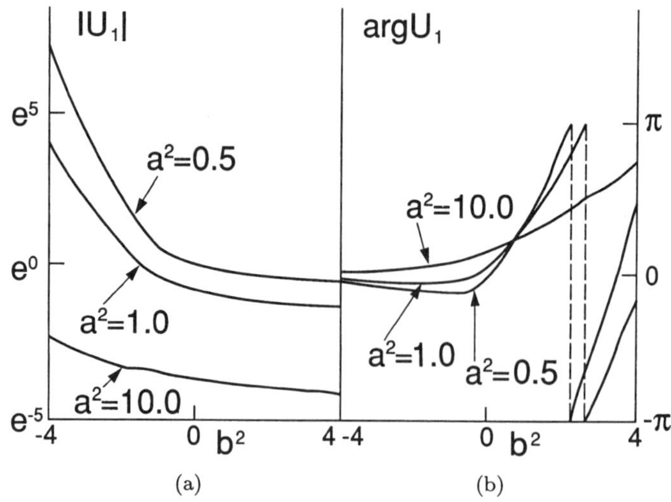

Fig. 5.3.   (a) Modulus ad (b) argument of the Stokes constant U_1 in the NT-case. (Taken from Ref. [51] with permission.)

equations:

$$Q_0 = \frac{1}{2}b^2(9a^2)^{-1/3}e^{i\pi/3}, \tag{5.66}$$

$$Q_1 = \frac{1}{3}, \tag{5.67}$$

$$Q_2 = \frac{1}{4}(1 - b^4)(9a^2)^{-2/3}e^{2i\pi/3}. \tag{5.68}$$

Some numerical values are presented in Fig. 5.3 for $|U_1|$ and $\arg U_1$ as a function of b^2 for some values of a^2.

5.2. Complete Semiclassical Solutions of General Curve Crossing Problems

On the basis of the exact analytical solutions of linear potential models obtained in the previous section, we can derive compact and yet accurate analytical solutions for the linear potential models [52–54, 60], and furthermore practically useful compact expressions of scattering matrices for *general* two-state curve crossing problems can be finally derived [61–63]. These are available for both LZ (Landau–Zener) and NT (Nonadiabatic Tunneling) cases of general curved potentials, whatever the coupling strength and the energy are. These do not require any sophisticated calculus such as complex contour integrals like in Eq. (3.22) and complicated special functions, and thus are very convenient even for general non-specialist users. Furthermore, the compact and accurate formulas have been derived for the Landau–Zener transition probability p. These are as simple as, and yet far better than the famous Landau–Zener formula $p_{\text{LZ}}^{(0)}$ of Eq. (3.4). They are simply expressed in terms of the basic parameters a^2 and b^2 (see Eqs. (3.14) and (5.10)) for $E \geq E_X$ ($E \geq E_b$) in the LZ(NT) case (see Figs. 2.1), but are still applicable to general curved potentials. At $E \leq E_X$ in the LZ-case another compact and accurate formula has been obtained for the Landau–Zener transition probability. At $E \leq E_b$ in the NT-case the compact and accurate expression has been derived for the overall nonadiabatic tunneling (transmission) probability. It should also be noted that all necessary phases are expressed in simple and compact forms.

Finally, the basic parameters a^2 and b^2 can be re-expressed only in terms of the corresponding adiabatic potentials; thus non-unique diabatization procedure is not required anymore, when only adiabatic potentials are available. These expressions actually give better results than the original definitions in terms of the diabatic potentials. Thus the theory does not require any information about the non-adiabatic couplings. This new semiclassical theory of very nice character outlined above is summarized in this section. This presents a complete set of solutions for the general two-state curve crossing problems after more than 60 years since the pioneering works done by Landau [1], Zener [2] and Stueckelberg [3]. The final recommended formulas are summarized in Appendix A. Some empirical corrections which are not given here are introduced there so that readers can apply them directly even to any extreme cases, although the original formulas given here are already good enough and work well in a wide range of energy and coupling strength.

5.2.1.   Landau–Zener (LZ) type

As was described in Sec. 5.1.1, once we know the Stokes constant U_1, then we can obtain the reduced scattering matrix for the two-state problem (see Eq. (5.15)). In order to derive the approximate expression of U_1, we need to know, first of all, the distribution of the four transition points in the complex z-plane. The transition points are zero points of the following quartic polynomial, i.e. the coefficient of the basic differential equation (see Eq. (5.8)):

$$z^4 - 2\frac{b^2}{a^2}z^2 - i\frac{4}{a^2}z + \left(\frac{b^2}{a^2}\right)^2 + \frac{1}{a^4} = 0 \,. \tag{5.69}$$

The four zero points t_j $(j = 1 - 4)$ are given as follows:

$$t_{1,2} = \pm x_1 + iy, \qquad t_{3,4} = \mp x_2 - iy \,, \tag{5.70}$$

where

$$x_1^2 = y^2 + \frac{b^2}{a^2} + \frac{1}{a^2y} \qquad (x_1 > 0) \,, \tag{5.71}$$

$$x_2^2 = y^2 + \frac{b^2}{a^2} - \frac{1}{a^2y} \qquad (x_2^2 \colon \text{real}) \,, \tag{5.72}$$

and y is a positive solution (there is only one) of

$$y^6 + \frac{b^2}{a^2}y^4 - \frac{1}{4a^4}y^2 - \frac{1}{4a^4} = 0 \,. \tag{5.73}$$

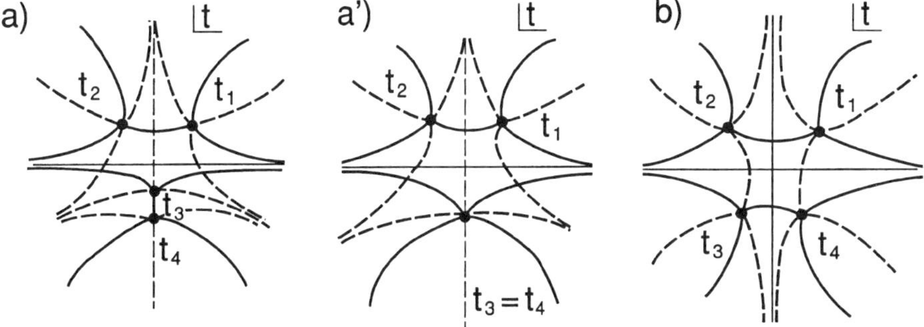

Fig. 5.4.   Distribution of transition points in the LZ-case. (a) $x_2^2 < 0$, (a') $x_2^2 = 0$, (b) $x_2^2 > 0$, - - -: Stokes line, ———: anti-Stokes line. (Taken from Ref. [51] with permission.)

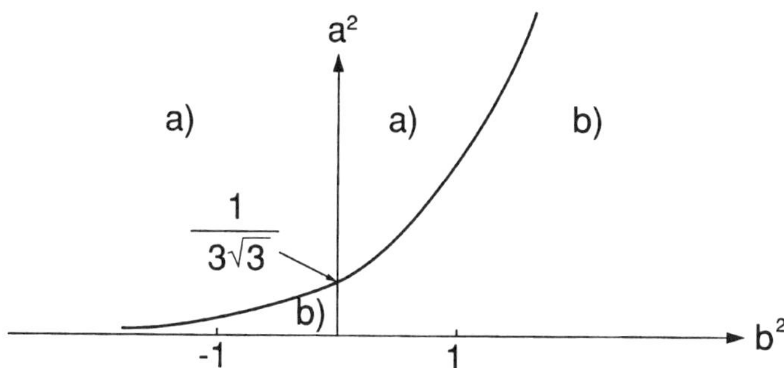

Fig. 5.5.   Regions of $x_2^2 < 0$ and $x_2^2 > 0$ in the (a^2, b^2) plane corresponding to the cases (a) and (b) of Fig. 5.4. (Taken from Ref. [51] with permission.)

The distribution of the transition points is shown in Figs. 5.4 and 5.5. The boundary between the regions (a) and (b) in Fig. 5.5 is

$$a^2 = \left[\frac{2}{3}\sqrt{b^2 + \frac{3}{4}} - \frac{b^2}{3}\right] \Big/ \left[2\sqrt{b^4 + \frac{3}{4}} - 2b^2\right]^2 . \tag{5.74}$$

Next, we consider the two limiting cases in order to derive compact analytical expressions of U_1. The first is the case that the two pairs of transition points (t_1, t_4) and (t_2, t_3) are well isolated from each other along the real axis and can be treated separately. Roughly speaking, this corresponds to $b^2 \gg 1$

in both LZ and NT-cases. The connection matrix L which connects the coefficients of the asymptotic solutions at $z \to \pm\infty$ and is directly related to scattering matrix can be well approximed by

$$L = F_1 F_0 F_2 , \qquad (5.75)$$

where F_j $(j = 1, 2)$ are obtained from the Weber equation with $F_1 (F_2)$ corresponding to $\mathrm{Re}(z) < 0$ (> 0) (see Eq. (4.43)), and F_0 represents the connection between the two pairs of transition points and is given by

$$F_0 = \begin{pmatrix} e^{-i\Phi} & 0 \\ 0 & e^{i\Phi} \end{pmatrix} , \qquad (5.76)$$

where

$$\Phi = \int_{-x_0}^{x_0} q^{1/2}(z)dz \qquad (5.77)$$

and

$$x_0 = \frac{1}{2}(x_1 + x_2) = \frac{1}{\sqrt{2a}}\sqrt{b^2 + \sqrt{b^4 + 1}} . \qquad (5.78)$$

From this procedure we can finally obtain the explicit expressions of U_1 (see Eq. (5.94)).

The second limiting case is that the two pairs (t_1, t_2) and (t_3, t_4) are well separated along the imaginary axis which is a Stokes line. This corresponds roughly to $b^2 \ll -1$ (see the case(a) in Figs. 5.4 and 5.5). The connection matrix G which connects the coefficients of the asymptotic solutions at $z \to \infty e^{\pm\pi i/2}$ can be well approximated as

$$G = H_1 H_0 H_2 , \qquad (5.79)$$

where H_j $(j = 1, 2)$ are again obtained from the Weber equation with H_1 (H_2) corresponding to $\mathrm{Im}\, z < 0$ (> 0) (see Eq. (4.46)), and H_0 represents the connection between the two pairs and is given by

$$H_0 = \begin{pmatrix} e^{-i\bar{\Phi}} & 0 \\ 0 & e^{i\bar{\Phi}} \end{pmatrix} , \qquad (5.80)$$

where

$$\tilde{\Phi} = \int_{-iy}^{iy} q^{1/2}(z)dz \,. \tag{5.81}$$

It should be noted that the above connection is made on the Stokes line and that one-half of the Stokes constants are assigned onto the Stokes line according to Jeffreys [47]. The above procedure leads to the explicit approximate expressionists of U_1 in this limiting case (see Eqs. (5.108) and (5.109)).

The first limiting case has been treated by using basically the same idea by many authors [4–8, 11–14]; but the present analysis is most accurate and thus can give much better results. In the treatments before, t_1 and t_4, and also t_2 and t_3, were assumed to be erroneously symmetric with respect to the real axis, for instance. The second limiting case has never been treated properly before. The formulas obtained in the above analysis can cover very wide range of a^2 and b^2 due to the careful analysis, and can be easily modified to cover practically the whole range of a^2 and b^2.

The reduced scattering matrix in the adiabatic representation is given by (cf. Eq. (5.15))

$$S_{\mathrm{LZ}}^{\mathrm{R}(a)} = \begin{pmatrix} (1 + U_1 U_2)e^{-2i\sigma} & -U_2 \\ -U_2 & (1 - U_1^* U_2)e^{2i\sigma} \end{pmatrix}, \tag{5.82}$$

where U_2 is defined by Eq. (5.16). The phase σ is given below in the subsections. The elastic scattering phases are

$$\eta_1^{(a)} = \lim_{R\to\infty} \left[\int_{t_1}^{R} k_1(R)dR - k_1(R)R + \frac{\pi}{4} \right] \tag{5.83}$$

and

$$\eta_2^{(a)} = \lim_{R\to\infty} \left[\int_{t_2}^{R} k_2(R)dR - k_2(R)R + \frac{\pi}{4} \right], \tag{5.84}$$

where t_j is the turning point on the adiabatic potential $E_j(R)$, (do not confuse with the zero points of Eq. (5.70)) (see Fig. 2.1(a)). The reduced scattering matrix elements are explicitly given as follows:

$$(S_{\mathrm{LZ}}^{\mathrm{R}(a)})_{11} = [p_{\mathrm{ZN}} + (1 - p_{\mathrm{ZN}})e^{2i\psi_{\mathrm{ZN}}}]e^{-2i\sigma_{\mathrm{ZN}}} \,, \tag{5.85}$$

$$(S_{\mathrm{LZ}}^{\mathrm{R}(a)})_{22} = [p_{\mathrm{ZN}} + (1 - p_{\mathrm{ZN}})e^{-2i\psi_{\mathrm{ZN}}}]e^{2i\sigma_{\mathrm{ZN}}} \tag{5.86}$$

and

$$(S_{\mathrm{LZ}}^{\mathrm{R}(a)})_{12} = (S_{\mathrm{LZ}}^{\mathrm{R}(a)})_{21} = -2i\sqrt{p_{\mathrm{ZN}}(1 - p_{\mathrm{ZN}})}\sin\psi_{\mathrm{ZN}}, \qquad (5.87)$$

where p_{ZN} is the Landau–Zener transition probability defined by Eq. (5.64) and

$$\psi_{\mathrm{ZN}} = \arg(U_1). \qquad (5.88)$$

The basic diabatic parameters a^2 and b^2 defined by Eqs. (3.14) and (5.10) can be re-expressed in terms of the adiabatic potentials as given below and thus non-unique diabatization procedure is not required anymore [62].

$$a^2 = \sqrt{d^2 - 1}\,\frac{\hbar^2}{\mu(T_2^{(0)} - T_1^{(0)})^2[E_2(R_0) - E_1(R_0)]} \qquad (5.89)$$

and

$$b^2 = \sqrt{d^2 - 1}\,\frac{E - [E_2(R_0) + E_1(R_0)]/2}{[E_2(R_0) - E_1(R_0)]/2} \qquad (5.90)$$

with

$$d^2 = \frac{[E_2(T_1^{(0)}) - E_1(T_1^{(0)})][E_2(T_2^{(0)}) - E_1(T_2^{(0)})]}{[E_2(R_0) - E_1(R_0)]^2}, \qquad (5.91)$$

where R_0 is the position at which $E_2(R) - E_1(R)$ becomes minimum (see Fig. 2.1(a)),

$$E_2(T_2^{(0)}) = E_1(T_1^{(0)}) = E_X \qquad (5.92)$$

and

$$E_X = \frac{1}{2}[E_2(R_0) + E_1(R_0)]. \qquad (5.93)$$

The formulas thus obtained are two sets: one is valid at $E \geq E_X$ and the other at $E \leq E_X$. These are explicitly presented in the following sub-subsections, although the final recomended formulas which contain empirical corrections and work well in any region of coupling strength and energy are summarized in Appendix A. As given above, the parameters a^2, b^2 and d^2 are well estimated from the adiabatic potentials with use of the reference points $T_j^{(0)}$ ($j = 1, 2$) [62]; the adiabatic phases, however, are found to be better estimated with use of the reference point R_0 which represents the minimum energy separation between the two adiabatic potentials [63]. The expressions given below are those based on this improvement. Only the final results are given in the following subsections. The derivation of these expressions can be found in the original references.

5.2.1.1. $E \geq E_X$ $(b^2 \geq 0)$

From the connection matrices the Stokes constant U_1 is simply given as

$$U_1 = \left(\frac{1}{p_{ZN}} - 1\right)^{1/2} e^{i\psi_{ZN}} \tag{5.94}$$

with

$$p_{ZN} = \exp\left[-\frac{\pi}{4ab}\left(\frac{2}{1+\sqrt{1+b^{-4}}}\right)^{1/2}\right], \tag{5.95}$$

and

$$\psi_{ZN} = \sigma_{ZN} + \phi_S, \tag{5.96}$$

where

$$\phi_S = -\frac{\delta}{\pi} + \frac{\delta}{\pi}\ln\left(\frac{\delta}{\pi}\right) - \arg\Gamma\left(i\frac{\delta}{\pi}\right) - \frac{\pi}{4}. \tag{5.97}$$

The famous Landau–Zener formula is equal to

$$p_{LZ}^{(0)} = \exp\left(-\frac{\pi}{4ab}\right), \tag{5.98}$$

and the formula (5.95) is as simple as, but yet far better than this.

Figure 3.2 demonstrates this fact. Since the exact p can be numerically calculated only in the linear potential model, Fig. 3.2 is the result for that model. The formula (5.95), however, can work well for general curved potentials, since nonadiabatic transition is very much well localized at the avoided crossing point. The phase integrals σ_{ZN} and δ_{ZN} are explicitly expressed as

$$\sigma_{ZN} = \sigma_0 + \int_{t_1}^{R_0} k_1(R)dR - \int_{t_2}^{R_0} k_2(R)dR \tag{5.99}$$

and

$$\delta_{ZN} = \delta_0 = \text{Im}\int_{R_0}^{R_*} [k_1(R) - k_2(R)]dR \simeq \frac{\sqrt{2}\pi}{4\sqrt{a^2}} \frac{F_+}{F_+^2 + F_-^2}, \tag{5.100}$$

where

$$\sigma_0 = \text{Re} \int_{R_0}^{R_*} [k_1(R) - k_2(R)]dR \simeq \frac{\sqrt{2}\pi}{4\sqrt{a^2}} \frac{F_-}{F_+^2 + F_-^2}, \tag{5.101}$$

$$F_\pm = \sqrt{\sqrt{(b^2 + \gamma_1)^2 + \gamma_2} \pm (b^2 + \gamma_1)}$$

$$\qquad + \sqrt{\sqrt{(b^2 - \gamma_1)^2 + \gamma_2} \pm (b^2 - \gamma_1)}, \tag{5.102}$$

$$\gamma_1 = 0.9\sqrt{d^2 - 1}, \tag{5.103}$$

$$\gamma_2 = \frac{7\sqrt{d^2}}{16}, \tag{5.104}$$

and R_0 is the position of minimum energy separation between the two adiabatic potentials and is found to be better as the reference point [63] than the turning points at the crossing energy which were used previously in Ref. [62].

These expressions might look messy at a glance, but actually do not involve any complex calculus and complicated special functions; and they are very convenient for practical applications. The essential points in deriving these expressions are as follows: The linear potential approximation was applied to the complex phase integral, but was minimized by introducing the real reference points R_0 instead of R_X (see Eq. (3.22) and Fig. 2.1(a)). The final expressions of σ_{ZN} and δ_{ZN} are obtained by applying the simple three-point quadrature to the integral. As before, the scattering matrix is expressed as a product of the adiabatic propagation matrices P and the nonadiabatic transition matrices I_X and O_X (see Eqs. (3.16)–(3.20)). The nonadiabatic transition matrix I_X is now given by

$$I_X = \begin{pmatrix} \sqrt{1 - p_{ZN}}\,e^{i(\psi_{ZN} - \sigma_{ZN})} & -\sqrt{p_{ZN}}\,e^{i\sigma_0} \\ \sqrt{p_{ZN}}\,e^{-i\sigma_0} & \sqrt{1 - p_{ZN}}\,e^{-i(\psi_{ZN} - \sigma_{ZN})} \end{pmatrix}. \tag{5.105}$$

The reduced scattering matrix $S_{\text{LZ}}^{\text{R}(a)}$ can also be expressed as

$$S_{\text{LZ}}^{\text{R}(a)} = I^t I \tag{5.106}$$

with

$$I = \begin{pmatrix} \sqrt{1-p_{ZN}}\,e^{i(\psi_{ZN}-\sigma_{ZN})} & -\sqrt{p_{ZN}}\,e^{i\sigma_{ZN}} \\ \sqrt{p_{ZN}}\,e^{-i\sigma_{ZN}} & \sqrt{1-p_{ZN}}\,e^{-i(\psi_{ZN}-\sigma_{ZN})} \end{pmatrix}. \tag{5.107}$$

5.2.1.2. $E \le E_X$ $(b^2 \le 0)$

The Stokes constant in this case is expressed as

$$\mathrm{Re}\,U_1 = \cos(\sigma_{ZN})\left\{ \sqrt{B(\sigma_{ZN}/\pi)}\,e^{\delta_{ZN}} - \sin^2(\sigma_{ZN})\frac{e^{-\delta_{ZN}}}{\sqrt{B(\sigma_{ZN}/\pi)}} \right\} \tag{5.108}$$

and

$$\mathrm{Im}\,U_1 = \sin(\sigma_{ZN})\left\{ B(\sigma_{ZN}/\pi)e^{2\delta_{ZN}} - \sin^2(\sigma_{ZN})\cos^2(\sigma_{ZN})\frac{e^{-2\delta_{ZN}}}{\sqrt{B(\sigma_{ZN}/\pi)}} \right.$$
$$\left. +\, 2\cos^2(\sigma_{ZN}) - 1 \right\}^{1/2}, \tag{5.109}$$

where

$$B(X) = \frac{2\pi X^{2X}}{X\Gamma^2(X)}. \tag{5.110}$$

The Landau–Zener transition probability p_{ZN} and the phase ψ_{ZN} can be explicitly given as

$$p_{ZN} = [1 + B(\sigma_{ZN}/\pi)e^{2\delta_{ZN}} - \sin^2(\sigma_{ZN})]^{-1} \tag{5.111}$$

and

$$\psi_{ZN} = \arg(U_1). \tag{5.112}$$

This is the first result ever obtained for $E \le E_X$ and works well over the whole range of coupling strength, as is demonstrated in Fig. 3.2. As is easily conjectured, this probability includes the effects not only of the nonadiabatic transition represented by σ_{ZN} but also of the tunneling from turning point to avoided crossing point represented by δ_{ZN}. The phase integrals σ_{ZN} and δ_{ZN} are explicitly given as follows:

$$\sigma_{ZN} = \sigma_0 \tag{5.113}$$

and

$$\delta_{\mathrm{ZN}} = \int_{t_1}^{R^0} |k_1(R)|dR - \int_{t_2}^{R^0} |k_2(R)|dR + \delta_0 , \qquad (5.114)$$

where σ_0 and δ_0 are given by Eqs. (5.101) and (5.100). The reduced scattering matrix $S_{\mathrm{LZ}}^{\mathrm{R}(a)}$ is still expressed by Eqs. (5.106) and (5.107) with the various parameters given above.

5.2.1.3.   *Numerical examples*

In order to demonstrate the validity and usefulness of the formulas presented in this section, numerical examples are shown in Figs. 5.6 and 5.7 for the inelastic transition probability $P_{12} = |S_{12}^{\mathrm{R}}|^2$ and the phase $\arg S_{11}$, respectively [9, 62]. The potentials employed in atomic units are

$$V_1(x) = 1.5e^{-\frac{3}{10}Kx} - 0.5, \qquad V_2(x) = e^{-\frac{5K}{K+1}x} \qquad (5.115)$$

and

$$V(x) = \frac{\sqrt{2K}}{\sqrt{K}+1}V_0 e^{-2.5x^2} \quad \text{with} \quad K = 1.2 . \qquad (5.116)$$

The coupling parameters V_0 in Eq. (5.116) is varied so that the parameters a^2 defined by Eq. (3.14) covers the three regimes: 10.0 (weak), 1.0 (intermediate) and 0.1 (strong). The adiabatic approximation means that a^2 and b^2 are estimated from adiabatic potentials by Eqs. (5.89)–(5.91). It should be noted that in these numerical calculations some empirical corrections have been introduced in the expressions of probabilty and phase, as is described in Appendix A, although these corrections play a role only in the regime of very weak or very strong diabatic coupling and in the very close vicinity of E_X. It should also be noted that the intermediate coupling strengh which appears in many practical cases of atomic and molecular processes corresponds to $a^2 \sim 1.0$. The complex phases given by Eqs. (5.100)–(5.104) and (5.113)–(5.114) fail, however, in a very strong coupling regime, $a^2 \lesssim 0.05$. The remedy for this can also be obtained and will be given in Appendix A.

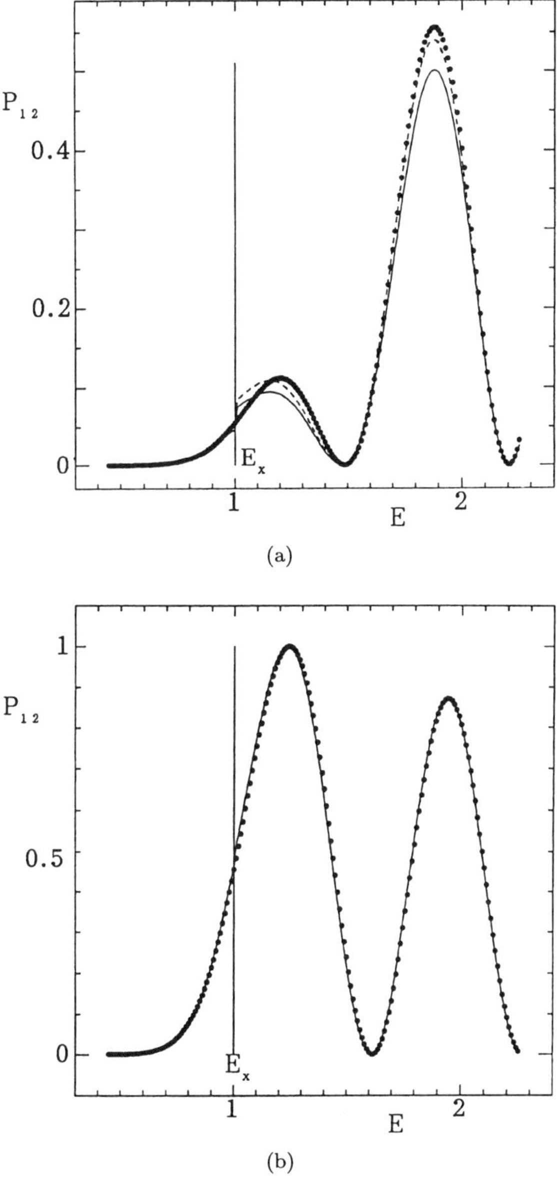

Fig. 5.6. Transition probability $P_{12} = |S_{12}|^2$ versus energy in the LZ case for the potential (5.115) and (5.116). $\cdots$: exact, ——: semiclassical(adiabatic), - - -: semiclassical (diabatic), (a) a^2(dia) = 0.1, (b) $a^2 = 1.0$, (c) $a^2 = 10.0$. (Taken from Ref. [9] with permission.)

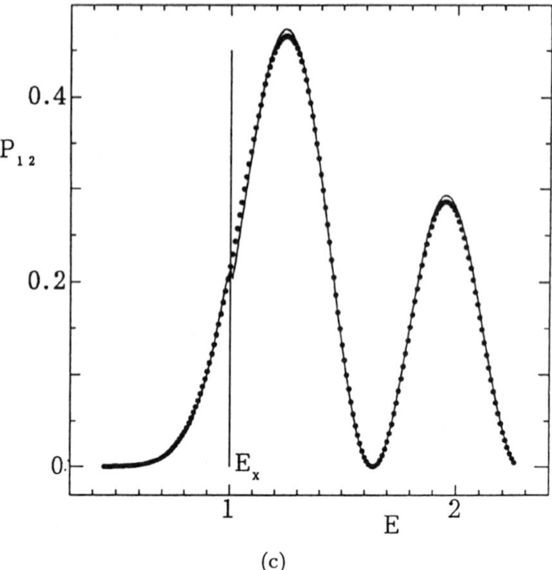

(c)

Fig. 5.6. (*Continued*).

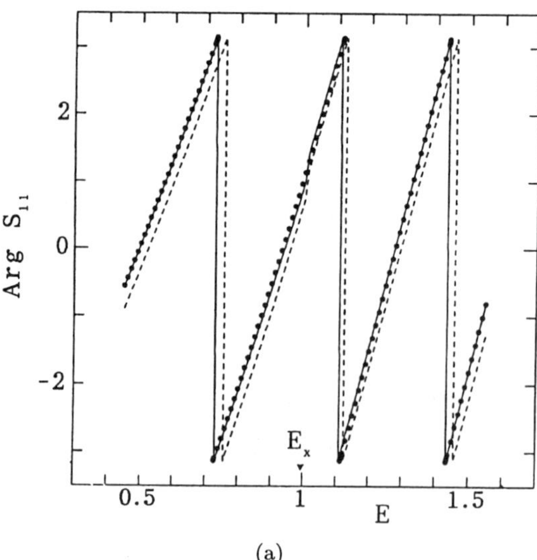

(a)

Fig. 5.7.   The same as Fig. 5.6 except for arg S_{11}.

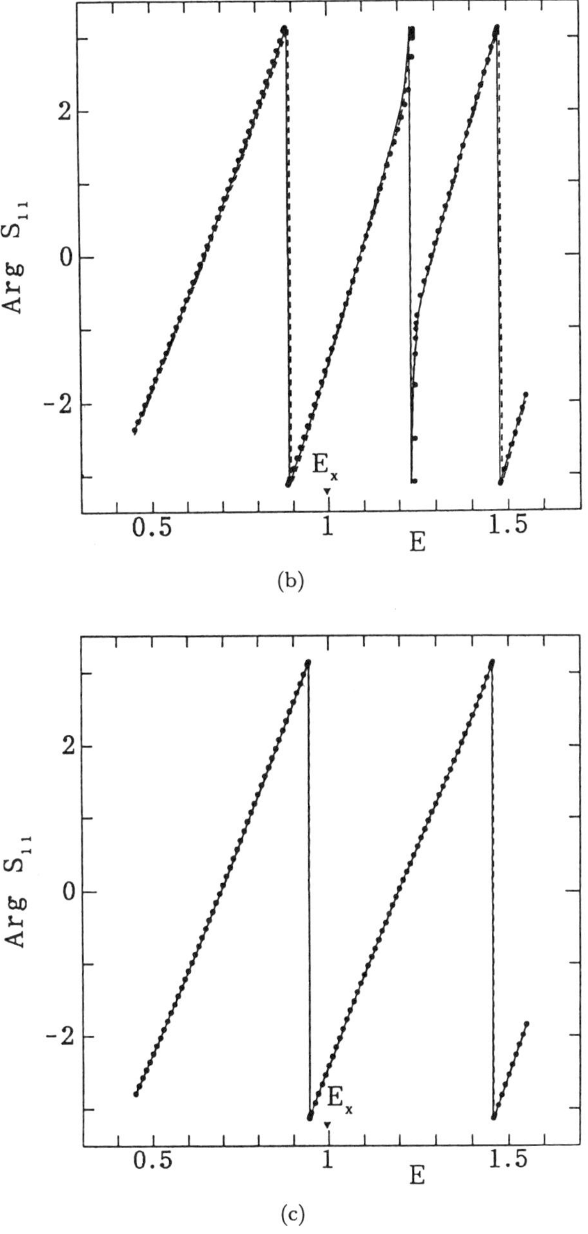

(b)

(c)

Fig. 5.7. (*Continued*).

5.2.2.   Nonadiabatic Tunneling (NT) Type

In the NT-case the situation is much more complicated than the LZ-case.
Eqs. (5.71)–(5.73) are replaced by

$$x_1^2 = y^2 + \frac{b^2}{a^2} - \frac{1}{a^2 y}, \tag{5.117}$$

$$x_2^2 = y^2 + \frac{b^2}{a^2} + \frac{1}{a^2 y}, \tag{5.118}$$

and

$$y^6 + \frac{b^2}{a^2} y^4 + \frac{1}{4a^4} y^2 - \frac{1}{4a^2} = 0, \tag{5.119}$$

where $y > 0$, but both x_1^2 and x_2^2 can be negative. The distribution of t_j's is
classified into six cases as shown in Fig. 5.8. The corresponding diagram in
(a^2, b^2) plane is shown in Fig. 5.9.

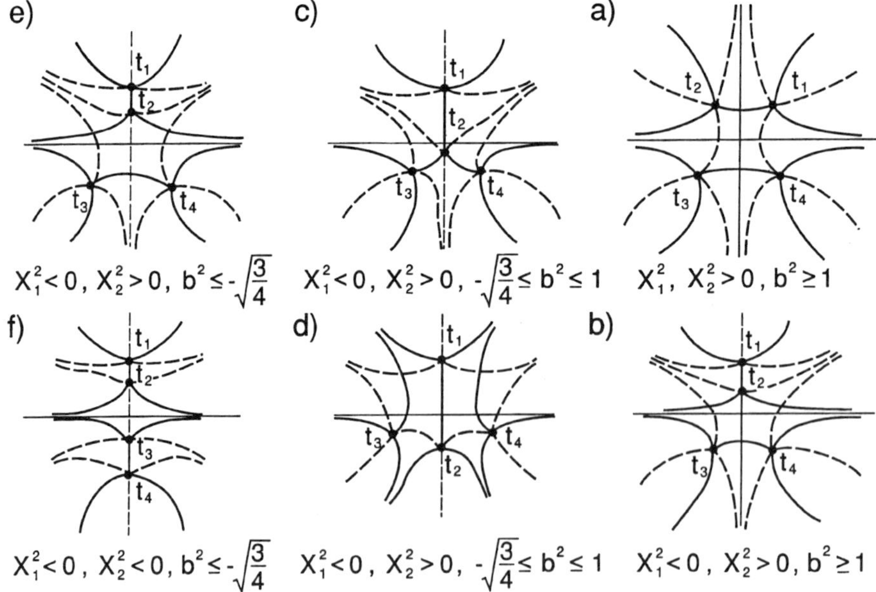

Fig. 5.8.   Distribution of transition points in the NT-case. - - -: Stokes line, ———: anti-Stokes line. (Taken from Ref. [51] with permission.)

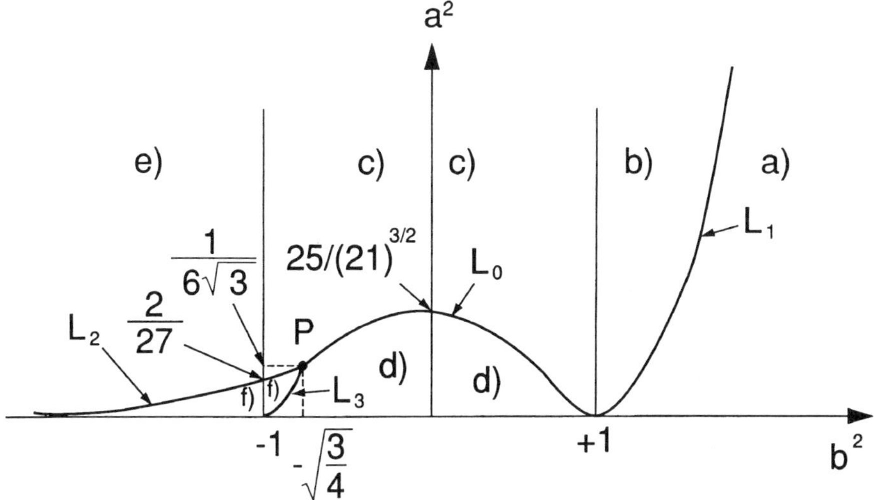

Fig. 5.9. Two-dimensional diagram in (a^2, b^2) plane to represent the various cases of Fig. 5.8. (Taken from Ref. [51] with permission.)

The reduced scattering matrix $S_{\rm NT}^{\rm R}$ given by Eqs. (5.60) and (5.61) is the one in the diabatic representation. In the adiabatic representation we have

$$S_{\rm NT}^{{\rm R}(a)} = \frac{1}{1 + U_1 U_2} \begin{pmatrix} e^{i\Delta_{11}} & U_2 e^{i\Delta_{12}} \\ U_2 e^{i\Delta_{12}} & e^{i\Delta_{22}} \end{pmatrix}, \qquad (5.120)$$

where U_2 is given by Eq. (5.61) and the additional phases Δ_{ij} are defined below. The elastic scattering phases $\eta_j^{(a)}$ in the adiabatic representation are defined as

$$\eta_1^{(a)} = \lim_{R \gg X_1} \left(\int_{X_1}^{R} k_1(R) dR - k_1(R) R + \frac{\pi}{4} \right) \qquad (5.121)$$

and

$$\eta_2^{(a)} = \lim_{R \ll X_2} \left(-\int_{X_2}^{R} k_1(R) dR + k_1(R) R + \frac{\pi}{4} \right), \qquad (5.122)$$

where the reference points $X_{1,2}$ are defined in each energy region. Another interesting and useful achievement is that the basic diabatic parameters a^2 and b^2 defied by Eq. (3.14) (see also Eq. (5.10)) can be re-expressed in terms of the adiabatic potentials as (see Fig. 2.1(b))

$$a^2 = \frac{(1 - \gamma^2)\hbar^2}{\mu(R_b - R_t)^2(E_b - E_t)} \tag{5.123}$$

and

$$b^2 = \frac{E - (E_b + E_t)/2}{(E_b - E_t)/2} \tag{5.124}$$

with

$$\gamma = \frac{E_b - E_t}{E_2(R_b + R_t/2) - E_1(R_b + R_t/2)} \qquad (0 < \gamma \le 1). \tag{5.125}$$

In the limiting case of $R_b = R_t (\gamma \to 1)$, the following formula should be used for a^2 instead of Eq. (5.123):

$$a^2 = \frac{\hbar^2(E_2'' - E_1'')}{4\mu(E_b - E_t)^2}, \tag{5.126}$$

where

$$E_1'' = \left[\frac{d^2 E_1(x)}{dx^2}\right]_{x=x_t} \qquad \text{and} \qquad E_2'' = \left[\frac{d^2 E_2(x)}{dx^2}\right]_{x=x_b}. \tag{5.127}$$

Another important matrix in the NT-case is the transfer matrix N which connects the waves on the right side of the barrier to those on the left side. This matrix is obtained from the reduced scattering matrix $S_{\mathrm{NT}}^{\mathrm{R}}$ as

$$N_{11} = \frac{1}{(S_{\mathrm{NT}}^{\mathrm{R}(a)})_{12}^*}, \qquad N_{22} = N_{11}^*$$

$$N_{12} = \frac{(S_{\mathrm{NT}}^{\mathrm{R}(a)})_{22}}{(S_{\mathrm{NT}}^{\mathrm{R}(a)})_{12}} \quad \text{and} \quad N_{22} = N_{12}^*. \tag{5.128}$$

In the following three subsections the Stokes constant U_1 and other basic quantities are presented. As in the LZ-case, the final recommended formulas which contain empirical corrections and ready for practical applications are summarized in Appendix A. Finally in Chaps. 11–13, the interesting unusual phenomenon of complete reflection will be further discussed. The derivation of various quantities can be found in the original references.

5.2.2.1. $E \le E_t$ $(b^2 \le -1)$

The Stokes constant U_1 is expressed as

$$\text{Re } U_1 = \sin(2\sigma_{\text{ZN}}) \left\{ \frac{1}{4}[B(\sigma_{\text{ZN}}/\pi)]^{1/2} e^{-\delta_{\text{ZN}}} + \frac{e^{\delta_{\text{ZN}}}}{[B(\sigma_{\text{ZN}}/\pi)]^{1/2}} \right\} \tag{5.129}$$

and

$$\text{Im } U_1 = \cos(2\sigma_{\text{ZN}}) \left\{ \frac{(\text{Re } U_1)^2}{\sin^2(2\sigma_{\text{ZN}})} + \frac{1}{\cos^2(\sigma_{\text{ZN}})} \right\}^{1/2}$$
$$- \frac{1}{2\sin(\sigma_{\text{ZN}})} \left| \frac{\text{Re } U_1}{\cos(\sigma_{\text{ZN}})} \right|, \tag{5.130}$$

where

$$\sigma_{\text{ZN}} = \frac{\pi}{16a|b|} \frac{[6 + 10(1 - b^{-4})^{1/2}]^{1/2}}{1 + (1 - b^{-4})^{1/2}}, \tag{5.131}$$

$$B(X) = \frac{2\pi X^{2X} e^{-2X}}{X\Gamma^2(X)}, \tag{5.132}$$

and

$$\delta_{\text{ZN}} = \int_{t_1^L}^{t_1^R} |k_1(R)| dR. \tag{5.133}$$

The additional phases Δ_{ij} in Eq. (5.120) are given by

$$\Delta_{12} = \Delta_{11} = \Delta_{22} = -2\sigma_{\text{ZN}}. \tag{5.134}$$

The reference points $X_{1,2}$ to define the elastic scattering phases (Eqs. (5.121) and (5.122)) are $X_1 = t_1^R$ and $X_2 = t_1^L$ (see Fig. 2.1(b)). The overall transmission (nonadiabatic tunneling) probability P_{12} is explicitly expressed as (see Eq. (5.62))

$$P_{12} = |(S_{\text{NT}}^{\text{R}(a)})_{21}|^2 = \frac{B(\sigma_{\text{ZN}}/\pi)e^{-2\delta_{\text{ZN}}}}{[1 + \frac{1}{4}B(\sigma_{\text{ZN}}/\pi)e^{-2\delta_{\text{ZN}}}]^2 + B(\sigma_{\text{ZN}}/\pi)e^{-2\delta_{\text{ZN}}}}. \tag{5.135}$$

The factor $e^{-2\delta_{\text{ZN}}}$ is nothing but the Gamov factor for tunneling, and the function $B(\sigma_{\text{ZN}}/\pi)$ clearly indicates the effect of nonadiabatic coupling on the tunneling.

5.2.2.2. $E_t \leq E \leq E_b$ $(|b^2| \leq 1)$

The Stokes constant U_1 is given by

$$U_1 = i[(1 + W^2)^{1/2}e^{i\phi} - 1]/W \tag{5.136}$$

with

$$\phi = \sigma_{\text{ZN}} + \arg \Gamma \left(\frac{1}{2} + i\frac{\delta_{\text{ZN}}}{\pi} \right) - \frac{\delta_{\text{ZN}}}{\pi} \ln \left(\frac{\delta_{\text{ZN}}}{\pi} \right) + \frac{\delta_{\text{ZN}}}{\pi} \tag{5.137}$$

and

$$W = \frac{1}{a^{2/3}} \int_0^\infty \cos \left(\frac{t^3}{3} - \frac{b^2}{a^{2/3}}t - \frac{1}{2a^{2/3}}\frac{1}{1 + a^{1/3}t} \right) dt \,, \tag{5.138}$$

where

$$\sigma_{\text{ZN}} = \text{Im} \left(\int_{R_1^{(2)}}^{R_2^{(2)}} k_2(R)dR \right) \tag{5.139}$$

and

$$\delta_{\text{ZN}} = \text{Im} \left(\int_{R_1^{(1)}}^{R_2^{(1)}} k_1(R)dR \right) . \tag{5.140}$$

Here $R_{1,2}^{(j)}$ are the complex turning points on the adiabatic potential $E_j(R)$. The quantities σ_{ZN} and δ_{ZN} can be further simplified [61] as shown in Appendix A. The expression of W is also improved for practical use by introducing empirical modifications (see Appendix A). The additional phases Δ_{ij} in Eq. (5.120) are

$$\Delta_{12} = \sigma_{\text{ZN}} \,, \tag{5.141}$$
$$\Delta_{11} = \sigma_{\text{ZN}} - 2\sigma_0^{\text{ZN}} \tag{5.142}$$

and

$$\Delta_{22} = \sigma_{\text{ZN}} + 2\sigma_0^{\text{ZN}} \tag{5.143}$$

with

$$\sigma_0^{\text{ZN}} = -\frac{1}{3}(R_t - R_b)k_1(R_t)(1 + b^2) \,, \tag{5.144}$$

where $R_t(R_b)$ is the position of the top (bottom) of the lower (upper) adiabatic potential (see Fig. 2.1(b)). The reference points $X_{1,2}$ to define $\eta_j^{(a)}$'s of

Eqs. (5.121) and (5.122) are $X_{1,2} = R_t$. The overall nonadiabatic tunneling (transmission) probability is equal to

$$P_{12} = |(S_{\text{NT}}^{\text{R}(a)})_{21}|^2 = \frac{W^2}{1 + W^2} \, . \tag{5.145}$$

5.2.2.3. $E \geq E_b$ $(b^2 \geq 1)$

We can obtain the following expression for the Stokes constant:

$$U_1 = i(1 - p_{\text{ZN}})^{1/2} e^{i\psi_{\text{ZN}}} \, , \tag{5.146}$$

where p_{ZN} is the Landau–Zener transition probability given by

$$p_{\text{ZN}} = \exp\left[-\frac{1}{4ab} \left(\frac{2}{1 + (1 - b^{-4})^{1/2}} \right)^{1/2} \right] \tag{5.147}$$

and ψ_{ZN} is the phase corresponding to ψ_{LS} of Eq. (3.32) and is given by

$$\psi_{\text{ZN}} = \sigma_{\text{ZN}} - \phi_{\text{S}}(\delta_{\text{ZN}}) \tag{5.148}$$

with

$$\sigma_{\text{ZN}} = \gamma_2(T_2^L, T_2^R) \, . \tag{5.149}$$

The phase integral γ_2 and the Stokes phase correction ϕ_{S} are defined by Eqs. (3.33) and (3.23), respectively. The nonadiabatic transition parameter δ_{ZN} is the counter part of σ_{ZN} at $E \leq E_t$ and is given by

$$\delta_{\text{ZN}} = \frac{\pi}{16ab} \frac{\{6 + 10[1 - b^{-4}]^{1/2}\}^{1/2}}{1 + [1 - b^{-4}]^{1/2}} \, . \tag{5.150}$$

The Landau–Zener transition probability p_{ZN} is as simple as the famous Landau–Zener formula $p_{\text{LZ}}^{(0)}$ of Eq. (3.4) and yet much better than the latter. It should be recalled that $p_{\text{LZ}}^{(0)}$ is expressed in terms of a^2 and b^2 as

$$p_{\text{LZ}}^{(0)} = \exp\left(-\frac{\pi}{4ab}\right) \, . \tag{5.151}$$

The additional phases Δ_{ij} in Eq. (5.120) are

$$\Delta_{12} = \gamma_2(T_2^L, T_2^R) \, , \tag{5.152}$$

$$\Delta_{11} = 2\gamma_2(T_2^L, R_b) - 2\sigma_0^{\text{ZN}} \tag{5.153}$$

and

$$\Delta_{22} = 2\gamma_2(R_b, T_2^R) + 2\sigma_0^{ZN} \tag{5.154}$$

with

$$\sigma_0^{ZN} = \left(\frac{R_b - R_t}{2}\right)\left\{k_1(R_t) + k_2(R_b) + \frac{1}{3}\frac{[k_1(R_t) - k_2(R_b)]^2}{k_1(R_t) + k_2(R_b)}\right\}. \tag{5.155}$$

The reference points $X_{1,2}$ in the definition of $\eta_j^{(a)}$ of Eqs. (5.121) and (5.122) are the same as in the case $E_t \leq E \leq E_b$, i.e., $X_{1,2} = R_t$. The overall nonadiabatic tunneling (transmission) probability P_{12} is equal to

$$P_{12} = |S_{21}^{R(a)}|^2 = \frac{4\cos^2\psi_{ZN}}{4\cos^2\psi_{ZN} + p_{ZN}^2/(1 - p_{ZN})}. \tag{5.156}$$

It should be noted that I-matrix for the NT-case can be defined only for $b^2 \geq 1$ (i.e., $E \geq E_b$), and is given by

$$I = \begin{pmatrix} \sqrt{1 - p_{ZN}}\,e^{i\phi_S} & \sqrt{p_{ZN}}\,e^{i\sigma_{ZN}} \\ -\sqrt{p_{ZN}}\,e^{-i\sigma_{ZN}} & \sqrt{1 - p_{ZN}}\,e^{-i\phi_S} \end{pmatrix}. \tag{5.157}$$

Here it should be noted that the sign in off-diagonal elements is different from the LZ-case. The I_X-matrix is given in the same way as before as

$$I_X = \begin{pmatrix} \sqrt{1 - p_{ZN}}\,e^{i\phi_S} & \sqrt{p_{ZN}}\,e^{i\sigma_0^{ZN}} \\ -\sqrt{p_{ZN}}\,e^{-i\sigma_0^{ZN}} & \sqrt{1 - p_{ZN}}\,e^{-i\phi_S} \end{pmatrix}. \tag{5.158}$$

The reduced scattering matrix derived before 1991, i.e. Eqs. (3.28)–(3.31), has essentially the same form as the new one here except that ψ_{ZN}, p_{ZN} and σ_0^{ZN} are improved.

5.2.2.4. *Complete reflection*

As is clear from Eq. (5.156), the transmission probability P_{12} becomes zero when the following condition is satisfied:

$$\psi_{ZN} = \gamma_2(T_2^L, T_2^R) - \phi_S = \left(\ell + \frac{1}{2}\right)\pi, \qquad \ell = 0, 1, 2, \ldots. \tag{5.159}$$

This is a very interesting phenomenon, because complete reflection of wave occurs in a simple potential like the one shown in Fig. 2.1(b) at certain discrete energies which satisfy Eq. (5.159). This phenomenon was proved to exist quantum mechanically exactly by the theory outlined in Sec. 5.1 in the linear potential model [51]. This occurs at $\mathrm{Im}(U_1) = 0$. The condition Eq. (5.159) is similar to the simple Bohr–Sommerfeld quantization condition with the extra correction term ϕ_S due to nonadiabatic coupling which takes a value in the range $(0, \pi/4)$.

This complete reflection phenomenon implies the following intriguing possibilities [57–59]: (1) bound state in the continuum, (2) molecular switching of transmission in a periodic system, and (3) control of molecular processes by lasers. These will be discussed in more detail in Chaps. 11–13.

5.2.2.5. *Numerical examples*

In order to demonstrate the validity and usefulness of the formulas presented in this section, numerical examples are shown in Figs. 5.10 and 5.11.

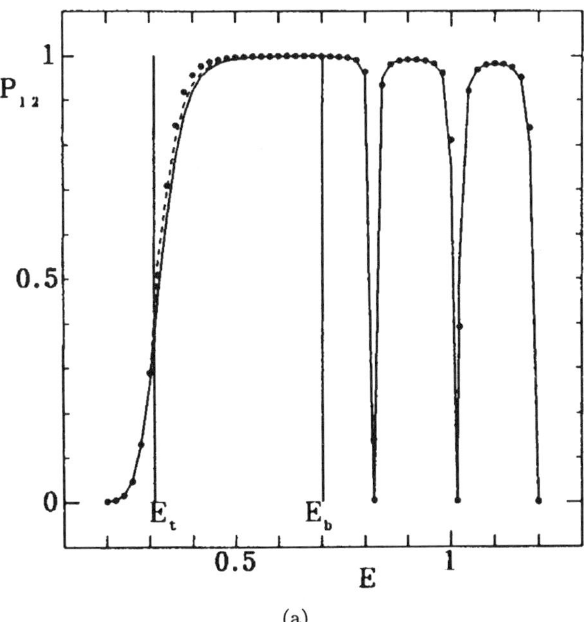

Fig. 5.10. Transition probability versus energy in the NT-case for the potential (5.160). $\cdots$: exact, ———: semiclassical (adiabatic), - - -: semiclassical (diabatic), (a) $a^2(dia) = 0.1$, (b) $a^2 = 1.0$, (c) $a^2 = 10.0$. (Taken from Ref. [9] with permission.)

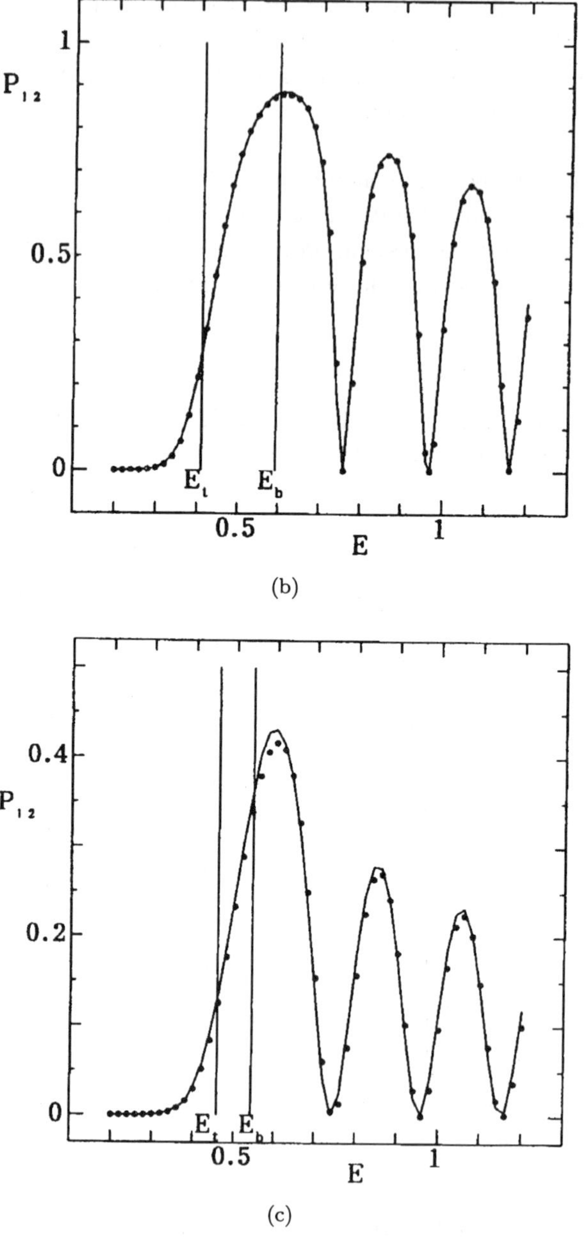

(b)

(c)

Fig. 5.10. (*Continued*).

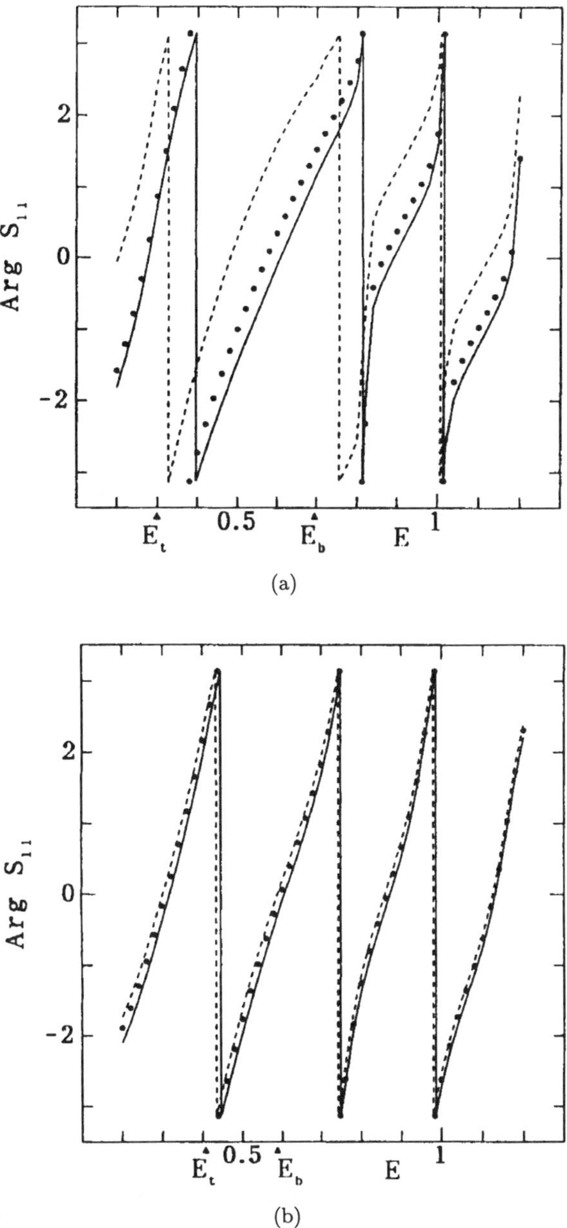

Fig. 5.11.   The same as Fig. 5.10 except for arg S_{12}. (Taken from Ref. [9] with permission.)

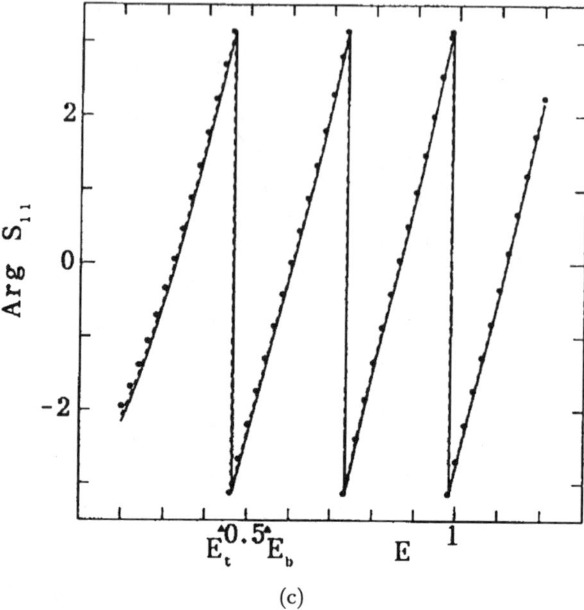

(c)

Fig. 5.11. (*Continued*).

These show the overall transmission probability $P_{12} = |(S^{\mathrm{R}}_{\mathrm{NT}})_{12}|^2$ and the phase $\arg(S^{\mathrm{R}(a)}_{\mathrm{NT}})_{12}$, respectively [61].

The potentials employed are given in atomic units by

$$V_1(x) = 0.5e^{7.5x},$$

$$V_2(x) = 0.5e^{-3.75x}, \tag{5.160}$$

$$V(x) = \frac{3}{2\sqrt{2}} V_0 e^{-2.5x^2}.$$

The coupling parameter V_0 is varied so that the parameter a^2 defined by Eq. (3.14) covers the three regimes: $a^2 = 10.0$ (weak), $a^2 = 1.0$ (intermediate), and $a^2 = 0.1$ (strong). The "adiabatic approximation" means that a^2 and b^2 are estimated from the adiabatic potentials by Eqs. (5.123)–(5.125). It should be noted that in these numerical examples the final recommended formulas presented in Appendix A are employed.

5.3. Non-Curve-Crossing Case

5.3.1. *Rosen–Zener–Demkov model*

The simplest quantum mechanical model of noncrossing case is the following: the diabatic potentials are constant with $V_2 - V_1 = \Delta > 0$ and the diabatic coupling is $V_0 e^{-\beta(R-R_X)}$, where $\Delta = 2V_0$, namely, at $R = R_X$ the diabatic coupling is equal to one-half of the diabatic potential energy difference Δ. Recently, Osherov and Voronin obtained the quantum mechanically exact analytical solution for this model in terms of the Meijer's G-function [64, 65]. In the adiabatic representation this system presents a three-channel problem at $E > V_2 > V_1$, since there is no repulsive wall at $R \ll R_X$ in the lower adiabatic potential. They have obtained the analytical expression of 3×3 transition matrix. Adding a repulsive potential wall at $R \ll R_X$ for the lower adiabatic channel and using the semiclassical idea of independent events of nonadiabatic transition at R_X and adiabatic wave propagation elsewhere, they finally derived the overall inelastic nonadiabatic transition probability P_{12} as follows:

$$P_{12} = \frac{\sinh(\pi k_1/\beta)\sinh(\pi k_2/\beta)\sin^2\tau}{\cosh^2(\pi(k_1-k_2)/2\beta)[\cosh^2(\pi(k_1+k2)/2\beta) - \sin^2\tau]}, \qquad (5.161)$$

where $k_j = k_j\,(R = \infty)$ and τ is the phase difference between the two adiabatic potentials from the turning point to R_X, corresponding to Eq. (3.13). At high collision energy we have

$$k_1 - k_2 \simeq \frac{\Delta}{\hbar\beta v}, \qquad (5.162)$$

where v is the velocity. Then Eq. (5.161) leads to

$$P_{12} \simeq \mathrm{sech}^2\left(\frac{\pi\Delta}{2\hbar\beta v}\right)\sin^2\sigma, \qquad (5.163)$$

which agrees with the Rosen–Zener–Demkov (RZD) formula Eq. (3.34) (see also Eq. (3.42)).

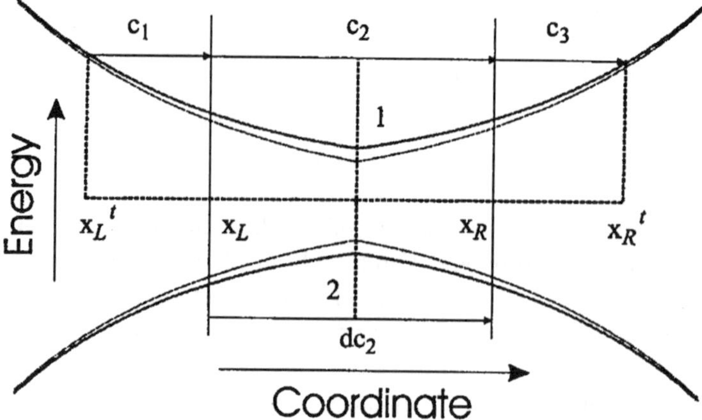

Fig. 5.12.   Diabatically avoided crossing model. Two diabatic potentials (thin lines) do not intersect and have a constant diabatic coupling. Adiabatic potentials are shown in bold. (Taken from Ref. [66] with permission.)

5.3.2.   *Diabatically avoided crossing model*

Another interesting non-curve-crossing problem is the following diabatically avoided crossing model (see Fig. 5.12) [66]:

$$V = \begin{pmatrix} U + Ve^{\alpha|x|} & C \\ C & U - Ve^{\alpha|x|} \end{pmatrix}, \tag{5.164}$$

where $U, V, C,$ and α are certain constants. Although the model can be solved exactly even for two different exponents, α_1 for $x < 0$ and α_2 for $x > 0$, we restrict ourselves to the case $\alpha_1 = \alpha_2$. It is interesting to note that a constant rotaion by $\pi/4$ reduces the above problem to a half-cut Rosen–Zener–Demkov model [64],

$$R\left(\frac{\pi}{4}\right) V R\left(-\frac{\pi}{4}\right) = \begin{pmatrix} U + C & Ve^{\alpha|x|} \\ -Ve^{\alpha|x|} & U - C \end{pmatrix}, \tag{5.165}$$

where $R(\theta)$ represents the rotation by angle θ. The adiabatic potentials are the same as those of the RZD model and are given by (see Fig. 5.12)

$$u^a_{1,2} = U \pm \sqrt{(Ve^{\alpha|x|})^2 + C^2}. \tag{5.166}$$

Introducing a new variable and the dimensionless parameters,

$$z = \frac{\mu^2 V^2}{4(\hbar\alpha)^4} e^{-2\alpha x} \quad \text{and} \quad q_i = \frac{\sqrt{2\mu(E - U_i)}}{\hbar\alpha} \qquad (i = 1, 2) \qquad (5.167)$$

with

$$U_1 = U + C \quad \text{and} \quad U_2 = U - C, \qquad (5.168)$$

it turns out that this model is one of those which can be solved exactly in terms of the Meijer's G-functions [65],

$$G_{pq}^{mn}\left(z \,\middle|\, \begin{matrix} \{a\} \\ \{b\} \end{matrix}\right), \qquad (5.169)$$

where a and b are sets of certain parameters. The four independent solutions in the present problem are given analytically by

$$\Psi_r = \frac{1}{\sqrt{2}}\left(\begin{matrix} G_{04}^{40}(e^{2\pi i r}z|\{b\}) - (-1)^{r+1}G_{04}^{40}(e^{2\pi i r}z|\{b'\}) \\ G_{04}^{40}(e^{2\pi i r}z|\{b\}) + (-1)^{r+1}G_{04}^{40}(e^{2\pi i r}z|\{b'\}) \end{matrix}\right),$$

$$\text{for } r = -1, \ldots, 2, \qquad (5.170)$$

where

$$\{b\} \equiv \{b_1, b_2, b_3, b_4\}, \qquad (5.171)$$

$$b_{1,2} = \pm i q_1/2, \qquad b_{3,4} = 1/2 \pm i q_2/2, \qquad \{b'\} = \{b\}|_{q_1 \leftrightarrow q_2}.$$

The total wave function can be expanded as

$$\begin{pmatrix} \psi_1 \\ \psi_2 \end{pmatrix} = \sum_{r=-1,0,1,2} c_r \Psi_r. \qquad (5.172)$$

Taking into account the asymptotic behaviour of the Meijer's G functions at $x \to -\infty$ $(z \to \infty)$,

$$G_{04}^{40}(e^{-2\pi i}z|\cdot) \sim e^{i\rho}/\sqrt{\rho},$$

$$G_{04}^{40}(z|\cdot) \sim e^{-\rho}/\sqrt{\rho},$$

$$G_{04}^{40}(e^{2\pi i}z|\cdot) \sim e^{-i\rho}/\sqrt{\rho}, \qquad (5.173)$$

$$G_{04}^{40}(e^{4\pi i}z|\cdot) \sim e^{\rho}/\sqrt{\rho},$$

$$\rho = 4z^{1/4},$$

and the boundary conditions, we have

$$\begin{pmatrix} \psi_1 \\ \psi_2 \end{pmatrix} = \alpha \Psi_{-1} + \beta \Psi_0 \qquad \text{for } x < 0 \tag{5.174}$$

and

$$\begin{pmatrix} \psi_1 \\ \psi_2 \end{pmatrix} = \gamma \Psi_{-1} + \delta \Psi_0 + \Psi_1 \qquad \text{for } x > 0 \,. \tag{5.175}$$

Here $|\alpha|^2(|\beta|^2)$ represents the transmission (reflection) coefficient. After some manipulations (see [66]) we can obtain the follwoing expressions for α and γ:

$$\alpha = -i \sin(\phi_1 + \phi_2) e^{i(\phi_2 - \phi_1)} \tag{5.176}$$

and

$$\gamma = -\cos(\phi_1 + \phi_2) e^{i(\phi_2 - \phi_2)} \,, \tag{5.177}$$

where ϕ_1 and ϕ_2 are defined by the following expressions:

$$\phi_1 = \arg[G^{(-1)}(b)G^{(0)}(b') + G^{(-1)}(b')G^{(0)}(b)] \tag{5.178}$$

and

$$\phi_2 = \arg[G^{(-1)},(b)G^{(0)},(b') + G^{(-1)},(b')G^{(0)},(b)] \,, \tag{5.179}$$

where

$$G^{(r)}(b) \equiv G_{04}^{40}(e^{2\pi i r} z_0 | b) \tag{5.180}$$

and

$$G^{(r)},(b) \equiv \frac{d}{dz} G_{04}^{40}(e^{2\pi i r} z_0 | b) \tag{5.181}$$

with $z_0 = z(x = 0)$. Unfotunately, the quantities $G^{(r)}(b)$ and $G^{(r)}_,(b)$ cannot be further simplified, because no analytical expressions are available for G-functions at finite arguments. The semiclassical analysis of this diabatically avoided crossing problem can, however, be carried out with use of the nonadiabatic transition matrix given by Eq. (3.36) and the adiabatic propagation matrices, when the two transition points, i.e. the real parts of the complex crossing points, are well separated from $x = 0$. The complete reflection and complete transmission phenomena in this model will be discussed in Chap. 11 by presenting numerical examples.

5.4. Exponential Potential Model

As was discussed in Sec. 3.3, the exponential potential model presents an interesting generalization of the Landau–Zener and the Rosen–Zener models and thus is worthwhile to investigate more thoroughly. Actually the model has attracted much attention recently. Osherov and Nakamura have solved certain special cases quantum mechanically exactly in terms of the Meijer's G-functions [67, 68]. The semiclassical solutions for a more general case have also been obtained [69, 70]. The first example of quantum mechanically exactly solvable ones is as follows: The diabatic model potentials are given by

$$V_{11}(x) = U_1 - V_1 e^{-\alpha x},$$
$$V_{22}(x) = U_2 - V_2 e^{-\alpha x}, \tag{5.182}$$
$$V_{12}(x) = V_{21}(x) = V e^{-\alpha x},$$

where

$$V_1 V_2 = V^2. \tag{5.183}$$

and $U_1 > U_2$ is assumed without losing generality. With use of the new variables,

$$z = -\frac{2m}{\hbar^2 \alpha^2}(V_1 + V_2)e^{-\alpha x}, \tag{5.184}$$

we can transform the original coupled Schrödinger equations into the following fourth-order single differential equation:

$$\left[\prod_{n=1}^{4}\left(z\frac{d}{dz} - b_n\right) - z\prod_{m=1}^{2}\left(z\frac{d}{dz} - a_m + 1\right)\right]\psi(z) = 0, \tag{5.185}$$

where

$$b_1 = iq_2, \quad b_2 = -iq_2, \quad b_3 = 1 + iq_1, \quad b_4 = 1 - iq_1, \tag{5.186}$$

$$a_1 = 1 + i\gamma, \quad a_2 = 1 - i\gamma, \tag{5.187}$$

$$q_j = \frac{\sqrt{2m}}{\hbar\alpha}\sqrt{E - U_j}, \quad (j = 1, 2, 3), \tag{5.188}$$

and

$$\gamma = \sqrt{\frac{V_1 q_1^2 + V_2 q_2^2}{V_1 + V_2}} \tag{5.189}$$

with $U_3 = \frac{U_1 V_2 + U_2 V_1}{V_1 + V_2}$. This equation can be solved exactly in terms of the Meijer's G functions [65].

Let us first consider the attractive potential case, V_j $(j = 1, 2) > 0$. We can obtain compact analytical expressions for the nonadiabatic transition matrix I_X (see Eq. (3.48)). When $E > U_1$, there are four open channels and the I_X-matrix becomes a 4×4 matrix. For the definition of the channel numbers, see Figs. 5.13.

Here we give the final expressions only for the nonadiabatic transition probabilities p_{ij} $(= |O_{ij}|^2)$,

$$p_{11} = \frac{\sinh(\pi(q_3 - q_1)) \sinh(\pi(q_2 + q_1))}{\sinh(\pi(q_3 + q_1)) \sinh(\pi(q_2 - q_1))} e^{-4\pi q_1}, \tag{5.190}$$

$$p_{12} = \frac{\sinh(\pi(q_3 - q_1)) \sinh(2\pi q_2) \sinh(2\pi q_1) \sinh(\pi(q_2 - q_3))}{\sinh^2(\pi(q_2 - q_1)) \sinh(\pi(q_3 + q_2)) \sinh(\pi(q_3 + q_1))}$$

$$\times e^{-2\pi q_2 - 2\pi q_1}, \tag{5.191}$$

$$p_{13} = 2 \frac{\sinh(\pi(q_3 - q_1)) \sinh(\pi(q_2 + q_1)) \sinh(2\pi q_1)}{\sinh(\pi(q_3 + q_1)) \sinh(\pi(q_2 - q_1))} e^{-2\pi q_2 - 2\pi q_1}, \tag{5.192}$$

$$p_{14} = \frac{\sinh(\pi(2\pi q_3) \sinh(\pi(q_2 + q_1)) \sinh(2\pi q_1) \sinh(\pi(q_2 - q_3))}{\sinh^2(\pi(q_1 + q_3)) \sinh(\pi(q_3 + q_2)) \sinh(\pi(q_2 - q_1))}$$

$$\times e^{-2\pi q_3 - 2\pi q_1}, \tag{5.193}$$

$$p_{22} = \left(\frac{\sinh(\pi(q_2 - q_3)) \sinh(\pi(q_2 + q_1))}{\sinh(\pi(q_2 + q_3)) \sinh(\pi(q_2 - q_1))} \right)^2 e^{-2\pi q_2}, \tag{5.194}$$

$$p_{23} = 2 \frac{\sinh(2\pi q_2) \sinh(\pi(q_2 + q_1)) \sinh(\pi(q_2 - q_3))}{\sinh(\pi(q_2 + q_3)) \sinh(\pi(q_2 - q_1))}$$

$$\times e^{2\pi q_3 - 2\pi q_2 - 2\pi q_1}, \tag{5.195}$$

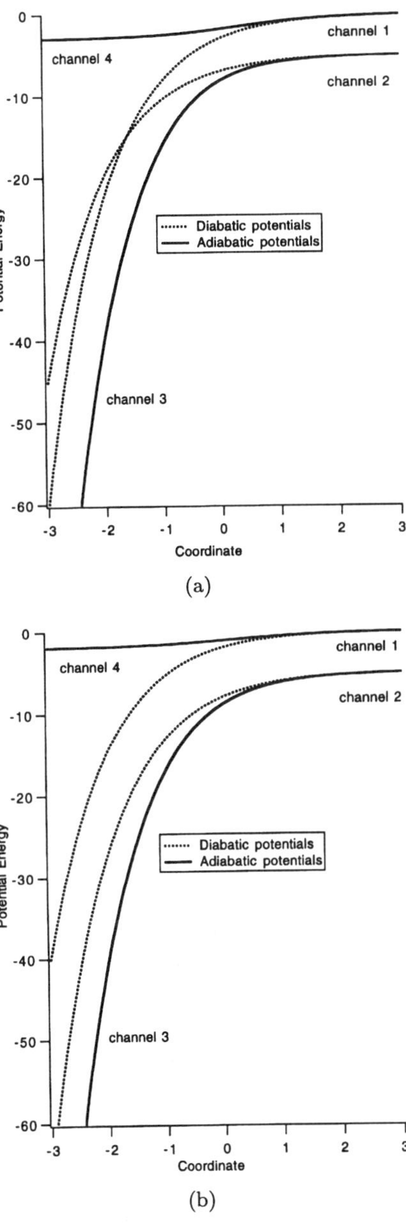

Fig. 5.13. (a) Schematic diabatic (- - -) and adiabatic (———) crossing potentials in the attractive case. The potential energies and the coordinate are scaled as $v_i = 2mV_i/(\hbar^2\alpha^2)$ and $z = \alpha x$. $u_1 = 0$, $u_2 = -5.0$, and $v_1 = 3.0$ and (b) The same as Fig. 5.13(a) except for the non-crossing case ($u_1 = 0$, $u_2 = -5.0$, and $v_1 = 2.0$). (Taken from Ref. [67] with permission.)

$$p_{24} = \frac{\sinh(2\pi q_2)\sinh(2\pi q_3)\sinh(\pi(q_2+q_1))\sinh(\pi(q_3-q_1))}{\sinh^2(\pi(q_2+q_3))\sinh(\pi(q_3+q_1))\sinh(\pi(q_2-q_1))}$$

$$\times\, e^{2\pi q_3 - 2\pi q_2}\,, \tag{5.196}$$

$$p_{33} = e^{4\pi(q_3-q_2-q_1)}\,, \tag{5.197}$$

$$p_{44} = \left[\frac{\sinh(\pi(q_3-q_1))\sinh(\pi(q_3-q_2))e^{2\pi q_3}}{\sinh(\pi(q_3+q_1))\sinh(\pi(q_3+q_2))}\right]^2 e^{4\pi q_3}\,, \tag{5.198}$$

$$p_{43} = 2\frac{\sinh(2\pi q_3)\sinh(\pi(q_3-q_1))\sinh(\pi(q_2-q_3))}{\sinh(\pi(q_2+q_3))\sinh(\pi(q_1+q_3))}$$

$$\times\, e^{2\pi q_3 - 2\pi q_1 - 2\pi q_2}\,. \tag{5.199}$$

Semiclassically, $p_{13}\,(\simeq p_{24})$ represents the nonadiabatic transition probability p for one passage of the transition region; and $p_{14}\,(\simeq p_{23})$ corresponds to $1-p$. It can be easily shown that the following simple expression holds except near the threshold region:

$$p = e^{-\pi(q_2-q_3)}\frac{\sinh(\pi(q_3-q_1))}{\sinh(\pi(q_2-q_1))} \xrightarrow{E\to\infty} \frac{V_1}{V_1+V_2}\,. \tag{5.200}$$

When $U_1 > E > U_3$, the channel 1 is closed and q_1 should be replaced by iq_1 in the above expressions. Naturally, $I_{1j} = I_{j1} = 0\,(j = 1-4)$. When $U_3 > E > U_2$, the channel 4 is also closed, and thus $I_{j4} = I_{4j} = 0\,(j = 1-4)$. Accordingly, q_3 should be further replaced by iq_3. When energy is further lower than U_2, only channel 3 is open and $q_j\,(j = 1-3)$ should be replaced by iq_j.

The repulsive case in which V_1 and V_2 are negative can also be solved by the G-functions: In the three channel case $(E > U_1)$, we obtain the following expressions:

$$p_{11} = \left[\frac{\sinh(\pi(q_2+q_1))\sinh(\pi(q_3-q_1))}{\sinh(\pi(q_2-q_1))\sinh(\pi(q_3+q_1))}\right]^2\,, \tag{5.201}$$

$$p_{12} = \frac{\sinh(\pi(q_3-q_1))\sinh(2\pi q_2)\sinh(2\pi q_1)\sinh(\pi(q_2-q_3))}{\sinh^2(\pi(q_2-q_1))\sinh(\pi(q_3+q_2))\sinh(\pi(q_3+q_1))}\,, \tag{5.202}$$

$$p_{13} = \frac{\sinh(2\pi q_1)\sinh(2\pi q_3)\sinh(\pi(q_2+q_1))\sinh(\pi(q_2-q_3))}{\sinh^2(\pi(q_3+q_1))\sinh(\pi(q_2+q_3))\sinh(\pi(q_2-q_1))}\,, \tag{5.203}$$

$$p_{22} = \left(\frac{\sinh(\pi(q_2 + q_1)) \sinh(\pi(q_2 - q_3))}{\sinh(\pi(q_2 + q_3)) \sinh(\pi(q_2 - q_1))} \right)^2 , \tag{5.204}$$

$$p_{23} = \frac{\sinh(2\pi q_2) \sinh(2\pi q_3) \sinh(\pi(q_2 + q_1)) \sinh(\pi(q_3 - q_1))}{\sinh^2(\pi(q_2 + q_3)) \sinh(\pi(q_3 + q_1)) \sinh(\pi(q_2 - q_1))} , \tag{5.205}$$

$$p_{33} = \left(\frac{\sinh(\pi(q_3 - q_1)) \sinh(\pi(q_2 - q_3))}{\sinh(\pi(q_1 + q_3)) \sinh(\pi(q_2 + q_3))} \right)^2 . \tag{5.206}$$

Semiclassically, p_{13} represents, as before, the nonadiabatic transition probability for one passage of the transition region, and is nicely given by the following formula except at the threshold:

$$p = e^{-\pi(q_3 - q_1)} \frac{\sinh(\pi(q_2 - q_3))}{\sinh(\pi(q_2 - q_1))} \xrightarrow{E \to \infty} \frac{V_2}{V_1 + V_2} . \tag{5.207}$$

Other probabilities can be nicely interpreted as follows:

$$p_{23} = 1 - p , \tag{5.208}$$

$$p_{11} = (1 - p)^2 , \tag{5.209}$$

$$p_{12} = (1 - p)p , \tag{5.210}$$

$$p_{22} = p^2 , \tag{5.211}$$

$$p_{33} = 0 . \tag{5.212}$$

The expressions in the two-channel ($U_1 > E > U_3$) and single-channel ($U_3 > E > U_2$) cases can be obtained in the same way as before. Some numerical examples are shown in Figs. 5.14(a) and 5.14(b). Figure 5.14(a) shows the exact results of p_{1j} ($j = 1-4$) for the attarctive case with $\epsilon_1 = 0, \epsilon_2 = -5.0$, and $\epsilon_3 = -3.0$, where ϵ_j are dimensionless parameters defined as $\epsilon_j = (2m/\hbar^2 \alpha^2)U_j$ and the abscissa is the dimensionless enegy defined by $\epsilon = (2m/\hbar^2 \alpha^2)E$. The interesting quantum mechanical threshold effects can be seen at $\epsilon \sim 0$. The complete reflection effect dies out very quickly, however. At energies off the threshold region, only p_{13} and p_{14} survive. Figure 5.14(b) demonstrates the accuracy of the simple semiclassical expressions given by Eq. (5.200). Except at the narrow threshold region, they agree with the exact ones perfectly.

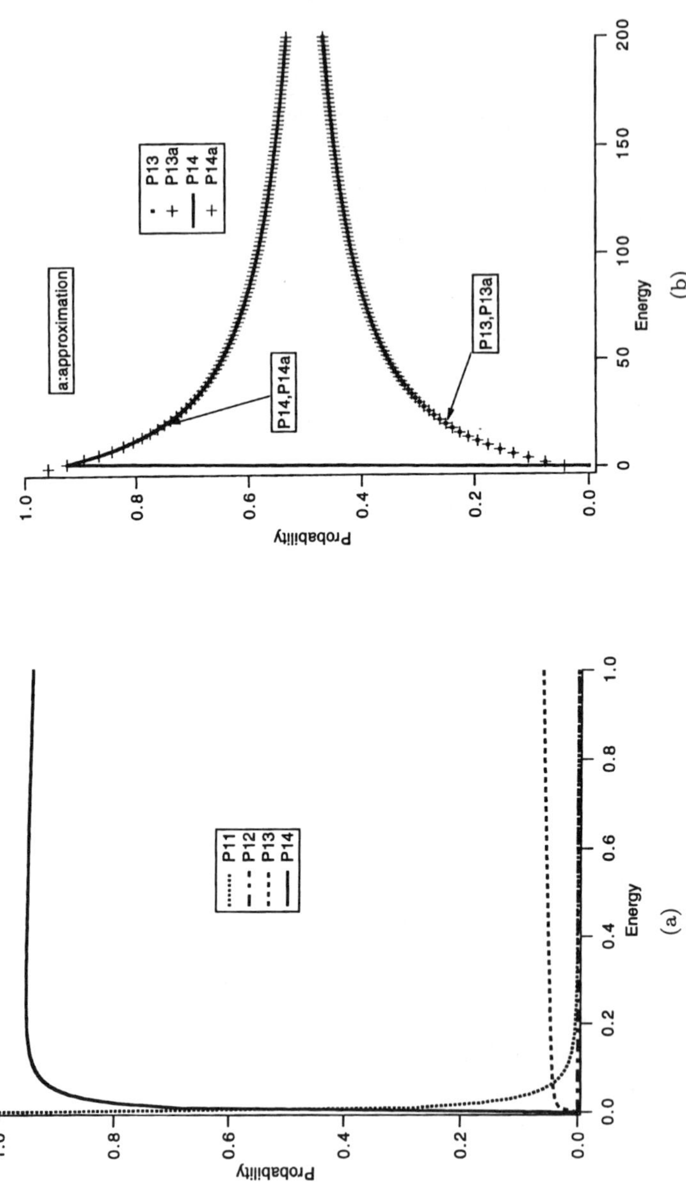

Fig. 5.14.   (a) Nonadiabatic transition probality p_{1j} ($j = 1 - 4$) for the transition from channel 1 to channel j in the case of Fig. 5.13(a). (b) The same as Fig. 5.14(a) for p_{1j} ($j = 3, 4$) together with the corresponding approximations. (Taken from Ref. [67] with permission.)

Osherov and Nakamura solved also a two-state problem quantum mechanically exactly for the following potential matrix [68]:

$$\hat{V}(x) = \begin{pmatrix} U_1 & Ve^{-\alpha x} \\ Ve^{-\alpha x} & U_2 + V_2 e^{-2\alpha x} \end{pmatrix}, \tag{5.213}$$

where $U_1 > U_2$ and $V_2 > 0$. It should be noted that the exponent of the second diabatic potential is two times of that of the diabatic coupling. The potentials are shown in Fig. 5.15. The lower adiabatic potential tends to a constant U_3 in the limit $x \to -\infty$, which is called channel 3. This problem can be solved again in terms of the Meijer's G-functions.

Only the final expressions of transition probabilities are given here.

$$p_{11} = \left(\frac{\cosh(\pi(q_1 + q_2)/2)\sinh(\pi(q_3 - q_1)/2)}{\sinh(\pi(q_1 + q_3)/2)\cosh(\pi(q_2 - q_1)/2)} \right)^2, \tag{5.214}$$

$$p_{12} = \frac{\sinh(\pi q_1)\sinh(\pi q_2)\sinh(\pi(q_3 - q_1)/2)\cosh(\pi(q_2 - q_3)/2)}{\cosh^2(\pi(q_2 - q_1)/2)\cosh(\pi(q_2 + q_3)/2)\sinh(\pi(q_1 - q_3)/2)}, \tag{5.215}$$

$$p_{13} = \frac{\sinh(\pi q_1)\sinh(\pi q_3)\cosh(\pi(q_2 - q_3)/2)\cosh(\pi(q_1 + q_2)/2)}{\cosh(\pi(q_2 + q_3)/2)\sinh^2(\pi(q_1 + q_3)/2)\cosh(\pi(q_2 - q_1)/2)}, \tag{5.216}$$

$$p_{22} = \left(\frac{\cosh(\pi(q_1 + q_2)/2)\cosh(\pi(q_2 - q_3)/2)}{\cosh(\pi(q_2 + q_3)/2)\cosh(\pi(q_2 - q_1)/2)} \right)^2, \tag{5.217}$$

$$p_{23} = \frac{\sinh(\pi q_2)\sinh(\pi q_3)\sinh(\pi(q_3 - q_1)/2)\cosh(\pi(q_1 + q_2)/2)}{\cosh^2(\pi(q_2 + q_3)/2)\sinh(\pi(q_1 + q_3)/2)\cosh(\pi(q_2 - q_1)/2)}, \tag{5.218}$$

$$p_{33} = \left(\frac{\sinh(\pi(q_3 - q_1)/2)\cosh(\pi(q_2 - q_3)/2)}{\sinh(\pi(q_1 + q_3)/2)\cosh(\pi(q_2 + q_3)/2)} \right)^2, \tag{5.219}$$

where

$$q_j = \frac{\sqrt{2m(E - U_j)}}{\hbar\alpha}, \qquad (j = 1 - 3). \tag{5.220}$$

In the special case of $V_2 = 0$, we have the Rosen–Zener case and

$$p_{13} \xrightarrow{V_2 \to 0} 2e^{-\pi(q_1 + q_2)} \frac{\sinh(\pi q_1)\cosh(\pi(q_1 + q_2)/2)}{\cosh(\pi(q_2 - q_1)/2)}, \tag{5.221}$$

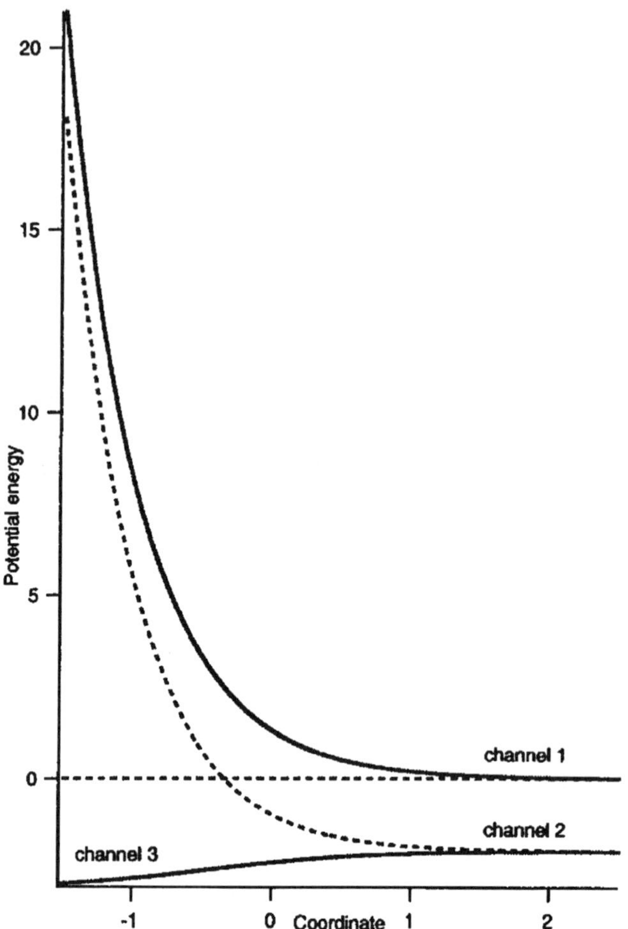

Fig. 5.15. Diabatic (dash line) and adiabatic (solid line) potentials. The energies and coordinate are scaled as $u = 2mU/(\hbar^2\alpha^2)$ and $z = \alpha x$. The dimensionless energy parameters are $u_1 = 0.0$, $u_2 = -2.0$, and $v = \sqrt{3}$. (Taken from Ref. [68] with permission.)

which coincides with the exact solution obtained by Osherov and Voronin [64]. It can be easily seen that

$$\text{Eq. } (5.221) = [1 + e^{\pi(q_2 - q_1)}]^{-1}(1 - e^{-2\pi q_2})(1 + e^{-\pi(q_1 + q_2)})$$

$$\equiv p_{\text{RZ}}(1 - e^{-2\pi q_2})(1 + e^{-\pi(q_1 + q_2)}), \tag{5.222}$$

where p_{RZ} is the Rosen–Zener probability and the residual two factors represent the quantum mechanical threshold effect. In the same way as before, we can obtain the accurate semiclassical expressions as follows:

$$p_{13} \simeq e^{\pi(q_1 - q_3)/2} \frac{\cosh(\pi(q_2 - q_3)/2)}{\cosh(\pi(q_2 - q_1)/2)},$$

$$p_{11} \simeq (1-p)^2, \qquad p_{12} \simeq (1-p)p, \tag{5.223}$$

$$p_{22} \simeq p^2, \qquad p_{23} \simeq 1-p, \qquad p_{33} \simeq 0.$$

It is interesting to note that cosh-function appears instead of sinh in Eq. (5.221).

The generalization of the potential matrix given by Eq. (5.182) without the restriction of Eq. (5.183) can be made in the semiclassical approximation [69, 70]. For simplicity we consider first the attractive potential case, i.e. $V_1 > V_2 > 0$ and $V > 0$. Thus the two potentials are open at $x \to -\infty$. First we introduce the following dimensionless quantities:

$$[E, U_1, U_2, V, V_1, V_2] \Leftarrow \frac{2M}{\hbar^2 \alpha^2} [E, U_1, U_2, V, V_1, V_2] \tag{5.224}$$

$$x \Leftarrow \alpha x.$$

Then the coupled differential equations are given by

$$-\psi_1''(x) + (U_1 - V_1 e^{-x} - E)\psi_1(x) + V e^{-x}\psi_2(x) = 0,$$

$$-\psi_2''(x) + (U_2 - V_2 e^{-x} - E)\psi_2(x) + V e^{-x}\psi_1(x) = 0. \tag{5.225}$$

Further introducing the new variable and parameters,

$$\rho^2 = 4V e^{-x/2}, \tag{5.226}$$

$$\nu^2 = 4(E - U_1), \qquad \mu^2 = 4(E - U_2), \tag{5.227}$$

and

$$\beta_j = \frac{V_j}{|V|}, \tag{5.228}$$

and using the transformations,

$$\psi_j(\rho) = \int_C p F_j(p) Z_a(p\rho) dp \quad \text{[Bessel transformation]} \tag{5.229}$$

and

$$F_1(p) = \frac{f_1(p)}{p^{1/2}(p^2 - a_1)(p^2 - a_2)}\,, \tag{5.230}$$

we obtain

$$\left[\frac{d^2}{dp^2} + \frac{1 + 4\mu^2}{4} \frac{p^4 - 4\epsilon p^2 + \lambda}{p^2(p^2 - a_1)(p^2 - a_2)}\right] f_1(p) = 0\,, \tag{5.231}$$

where

$$\epsilon = \frac{(\beta_1 + \beta_2)/4 + (\beta_1\mu^2 + \beta_2\nu^2)}{1 + 4\mu^2}\,, \tag{5.232}$$

$$\lambda = \frac{(\beta_1\beta_2 - 1)(1 + 4\nu^2)}{1 + 4\mu^2}\,, \tag{5.233}$$

$$a_{1,2} = \frac{1}{2}(\beta_1 + \beta_2) \mp \left[\frac{(\beta_1 - \beta_2)^2}{4} + 1\right]^{1/2}\,, \tag{5.234}$$

and $Z_a(z)$ is an appropriately chosen Bessel function (the Hankel function $H_{\pm\nu}^{(1,2)}$) which satisfies

$$\left[z^2 \frac{d^2}{dz^2} + z\frac{d}{dz} + z^2 + \nu^2\right] Z_a(z) = 0\,. \tag{5.235}$$

The function $F_2(p)$ is simply obtained from $F_1(p)$ by

$$F_2(p) = \mathrm{sgn}(V)(\beta_1 - p^2)F_1(p)\,. \tag{5.236}$$

The integral contour C should be chosen so that the following conditions are satisfied:

$$F_j(p)\frac{dZ_{i\nu}}{dp}p^n\bigg|_C = 0 \qquad (n = 1, 3)\,,$$

$$\frac{d}{dp}(F_j(p)p^n)Z_{i\nu}\bigg|_C = 0 \qquad (n = 1, 3)\,, \tag{5.237}$$

$$F_j(p)p^n Z_{i\nu}|_C = 0 \qquad (n = 0, 2)\,.$$

Under the conditions of $\mu^2, \nu^2 \gg 1$, we can simplify Eq. (5.231) as

$$\left[\frac{d^2}{dp^2} + P_0(p)\right] f_1(p) = 0, \qquad (5.238)$$

where

$$P_0(p) = \mu^2 \frac{(p^2 - c_1)(p^2 - c_2)}{p^2(p^2 - a_1)(p^2 - a_2)} \qquad (5.239)$$

with

$$c_{1,2} = \frac{1}{2}(\beta_1 + \beta_2\nu^2/\mu^2) \mp \left[(\beta_1 - \beta_2\nu^2/\mu^2)^2/4 + \frac{\nu^2}{\mu^2}\right]^{1/2}. \qquad (5.240)$$

Now, we apply the WKB approximation to Eq. (5.231) and use $H^{(2)}_{i\nu}(p\rho)$, the second kind of Hankel function, for $Z_a(p\rho)$. Then we have

$$F_1(p) \simeq \frac{\exp[i \int^p \sqrt{P_0}\,dp]}{p^{1/2}(p^2 - a_1)(p^2 - a_2)P_0^{1/4}} \qquad (5.241)$$

and

$$\psi_1(p) \simeq \int_C \frac{e^{iS(p,\rho)}p^{1/2}}{(p^2 - a_1)(p^2 - a_2)(p^2\rho^2 + \nu^2)^{1/4}P_0^{1/4}}\,dp, \qquad (5.242)$$

where

$$S(p,\rho) = \int^p \sqrt{P_0(p)}\,dp - \int^{p\rho} \sqrt{1 + \nu^2/\xi^2}\,d\xi. \qquad (5.243)$$

The integral contour C is taken as shown in Fig. 5.16 in the attractive case.

In order to evaluate Eq. (5.242) we use the saddle-point method. The details of the derivation are not provided here, but the final expressions of the asymptotic behavior of the wave functions are given as

$$\psi_1(x) \xrightarrow{x \to \infty} A_1 \frac{\rho^{-i\nu}}{\sqrt{\nu}}, \qquad (5.244)$$

$$\psi_1(x) \xrightarrow{x \to -\infty} B_1 \frac{e^{-i\sqrt{a_2}\rho}}{a_2^{1/4}\rho^{1/2}}, \qquad (5.245)$$

$$\psi_2(x) \xrightarrow{x \to \infty} A_2 \frac{\rho^{-i\mu}}{\sqrt{\mu}}, \qquad (5.246)$$

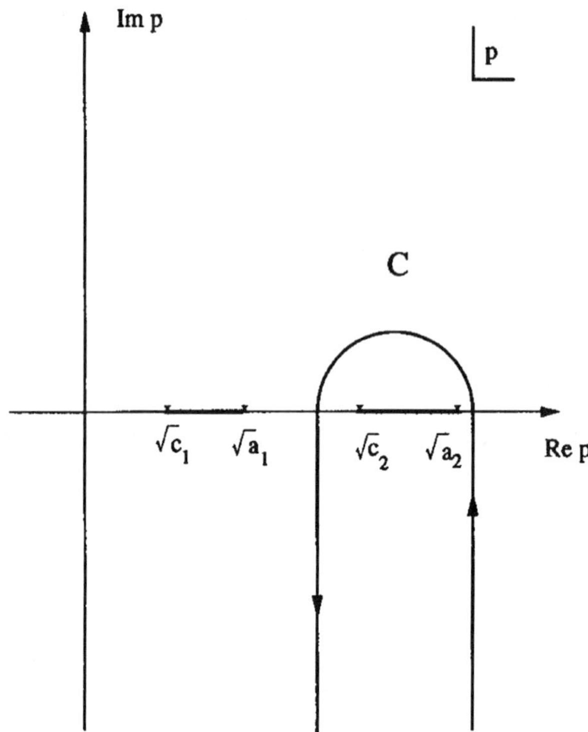

Fig. 5.16.   The integral contour C to define the wave function ψ_1. (Taken from Ref. [69] with permission.)

where

$$A_1 = \frac{\pi i}{\sqrt{\nu}} e^{-i\phi_0} (a_2 - a_1)^{i\delta - 1} e^{-\pi\delta_2} \frac{\Gamma(1-\delta)}{\Gamma(1-i\delta_1)\Gamma(1-i\delta_2)} \,, \qquad (5.247)$$

$$B_1 = \frac{\pi i}{\sqrt{\mu}} \left(\frac{\delta}{\sqrt{\delta_1\delta_2}} \right)^{i\delta - 1} a_2^{i\nu/2} \left(\frac{a}{\mu} \right)^{i\delta_2} \frac{e^{-\pi\delta_2/2}}{\Gamma(1-\delta_2)} \,, \qquad (5.248)$$

$$A_2 = \frac{1}{\sqrt{\nu}} e^{-\pi\delta_2 + \pi\delta/2} \sinh(\pi\delta_2)\Gamma(i\delta) e^{-i\phi_0 + i\delta \ln(4\nu)} \,, \qquad (5.249)$$

$$\phi_0 = \nu - \nu \ln(2\nu) \,. \qquad (5.250)$$

The parameters δ_1, δ_2 and δ are given by

$$\delta_1 = \frac{\mu - \nu}{4}\left(1 + \cos(2g_0)\right), \tag{5.251}$$

$$\delta_2 = \frac{\mu - \nu}{4}\left(1 - \cos(2g_0)\right), \tag{5.252}$$

$$\delta = \delta_1 + \delta_2 = \frac{\mu - \nu}{2}, \tag{5.253}$$

$$\cos(2g_0) = \frac{(\beta_1 - \beta_2)/2}{\sqrt{1 + (\beta_1 - \beta_2)^2/4}}. \tag{5.254}$$

The nonadiabatic transition probability p for one passage of the transition region is finally obtained as

$$p \equiv \left|\frac{\phi_2(\infty)}{\phi_1(-\infty)}\right|^2 = \left|\frac{\psi_1(\infty)}{\psi_1(-\infty)/\cos(2g_0)}\right|^2 = e^{-\pi\delta_2}\frac{\sinh(\pi\delta_1)}{\sinh(\pi\delta)}. \tag{5.255}$$

The parameters δ_1, δ_2 and δ given by Eqs. (5.251)–(5.253) are actually defined more generally by the following contour intergrals:

$$\delta_j = \frac{1}{2\pi i}\int_{L_j}\sqrt{P_0(p)}\,dp \qquad (j = 1, 2) \tag{5.256}$$

and

$$\delta = \delta_1 + \delta_2 = \frac{1}{2}(\mu - \nu), \tag{5.257}$$

where the contours L_j are defined in Fig. 5.17. The expressions given in Eq. (5.251)–(5.253) are actually the high energy approximation $(\nu^2/\mu^2 \to 1)$ to the above general definition.

Equation (5.255) has the similar form as that obtained before in the Nikitin's model (see Eq. (3.49)) and in the special exponential model (see Eq. (5.207)). The present formula gives a generalization, and the parameter correspondence is as follows:

$$\begin{aligned}
\delta_1 &\leftrightarrow \xi - \xi_p \quad \text{[Nikitin model]} \leftrightarrow q_3 - q_1 \quad \text{[special case]} \\
\delta_2 &\leftrightarrow \xi_p \quad \text{[Nikitin model]} \leftrightarrow q_2 - q_3 \quad \text{[special case]},
\end{aligned} \tag{5.258}$$

Our next task is to derive expressions for the dynamical phases ϕ and ψ, which are defined as the phases of the nonadiabatic transition matrix elements in the

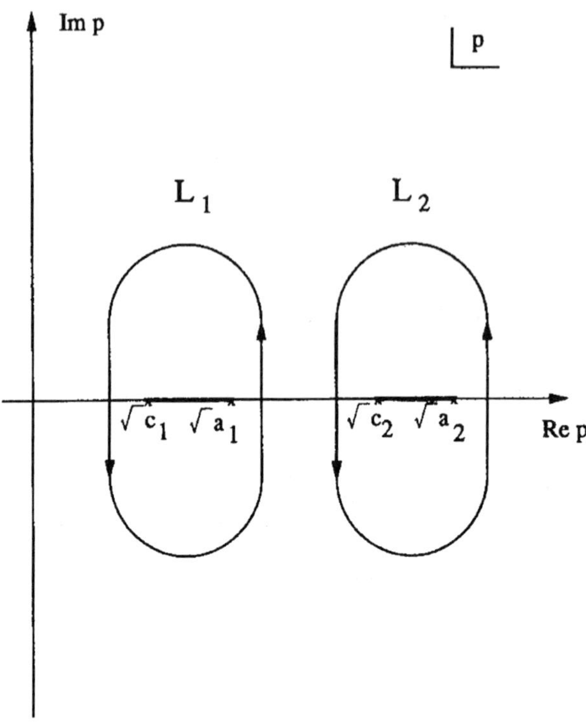

Fig. 5.17.   The integral contours L_1 and L_2 to define δ_1 and δ_2. (Taken from Ref. [69] with permission.)

adiabatic representation as (cf. Eq. (3.48)),

$$I_X = \begin{pmatrix} \sqrt{1-p}\,e^{i\phi} & -\sqrt{p}\,e^{-i\psi} \\ \sqrt{p}\,e^{i\psi} & \sqrt{1-p}\,e^{-i\phi} \end{pmatrix}. \tag{5.259}$$

This matrix represents the nonadiabatic transition amplitude for one passage of the transition region. In order to obtain these phases we utilize the exact solutions in the special case $(V_1 V_2 = V^2)$ [67]. Using the semiclassical expressions of phases along the adiabatic potentials and the exact total phases in the special case, we can finally obtain the following formulas for ϕ and ψ:

$$\phi = \gamma(\delta_2) - \gamma(\delta), \tag{5.260}$$

and

$$\psi = \gamma(\delta_1) - \gamma(\delta) - 2\left[\sqrt{\delta\delta_2} + \frac{\delta_1}{2}\ln\frac{\sqrt{\delta} - \sqrt{\delta_2}}{\sqrt{\delta} + \sqrt{\delta_2}}\right], \qquad (5.261)$$

where

$$\gamma(X) = X\ln(X) - X - \arg\Gamma(iX). \qquad (5.262)$$

The parameters δ_1 and δ_2 here are those obtained from Eqs. (5.251)–(5.253) under the condition $V_1 V_2 = V^2$, and thus the above formulas for the dynamical phases can be generalized with use of the general definition of Eq. (5.256). The functionality of Eqs. (5.260) and (5.261) is the same as that of Nikitin's formulas (cf. Eqs. (3.50) and (3.51)), but the present theory is naturally more general than the latter because of the generalization of the parameters. If we take a limit $\delta \to \infty$ or $\delta_2 \to 0$, then we can obtain the Landau–Zener formula. In the limit $\delta_1 = \delta_2$, on the other hand, we can cover the Rosen–Zener theory (see Eq. (3.35)).

The parameters δ_1 and δ_2 can be actually further generalized as

$$\delta_1 = \frac{1}{\pi}\mathrm{Im}\left(\int_{x_1}^{x_*} k_1(x)dx - \int_{x_2}^{x_*} k_2(x)dx\right) \qquad (5.263)$$

$$\delta_2 = \frac{1}{\pi}\mathrm{Im}\left(\int_{\mathrm{Re}(x_*)}^{x_*} [k_1(x) - k_2(x)]dx\right), \qquad (5.264)$$

where x_j $(j = 1, 2)$ are the complex turning points and x_* is the complex crossing point. Then the final expressions of p and the dynamical phases obtained above are free from the parameters of the original exponential potential models.

In the repulsive potential case, the above formulas for δ_1 and δ_2 cannot be used, because the the turning points become real and can be moved to the reference point $\mathrm{Re}(x_*)$ without affecting the imaginary part. This would imply $\delta_1 = \delta_2$, which is not correct. Even in the attractive case, the above expression of δ_1 is not convenient, because analytical property around the complex turning point is required despite the fact that the nonadiabatic transition occurs locally around the crossing region. These defects can be removed by considering the following symmetry of the problem [70]. For a given set of repulsive potentials with $V_{1,2} < 0$, we define the corresponding attractive potentials by changing the constants V_1 and V_2 to $-V_1$ and $-V_2$. This change of sign is formally equivalent to

$$V_{ii}(x) \to V_{ii}(\infty) - [V_{ii}(x) - V_{ii}(\infty)], \qquad (5.265)$$

namely the potentilas are inverted around the respective asymptotic enery levels. Then we have

$$\delta_1 = \frac{1}{\pi}\mathrm{Im}\left\{\int_{\mathrm{Re}(x_*)}^{x_*} [k_2^{(a)}(x) - k_1^{(a)}(x)]dx\right\}, \qquad (5.266)$$

$$\delta_2 = \frac{1}{\pi}\mathrm{Im}\left\{\int_{x_2}^{x_*} k_2^{(a)}(x)dx - \int_{x_1}^{x_*} k_1^{(a)}(x)dx\right\}, \qquad (5.267)$$

where $k_j^{(a)}(x)\,(j = 1,2)$ denotes the local adiabatic momenta obtained from the attractive potentials defined above. Equation (5.267) is not useful for practical applications, since this contains the complex turning points and requires the global analytical structure of the potentilal matrix elements. Actually, we can choose the two generalized parameters expressed in terms of contour integrals, δ_1 given in Eq. (5.266) and δ_2 expressed as

$$\delta_2 = \frac{1}{\pi}\mathrm{Im}\left\{\int_{\mathrm{Re}(x_*)}^{x_*} [k_2^{(r)}(x) - k_1^{(r)}(x)]dx\right\}, \qquad (5.268)$$

where $k_j^{(r)}(x)\,(j = 1,2)$ are the local adiabatic momenta of the original jth repulsive adiabatic potentials, and x_* is the corresponding complex crossing point. Similarly, in the case of attractive potentials, we can use Eq. (5.264) for δ_2 and Eq. (5.268) for δ_1 instead of Eq. (5.263). In this case $k_j^{(r)}(x)\,(j = 1,2)$ in Eq. (5.268) are the local momenta corresponding to the repulsive potentials obtained from the original attractive ones according to the transformation given in Eq. (5.265).

Some numerical examples in comparison with the exact numerical solutions of coupled differential equations are given in Figs. 5.18 and 5.19.

5.5.   Mathematical Implications

The mathematical method used in Sec. 5.1 to derive the exact expressions of the Stokes constants for the differential equation Eq. (5.19) with the quartic

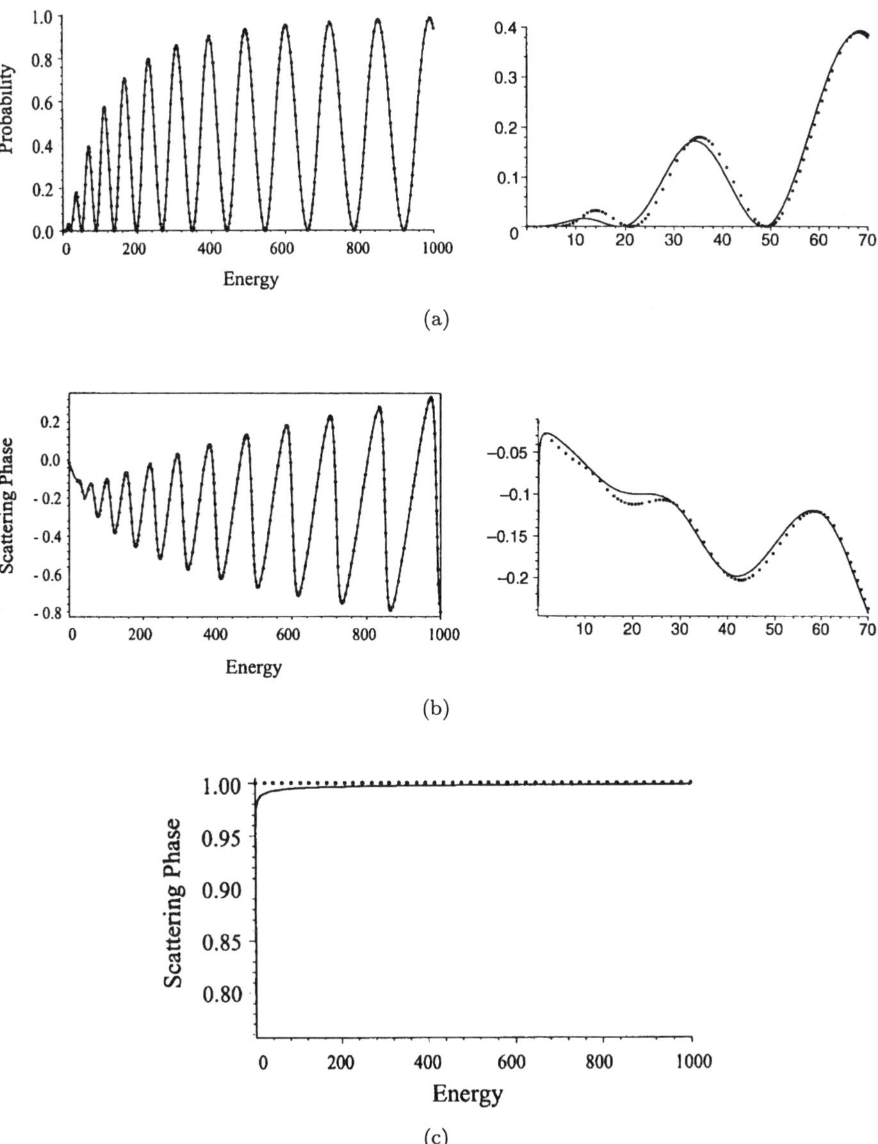

(a)

(b)

(c)

Fig. 5.18. Accuracy of the reduced semiclassical S^{R}-matrix in the case of model potential with $V_1 = -30$, $V_2 = -40$, $V = 20$, $U_1 = 0$ and $U_2 = -15$ (an example of large asymptotic energy separation). (a) $P = |S^{\mathrm{R}}_{12}(E)|^2$, (b) $\Phi = \arg\{S^{\mathrm{R}}_{11}(E)/S^{\mathrm{R}}_{22}(E)\}/\pi$, (c) $\Psi = \arg\{S^{\mathrm{R}}_{11}(E)S^{\mathrm{R}}_{22}(E)\}/\pi$. (Taken from Ref. [70] with permission.)

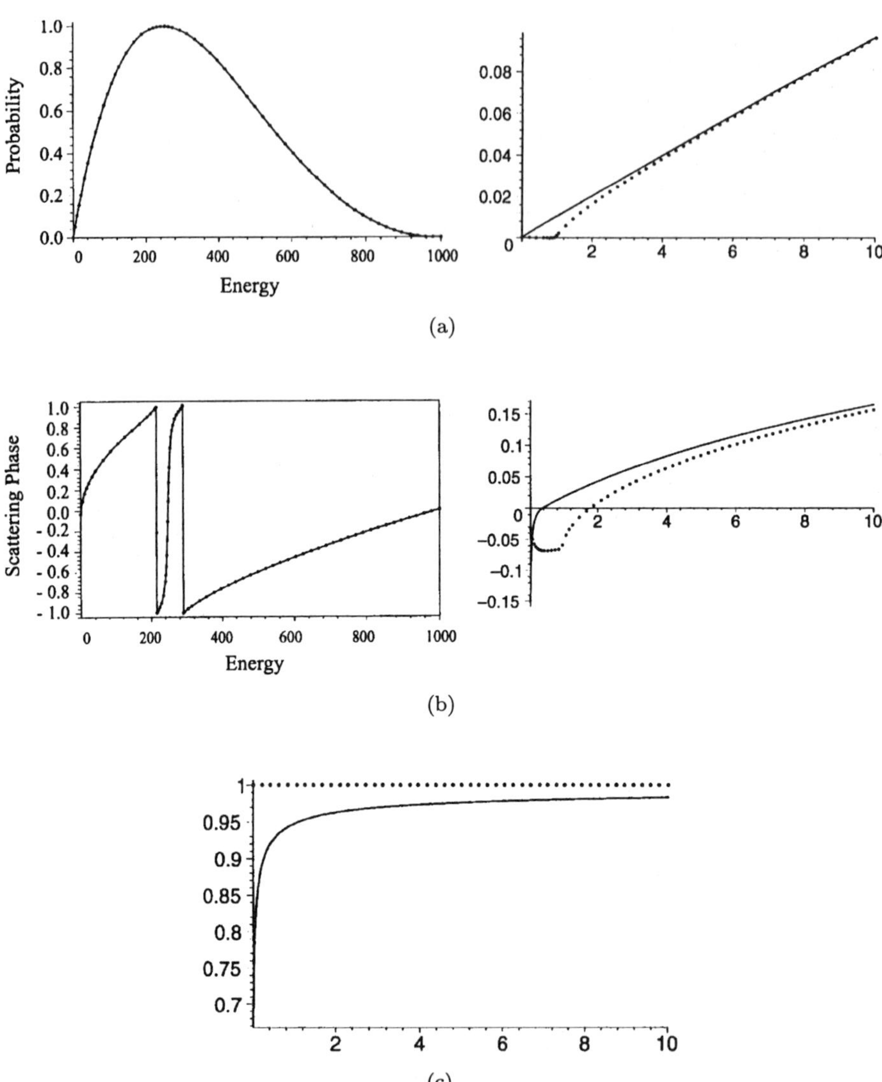

(a)

(b)

(c)

Fig. 5.19.　The same as Fig. 5.18 for the potential with $V_1 = -0.1$, $V_2 = -0.1002$, $V = 0.005$, $U_1 = 0$ and $U_2 = -0.1$. (An example of almost parallel potentials with small asymptotic energy separation). (Taken from Ref. [70] with permission.)

polynomial coefficient function $q(z)$ can also be applied to the same differential equation with the following three kinds of coefficient functions [50]:

$$\text{(i)} \qquad q(z) = \sum_{j=0}^{3} a_j z^j \quad \text{with } a_2 = 0\,, \tag{5.269}$$

$$\text{(ii)} \qquad q(z) = a_{2N-1} z^{2N-1} + \sum_{j=-\infty}^{N-2} a_j z^j\,, \tag{5.270}$$

$$\text{(iii)} \qquad q(z) = a_{2N} z^{2N} + \sum_{j=-\infty}^{N-1} a_j z^j\,, \tag{5.271}$$

where z is a complex variable, a_j are complex coefficients and N is an arbitrary positive integer. According to Hinton [55], the following transformation is introduced:

$$\phi(z) = z^{-M/4} u(\zeta) \tag{5.272}$$

and

$$\zeta = \frac{4}{M+2}(-a_M)^{1/2} z^{(M+2)/2}\,, \tag{5.273}$$

where M is the highest positive order of $q(z)$, i.e. $q(z) \to a_M z^M$ at $|z| \to \infty$. Then the differential equation for $u(\zeta)$ is

$$\frac{d^2}{d\zeta^2} u(\zeta) + Q(\zeta) u(\zeta) = 0\,, \tag{5.274}$$

where

$$Q(\zeta) = -\frac{1}{4} - \frac{q(z) - a_M z^M}{4 a_M z^M} - \frac{(M+2)^2 - 4}{64 a_M} z^{-(M+2)}\,. \tag{5.275}$$

There are $M + 2$ Stokes lines in the complex z-plane in the asymptotic region and all of these coincide with the x-axis in the complex ζ-plane. The Stokes constants U_j $(j = 1, M + 2)$ in the z-plane and those T_j $(j = 1, M + 2)$ in the ζ-plane are connected in simple relations, and the exact expressions for T_j can be obtained by generalizing the coupled wave integral equations method devised by Hinton [55] (see Ref. [50]). This means that new special functions, in principle, could be constructed, corresponding to the differential equations which have the four kinds of coefficient functions (the above three cases and

the quartic polynomial case discussed in Sec. 5.1), although the expressions
of the Stokes constants have rather complicated forms of series expansions.
Below, skipping the mathematical derivations, only the final expressions of
Stokes constants are given for the above three cases.

5.5.1. *Case* (i)

The distribution of the Stokes lines is shown in Fig. 5.20. The relations between
U_j and T_j are simply

$$U_j = T_j, \qquad (j = 1 - 5), \qquad (5.276)$$

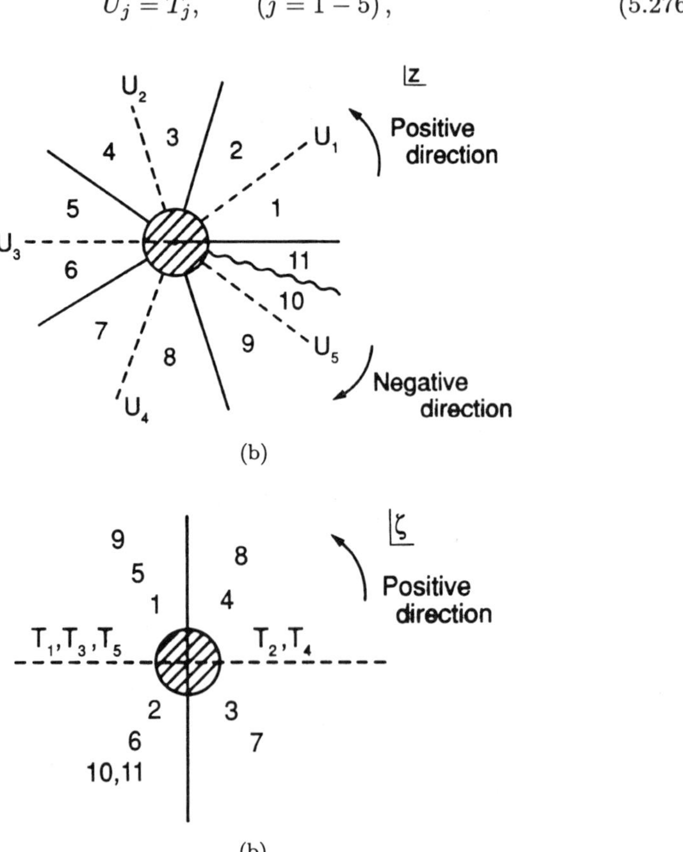

Fig. 5.20.   Stokes (- - -) and anti-Stokes (——) lines in the asymptotic region. ∿: branch
cut, and $\arg \zeta = 3 \arg(z + \pi/2)$. (Taken from Ref. [50] with permission.)

and T_j are related with each other by

$$T_{i+1}(Q_0, Q_2) = T_i(\omega^4 Q_0, \omega^6 Q_2) \quad \text{for } i = 1 - 4, \tag{5.277}$$

where

$$Q(\zeta) = -\frac{1}{4} + \frac{Q_0}{\zeta^{4/5}} + \frac{Q_2}{\zeta^{6/5}} + \frac{Q_3}{\zeta^2}, \tag{5.278}$$

and

$$\omega = e^{-i\pi/3}. \tag{5.279}$$

The Stokes constant T_1 is given by

$$T_1 = \sum_{s=0}^{\infty} \sum_{n=0}^{\infty} \frac{(10Q_0)^n}{n!} \Delta_s \frac{-2\pi i}{\Gamma(-(n-s-5)/5)} \exp(i\pi(n-s-5)/5), \tag{5.280}$$

where

$$\Delta_s = \sum_{n+m=s} B_{n+1}^{(1)} \beta_m^{(1)}, \tag{5.281}$$

$$\alpha_{n+5}^{(1)} = -\sum_{q+s=n} \delta_s W_{sq}, \tag{5.282}$$

$$\beta_{n+1}^{(1)} = -\frac{5}{n+1} \sum_{p+q=n} B_{p+1}^{(2)} \alpha_{q+1}^{(1)}, \tag{5.283}$$

$$B_n^{(1)} = \sum_{K=0}^{n-1} B_{n-K} T_K(d_1, d_2), \tag{5.284}$$

$$B_{K=0}^{(2)} = \sum_{K=0}^{n-1} B_{n-K} T_K(-d_1, -d_2), \tag{5.285}$$

$$T_{2K}(d_1, d_2) = \sum_{n=[K/2]}^{K} \Theta(3n - K) \frac{(10d_1)^{3n-K}(10d_2/3)^{K-n}}{(3n-K)!(K-n)!}, \tag{5.286}$$

$$T_{2K+1}(d_1, d_2) = \sum_{n=[(K-1)/3]}^{K} \Theta(3n + 1 - K)$$

$$\times \frac{(10d_1)^{3n+1-K}(10d_2/3)^{K-n}}{(3n + 1 - K)!(K - n)!} \,, \tag{5.287}$$

$$d_1 = -Q_2, \qquad d_2 = -Q_0^2 \,, \tag{5.288}$$

$$B_n = v_n + \sigma_n \,, \tag{5.289}$$

$$v_{2n} = 0, \tag{5.290}$$

$$v_{2n+1} = \frac{\sum_{m=2}^{2n} v_m v_{2n+2-m} + P_{2n+1}}{(2n + 5)/5 - 2v_1} \,, \tag{5.291}$$

$$\sigma_j(j = 1 - 4, 2n) = 0 \,, \tag{5.292}$$

$$\sigma_5 = 4Q_0/5 \,, \tag{5.293}$$

$$\sigma_7 = 6Q_2/5 \,, \tag{5.294}$$

$$\sigma_9 = 2Q_0\sigma_5 + 8Q_0^2/5 \,, \tag{5.295}$$

$$\sigma_{11} = 2Q_0\sigma_7 + 2Q_2\sigma_5, \tag{5.296}$$

$$\sigma_{2n+1} = 2Q_0\sigma_{2n-3} + 2Q_2\sigma_{2n-5} + 2Q_0^2\sigma_{2n-7} \,, \tag{5.297}$$

$$P_1 = 2Q_0Q_2 + 0.21, \qquad P_j = 0 \quad (j = 2, 4, 6), \qquad P_3 = Q_2^2 + 2Q_0^3 \,,$$
$$P_5 = 2Q_2Q_0^2, \qquad P_7 = Q_0^4 \,. \tag{5.298}$$

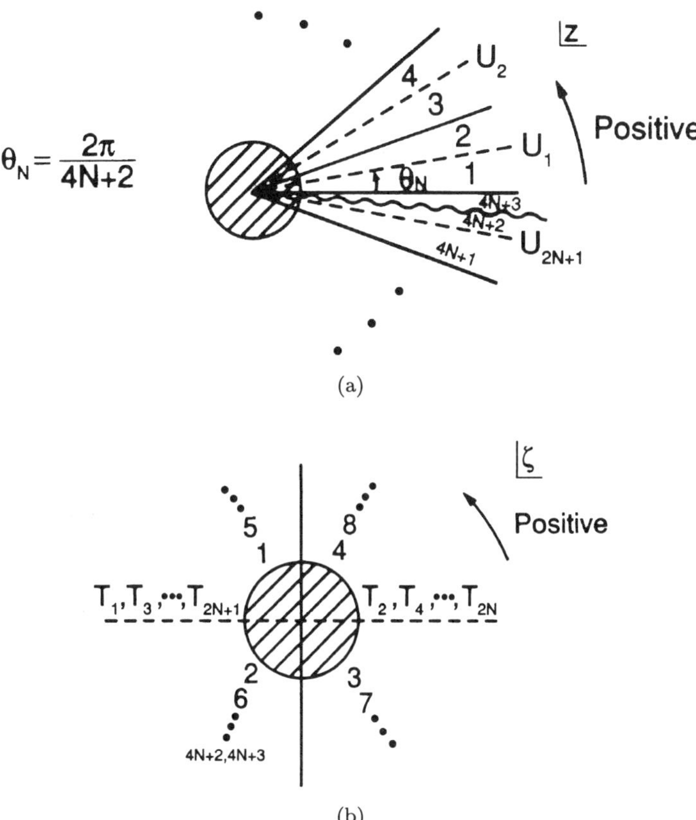

$$\theta_N = \frac{2\pi}{4N+2}$$

Fig. 5.21. Stokes (- - -) and anti-Stokes (———) lines in the asymptotic region. $\sim\!\sim$: branch cut, and $\arg \zeta = (N + 1/2) \arg z + \pi/2$. (Taken from Ref. [50] with permission.)

5.5.2. *Case* (ii)

The Stokes lines in the asymptotic region are shown in Fig. 5.21. The relations among the Stokes constants are simply given by

$$U_j = T_j \qquad (j = 1, \ldots, 2N + 1)$$

$$T_{j+1}(Q_1, Q_2, \ldots) = T_j(\omega^{2(N+1)}Q_1, \omega^{2(N+2)}Q_2, \ldots), \tag{5.299}$$

where

$$\omega = e^{-i\pi/(2N+1)} \tag{5.300}$$

and

$$Q(\zeta) = -\frac{1}{4} + \sum_{n=1}^{\infty} Q_n \zeta^{-2[(N+n)/(2N+1)]} \, . \tag{5.301}$$

The Stokes constant T_1 is given by

$$T_1 = -2\pi i \sum_{s=0}^{\infty} \left(\sum_{q=0}^{s} B_{s-q+1}^{(1)} \beta_q^{(1)} \right) \frac{e^{i\pi(-(2N+1+s)/(2N+1))}}{\Gamma[(2N+1+s)/(2N+1)]} \, , \tag{5.302}$$

where

$$\beta^{(1)} = -\frac{2N+1}{n} \sum_{n-m+1}^{(2)} \alpha_m^{(1)} \, , \tag{5.303}$$

$$\alpha_{2N+n}^{(1)} = \sum_{s=1}^{n} \left(\sum_{q=0}^{s-1} B_{s-q}^{(1)} \beta_q^{(1)} \right) (-1)^{1+(n-s)/(2N+1)}$$

$$\times \frac{\Gamma[(2N+n)/(2N+1)]}{\Gamma[(2N+s)/(2N+1)]} \gamma \left(\frac{n-s}{2N+1} \right) , \tag{5.304}$$

$$\alpha_j^{(1)} = 0, \qquad (j = 1, \ldots, 2N) \, , \tag{5.305}$$

$$\gamma(X) = \begin{cases} 1 & (\text{for } X = 0, 1, 2, \ldots) \\ 0 & (\text{otherwise}) \end{cases} , \tag{5.306}$$

$$B_n^{(1)} = B_n + \sum_{s=1}^{n-1} B_{n-s} T_s(d_1, d_2, \ldots, d_N) \, , \tag{5.307}$$

$$B_n^{(2)} = B_n + \sum_{s=1}^{n-1} B_{n-s} T_s(-d_1, -d_2, \ldots, d_N) \, , \tag{5.308}$$

$$d_n = -2Q_n \frac{2N+1}{2n-1} \, , \tag{5.309}$$

$$T_{2s}(d_1, d_2, \ldots, d_N) = \sum_{m=[s/(2N-1)]}^{s} \frac{\Theta[m(2N-1) - s]}{(2m)!}$$

$$\times \sum_{n_1+n_2+\cdots+n_{2m}=m+s} d_{n_1} \cdots d_{n_{2m}} \, , \tag{5.310}$$

$$T_{2s-1}(d_1, d_2, \ldots, d_N) = \sum_{m=[(s-N)/(2N-1)]}^{s-1} \frac{\Theta[m(2N-1) - s + N]}{(2m+1)!}$$

$$\times \sum_{n_1+n_2+\cdots+n_{2m+1}=m+s} d_{n_1} d_{n_2} \cdots d_{n_{2m+1}}, \quad (5.311)$$

$$B_n = v_n \quad \text{for } 1 \le n \le 2N + 2 \,,$$

$$B_{2n} = v_{2n} \quad \text{for } N + 2 \le n \,, \quad (5.312)$$

$$B_{2n+1} = v_{2n+1} + \sigma_{n-N} \quad \text{for } N + 1 \le n \,,$$

$$v_{2s}\left(-\frac{2N+2s}{2N+1}\right) + \sum_{m=1}^{2s} v_m v_{2s+1-m} = 0 \,,$$

$$v_{2s+1}\left(-\frac{2N+2s+1}{2N+1}\right) + \sum_{m=1}^{2s+1} v_m v_{2s+2-m} + P_{s+N+1} = 0 \,, \quad (5.313)$$

$$v_1^2 - v_1 + P_{N+1} = 0 \,,$$

$$\sigma_m = \begin{cases} Q_m \dfrac{2(N+m)}{2N+1} & \text{for } 1 \le m \le N \,, \\[2ex] 0 & \text{for } m = N + 1 \,, \\[2ex] \displaystyle\sum_{n=1}^{n-N-1} 2Q_n \sigma_{m-N-n} & \text{for } N + 2 \le m \le 2N + 1 \,, \\[2ex] \displaystyle\sum_{n=1}^{N} 2Q_n \sigma_{m-N-n} & \text{for } 2N + 2 \le m \,, \end{cases} \quad (5.314)$$

and

$$P_s = \begin{cases} Q_{N+1} & \text{for } s = N + 1 \,, \\[2ex] Q_s + \displaystyle\sum_{m=1}^{s-N-1} Q_m Q_{s-N-m} & \text{for } N + 2 \le s \le 3N \,, \\[2ex] Q_s & \text{for } 3N + 1 \le s \,. \end{cases} \quad (5.315)$$

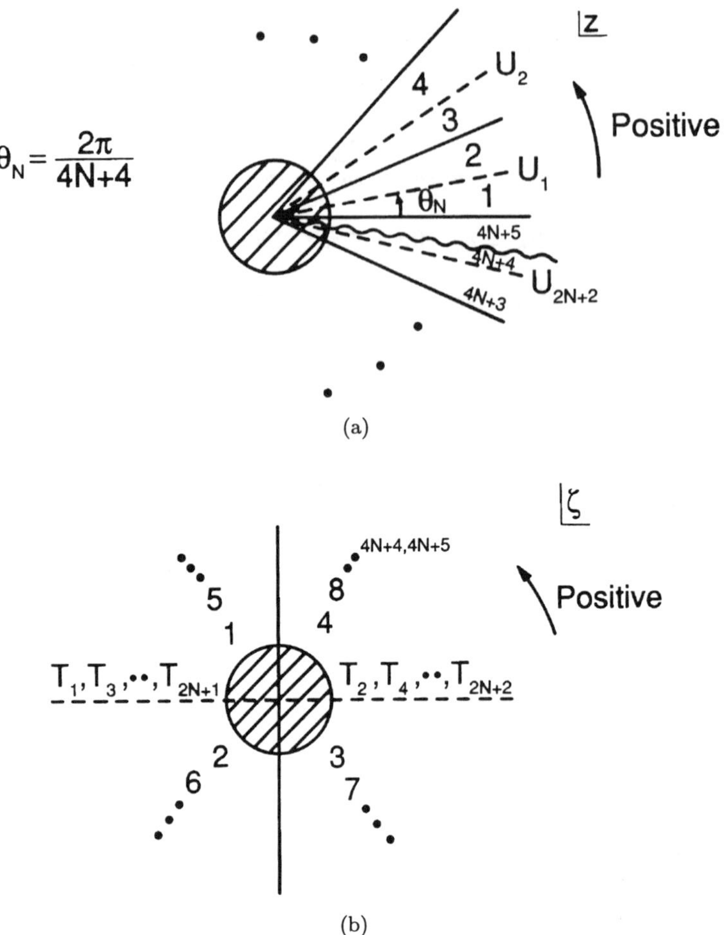

Fig. 5.22.　Stokes (- - -) and anti-Stokes (———) line in the asymptotic region. $\sim\!\sim$: branch cut, $\arg\zeta = (N+1)\arg(z+\pi/2)$. (Taken from Ref. [50] with permission.)

5.5.3.　*Case* (iii)

The Stokes line distribution is shown in Fig. 5.22, and the relations among the Stokes constants are given by

$$U_{2N+1} = T_{2N+1}\left[\frac{2}{3}i\sqrt{a_{2N}}\right]^{-2Q_1},$$

$$U_{2N+2} = T_{2N+2}\left[\frac{2}{3}i\sqrt{a_{2N}}\right]^{2Q_1},$$

$$(5.316)$$

and

$$T_{2N+1}(Q_1, Q_2, \ldots) = T_{2N}(Q_1 \omega^{N+1}, Q_2 \omega^{N+2}, \ldots) e^{2\pi i Q_1},$$
$$T_{2N+2}(Q_1, Q_2, \ldots) = T_{2N+1}(Q_1 \omega^{N+1}, Q_2 \omega^{N+2}, \ldots) e^{-2\pi i Q_1}, \tag{5.317}$$

where

$$Q(\zeta) = -\frac{1}{4} + \sum_{n=1}^{\infty} Q_n \zeta^{-(N+n)/(N+1)}. \tag{5.318}$$

Finally, the Stokes constant T_1 is given by

$$T_1 = -2\pi i e^{i\pi\nu} \sum_{n=0}^{\infty} \left(\sum_{p+q=n} B_{p+1}^{(1)} \beta_q^{(1)} \right) \frac{e-i\pi[(N+n+1)/(N+1)]}{\Gamma[-\nu + (N+n+1)/(N+1)]}, \tag{5.319}$$

where

$$\beta_n^{(1)} = -\frac{N+1}{n} \sum_{p+q=n-1} B_{p+1}^{(2)} \alpha_{q+1}^{(1)}, \tag{5.320}$$

$$\alpha_{n+N+1}^{(1)} = \sum_{s=0}^{n} \left(\sum_{p+q=s} B_{p+1}^{(1)} \beta_q^{(1)} \right) (-1)^{(n-s)/(N+1)+1}$$
$$\times \frac{\Gamma(-\nu + (N+n+1)/(N+1))}{\Gamma(-\nu_{(}N+s+1)/(N+1))} \gamma \left(\frac{n-s}{N+1} \right), \tag{5.321}$$

$$\alpha_j^{(1)} = 0 \qquad (j = 1, 2, \ldots, N), \tag{5.322}$$

$$\nu = 2Q_1, \tag{5.323}$$

$$B_n^{(1)} = B_n + \sum_{s=1}^{n-1} B_{n-s} T_s(d_1, d_2, \ldots, d_N), \tag{5.324}$$

$$B_n^{(2)} = B_n + \sum_{s=1}^{n-1} B_{n-s} T_s(-d_1, -d_2, \ldots, -d_N), \tag{5.325}$$

$$T_s(d_1, d_2, \ldots, d_N) = \sum_{m=[s/N]}^{s} \frac{1}{m!} \Theta(mN - s)$$

$$\times \sum_{n_1+n_2+\cdots+n_m=s} d_{n_1} d_{n_2} \cdots d_{n_m} , \qquad (5.326)$$

$$d_n = -2Q_{n+1} \frac{N+1}{n} , \qquad (5.327)$$

$$B - n = \begin{cases} v_n & \text{for } 1 \leq n \leq N+1 , \\ v_n + \sigma_{n-N-1} & \text{for } N+2 \leq n , \end{cases} \qquad (5.328)$$

$$\sigma_m = \begin{cases} Q_m \left(\dfrac{N+m}{N+1} \right) & \text{for } 1 \leq m \leq N+1 , \\ \displaystyle\sum_{n=1}^{m-N-1} 2Q_n \sigma_{m-N-n} & \text{for } N+2 \leq m \leq 2N+2 , \\ \displaystyle\sum_{n=1}^{N+1} 2Q_n \sigma_{m-N-n}, & \text{for } 2N+3 \leq m , \end{cases} \qquad (5.329)$$

$$v_n = \frac{\sum_{m=2}^{n-1} v_m v_{n+1-m} + P_{n+N+1}}{(N+n)/(N+1) - 2v_1} ,$$

$$v_2 = \frac{P_{N+3}}{(N+2)/(N+1) - 2v_1} ,$$

$$v_1^2 - v_1 + P_{N+2} = 0 , \qquad (5.330)$$

and

$$P_s = \begin{cases} Q_s + \displaystyle\sum_{m=1}^{s-N-1} Q_m Q_{s-N-m} & \text{for } N+2 \leq s \leq 3N+2 , \\ Q_s & \text{for } 3N+3 \leq s . \end{cases} \qquad (5.331)$$

Chapter 6

Basic Two-State Theory for Time-Dependent Processes

Time-dependent processes naturally obey the time-dependent Schrödinger equations, and there are two types of time dependencies. One is an approximation to the time-independent quantum mechanical scattering problems: namely, a certain time-dependence is *a priori* assumed for the adiabatic parameter $R(t)$. The most typical example is a high energy approximation of atomic collision processes in which the nuclear motion is assumed to obey a certain *a priori* given classical trajectory of $R(t)$. The other is an intrinsically time dependent problem: namely, quantum mechanical processes in a certain time-dependent external field. This becomes more and more important nowadays because of the remarkable progress of laser technology. Laser intensity and frequency can now be tailored to have various shapes, i.e. time-dependencies. Processes induced by the time-dependence of the external field are nothing but the time-dependent nonadiabatic processes. The external field can also be any other such as electric and magnetic fields, of course. Generally speaking, the time-dependent theory is easier than the time-independent theoy, simply because the Schrödinger equation of the former is the first order differential equation compared to the second order of the time-independent case. Present status of the basic theories of the two-state time-dependent nonadiabatic transition is summarized in this chapter.

6.1. Exact Solution of Quadratic Potential Problem

The famous Landau–Zener formula, Eq. (3.4), is the exact solution of the linear potential model in which the diabatic potentials are linear functions of time

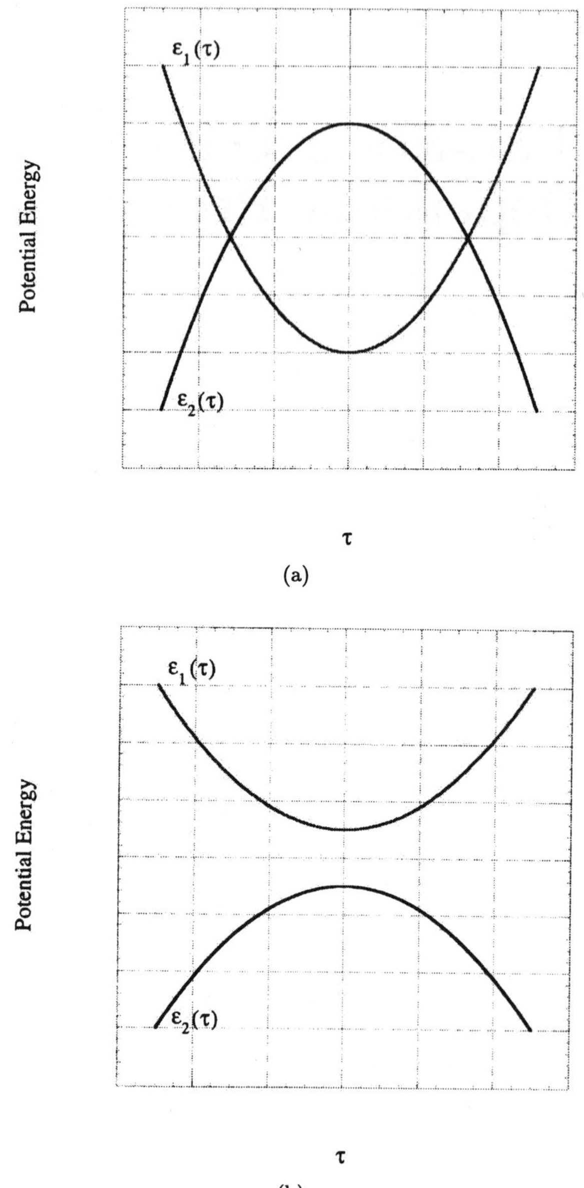

Fig. 6.1.   (a) Schematic crossing quadratic potentials. (b) Schematic non-crossing quadratic potentials. (Taken from Ref. [71] with permission.)

and the diabatic coupling is constant. This was solved by Zener [2]. This cannot present an exact solution, however, for the linear potential model in the coordinate space, as was explained in the previous chapter. The formula works well at high energies, where the turning points are well separated from the crossing points, i.e. transition point. However, when the energy becomes lower and the turning points come close to the crossing point, then the theory breaks down, because the two crossing points, one on the way-in and the other on the wayout, cannot be treated independently (see Figs. 2.1(a) and 6.1). This corresponds to a two-closely-lying-crossing-point problem in the time-dependent formalism. Furthermore, when the energy is lower than the crossing point, no real crossing appears even in the diabatic representation and the Landau–Zener formula does not work at all. All these cases were solved, however, by Zhu and Nakamura, as is described in the previous chapter. This suggests that we can formulate a new time-dependent theory which can solve all these cases [71].

Let us start with the two coupled time-dependent Schrödinger equations for a quadratic potential model in which the diabatic potentials are quadratic functions of time and the diabatic coupling is constant.

$$i\hbar \frac{d}{dt}\begin{pmatrix} c_1(t) \\ c_2(t) \end{pmatrix} = \begin{pmatrix} \hat{\epsilon}_1(t) & V_0 \\ V_0 & \hat{\epsilon}_2(t) \end{pmatrix} \begin{pmatrix} c_1(t) \\ c_2(t) \end{pmatrix}, \qquad (6.1)$$

where V_0 is a constant diabatic coupling and $\hat{\epsilon}_1(t)$ and $\hat{\epsilon}_2(t)$ are the quadratically time-dependent potentials given by

$$\begin{aligned} \hat{\epsilon}_1(t) &= \alpha_1 t^2 + \beta_1 , \\ \hat{\epsilon}_2(t) &= \alpha_2 (t - \gamma)^2 + \beta_2 . \end{aligned} \qquad (6.2)$$

Introducing the dimensionless variables,

$$\tau = \frac{2V_0(t)}{\hbar}\left(t + \frac{\alpha_2\gamma}{\alpha_1 - \alpha_2}\right), \qquad (6.3)$$

$$\Delta\epsilon(\tau) = \alpha\tau^2 - \beta, \qquad (6.4)$$

$$\alpha = \frac{(\alpha_1 - \alpha_2)\hbar^2}{8V_0^3}, \qquad (6.5)$$

and

$$\beta = \frac{1}{2V_0}\left(\beta_2 - \beta_1 + \frac{\alpha_1\alpha_2^2\gamma^2}{\alpha_1 - \alpha_2}\right), \tag{6.6}$$

we have

$$i\hbar\frac{d}{dt}\begin{pmatrix} c_1(\tau) \\ c_2(\tau) \end{pmatrix}$$

$$= \begin{pmatrix} 0 & \frac{1}{2}\exp\left[i\int_{\tau_0}^{\tau}\Delta\epsilon(\tau)d\tau\right] \\ \frac{1}{2}\exp\left[-i\int_{\tau_0}^{\tau}\Delta\epsilon(\tau)d\tau\right] & 0 \end{pmatrix}\begin{pmatrix} c_1(\tau) \\ c_2(\tau) \end{pmatrix}. \tag{6.7}$$

It is clearly seen that this problem can be described completely by the two parameters α and β. The parameter α can be assumed to be positive without loss of generality. The positive (negative) β corresponds to the crossing (non-crossing) case. The negative case is called "diabatically avoided crossing."

Now we try to obtain the properties and exact solutions of the transition matrix of this problem. First we introduce the evolution operator $F(z, z_0)$ which satisfies

$$\begin{pmatrix} c_1(z) \\ c_2(z) \end{pmatrix} = F(z, z_0)\begin{pmatrix} c_1(z_0) \\ c_2(z_0) \end{pmatrix}, \tag{6.8}$$

where z is a complex variable, the real part of which is τ. It can be shown from Eq. (6.7) that $F(z, z_0)$ satisfies the following properties:

$$\det F(z, z_0) = 1 \tag{6.9}$$

$$F(z, z_0) = \begin{pmatrix} 0 & -1 \\ 1 & 0 \end{pmatrix}F^*(z, z_0)\begin{pmatrix} 0 & 1 \\ -1 & 0 \end{pmatrix}. \tag{6.10}$$

The transition matrix T^{R} is defined by

$$\begin{pmatrix} c_1(\infty) \\ c_2(\infty) \end{pmatrix} = T^{\mathrm{R}}\begin{pmatrix} c_1(-\infty) \\ c_2(-\infty) \end{pmatrix}, \tag{6.11}$$

and satisfies $T^{\mathrm{R}} = F(\infty, -\infty)$. The overall transition probability P_{12} is equal to $|T_{12}^{\mathrm{R}}|^2$. The transition amplitude T^{R} naturally satisfies the unitarity and the following properties:

$$T_{11}^{\mathrm{R}} = (T_{22}^{\mathrm{R}})^* \quad \text{and} \quad T_{12}^{\mathrm{R}} = T_{21}^{\mathrm{R}} = \text{pure imaginary}. \tag{6.12}$$

In order to derive analytical expression of T^R we first transform Eq. (6.7) into the following single second order differential equation:

$$\frac{d^2 u(\tau)}{d\tau^2} + I(\tau)u(\tau) = 0 \,, \tag{6.13}$$

where

$$u(\tau) = c_1(\tau)\exp\left[-\frac{i}{2}\int_{\tau_0}^{\tau} \Delta\epsilon(\tau)d\tau\right] \tag{6.14}$$

and

$$I(\tau) = \frac{i}{2}\Delta\dot{\epsilon}(\tau) + \frac{1}{4}[\Delta\epsilon(\tau)]^2 + \frac{1}{4} = \frac{1}{4} - i\alpha\tau + \frac{1}{4}(\alpha\tau^2 - \beta)^2 \,. \tag{6.15}$$

It is interesting to note that Eq. (6.13) has exactly the same form as Eq. (5.7) with the correspondence:

$$a^2 \Longleftrightarrow \alpha \quad \text{and} \quad b^2 \Longleftrightarrow \beta\,. \tag{6.16}$$

This indicates that the present time-dependent quadratic potential model can be solved in the same way as that of the Landau–Zener type time-independent linear potential model and that the transition matrix T^R corresponds to the reduced scattering matrix S^R in the time-independent problem. Furthermoe, $\beta > 0\,(\beta < 0)$ corresponds to $b^2 > 0\,(b^2 < 0)$. This is exactly what was explained above. It is interesting to note that the NT (nonadiabatic tunneling) case does not appear in the time-dependent problem. In the NT-case the wave bifurcates into tunneling and reflection portions and the process cannot be described by a single time dependence.

The derivation of T^R required in the present discussion is the same as that of the LZ-case described in Sec. 5.1.1. Finally the transition matrix T^R is given by

$$T^R = \begin{pmatrix} 1 + U_1 U_2 & -U_2 \\ -U_2 & 1 - U_1^* U_2 \end{pmatrix} \tag{6.17}$$

with

$$U_2 = \frac{2i\,\mathrm{Im}\,U_1}{1 + |U_1|^2}\,, \tag{6.18}$$

where U_1 and U_2 are the corresponding Stokes constants which can be given by the expressions in Sec. 5.1.1 with the appropriate replacements of parameters corresponding to Eq. (6.16). The nonadiabatic transition probability P_{12} is as usual given by

$$P_{12} = 4p(1-p)\sin^2(\arg U_1) \quad \text{with} \quad p = \frac{1}{1+|U_1|^2}. \tag{6.19}$$

The exact expression of the Stokes constant U_1 is given before (Eqs. (5.30) and (5.34)).

6.2.   Semiclassical Solution in General Case

The above discussion together with the contents of Sec. 5.2 suggests us that we can derive compact and accurate semiclassical formulas for the two-state quadratic potential model and also for such a more general case that has the same analytical structure as that of the quadratic potential model. The transition matrix T which is equivalent to the scattering matrix in the time-independent scheme can also be expressed as a product of propagation matrices and nonadiabatic transition matrices as in Eq. (3.16).

6.2.1.   *Two-crossing case:* $\beta \geq 0$ *(see Fig. 6.1(a))*

This corresponds to the case of $E \geq E_X$ in the time-independent Landau–Zener type. The overall transition matrix from A to B is given by

$$T = P_{BX_2} I_{X_2} P_{X_2 X_1} I_{X_1} P_{X_1 A}, \tag{6.20}$$

where the nonadiabatic transition matrix I_X and the adiabatic propagation matrix P_{BA} are given by

$$I_X = \begin{pmatrix} \sqrt{1-p}\,e^{i\phi_S} & -\sqrt{p}\,e^{i\sigma_0} \\ \sqrt{p}\,e^{-i\sigma_0} & \sqrt{1-p}\,e^{-i\phi_S} \end{pmatrix}, \tag{6.21}$$

$$P_{BA} = \begin{pmatrix} \exp\left[\dfrac{i}{2\hbar}\displaystyle\int_{t_A}^{t_B} \Delta E(t)dt\right] & 0 \\ 0 & \exp\left[-\dfrac{i}{2\hbar}\displaystyle\int_{t_A}^{t_B} \Delta E(t)dt\right] \end{pmatrix}. \tag{6.22}$$

Here $\Delta E(t) > 0$ is the adiabatic energy difference, $E_2(t) - E_1(t)$, and the matrix indices 1 and 2 correspond to the state suffix. The other parameters (p and ϕ_S) are the same as before given in Sec. 5.2.1.1. The basic parameters a^2 and b^2

are replaced by α and β according to the correspondence given by Eq. (6.16). When the two avoided crossings are well separated, these parameters can be estimated at each avoided crossing by

$$\alpha = \frac{\sqrt{d^2 - 1}\hbar^2}{2V_0^2(t_t^2 - t_b^2)} \tag{6.23}$$

and

$$\beta = -\sqrt{d^2 - 1}\left(\frac{t_b^2 + t_t^2}{t_t^2 - t_b^2}\right) \tag{6.24}$$

with

$$V_0 = \frac{1}{2}(E_2(t_0) - E_1(t_0)) \tag{6.25}$$

and

$$d^2 = \frac{[E_2(t_b) - E_1(t_b)][E_2(t_t) - E_1(t_t)]}{[E_2(t_0) - E_1(t_0)]^2}, \tag{6.26}$$

where t_0 is the real position at which $E_2(t) - E_1(t)$ becomes minimum, and $t_b(t_t)$ is the bottom (top) of the potential $E_2(t)(E_1(t))$. When the two avoided crossings are not well separated, it is better to fit the curve of $\Delta E(t)$ by a quartic polynomial and then the two parameters are given by the following expressions:

$$\alpha = \frac{\hbar^2\sqrt{(\Delta E_M - \Delta E_m)(\Delta E_M + \Delta E_m)}}{(\Delta E_m)^3 t_m^2}, \tag{6.27}$$

$$\beta = \frac{\sqrt{(\Delta E_M - \Delta E_m)(\Delta E_M + \Delta E_m)}}{\Delta E_m} \tag{6.28}$$

with

$$\Delta E_m = \frac{1}{2}[\Delta E(t_m^{(+)}) + \Delta E(t_m^{(-)})], \tag{6.29}$$

$$\Delta E_M = \Delta E(t_M), \tag{6.30}$$

and

$$t_m = \frac{1}{2}(t_m^{(+)} + t_m^{(-)}), \tag{6.31}$$

where $t_m^{(\pm)}$ are the positions at which $\Delta E(t) = E_2(t) - E_1(t)$ becomes minimum, and t_M is the position at which ΔE becomes maximum. The quantities σ_0 and δ in the time-independent formulation defined by the integral of $k_1(R) - k_2(R)$ from R_0 to R_* (see Eqs. (3.22), (5.99)–(5.101)) are replaced by

$$\sigma_0 + i\delta = \frac{1}{\hbar} \int_{t_0}^{t_*} \Delta E(t) dt \,, \tag{6.32}$$

where t_* represents the complex adiabatic crossing point corresponding to the minimum position t_0 which is either $t_m^{(+)}$ or $t_m^{(-)}$ in the case of the above quartic polynomial. When ΔE is symmetric, then $t_m^{(\pm)}$ give the same results. When the two crossings are far apart from each other, then they are treated separately as mentioned above. With σ_0 and δ given by Eq. (6.32), and with the expression of the Stokes constant U_1 given before, we can obtain an accurate formula for the transition matrix. However, as in the time-independent case, we can use the quadratic potential model and avoid the complex calculus in Eq. (6.32) to obtain compact explicit expressions for σ_0 and δ. The best usage of the quadratic potential approximation is equivalent to the best usage of the linear potential approximation in the time-independent framework; and thus σ_0 and δ are finally given by Eqs. (5.100)–(5.104), where a^2, b^2, and d^2 should naturally be replaced by α, β, and d^2 given above.

Now, the reason why $\Delta E(t)/2$ appears in the propagation matrix may be understood as follows. First of all, it should be noted that the time-dependent coupled equations always depend only on the energy difference (see Eq. (6.7)). If we transform the original coupled equations in the diabatic representation, Eq. (6.1), into the adiabatic representation, then we have

$$i\hbar \frac{d}{dt} \begin{pmatrix} d_1(t) \\ d_2(t) \end{pmatrix} = \begin{pmatrix} E_1(t) & \dot{\theta} \\ -\dot{\theta} & E_2(t) \end{pmatrix} \begin{pmatrix} d_1(t) \\ d_2(t) \end{pmatrix}, \tag{6.33}$$

where $d_j(t)$ $(j = 1, 2)$ and $E_j(t)$ $(j = 1, 2)$ are the coefficients and the adiabatic energies, respectively, and $\dot{\theta}$ represents the nonadiabatic coupling which is a time-derivative of the transformation angle θ given by

$$\theta = \frac{1}{2} \tan^{-1} \left(\frac{2V_0}{\epsilon_2 - \epsilon_1} \right). \tag{6.34}$$

Since the nonadiabatic coupling is very much localized at the avoided crossing t_X, at $t \le t_X - 0$ the matrix in the right hand side of Eq. (6.33) is diagonal and the solution is simply given by the adiabatic propagation,

$$d_j(t) = d_j(t=0)\exp\left[-\frac{i}{\hbar}\int_0^t E_j(t)dt\right].$$

(6.35)

This is true also for $t \geq t_X + 0$. Thus, in general, we can have the following adiabatic propagation formula for $t > t_X$ starting from $t = 0$,

$$\begin{pmatrix} d_1(t) \\ d_2(t) \end{pmatrix} = \begin{pmatrix} \exp\left[-\frac{i}{\hbar}\int_{t_X}^t E_1(t)dt\right] & 0 \\ 0 & \exp\left[-\frac{i}{\hbar}\int_{t_X}^t E_2(t)dt\right] \end{pmatrix}$$

$$\times I_X \begin{pmatrix} \exp\left[-\frac{i}{\hbar}\int_0^{t_X} E_1(t)dt\right] & 0 \\ 0 & \exp\left[-\frac{i}{\hbar}\int_0^{t_X} E_2(t)dt\right] \end{pmatrix}$$

$$\times \begin{pmatrix} d_1(t=0) \\ d_2(t=0) \end{pmatrix},$$

(6.36)

where I_X represents the nonadiabatic transition matrix at t_X. The adiabatic energies $E_{1,2}(t)$ are, as usual, given by

$$E_{1(2)}(t) = \frac{1}{2}[\epsilon_1(t) + \epsilon_2(t)] - (+)\frac{1}{2}\Delta E(t).$$

(6.37)

The first common term can be factored out and thus finally the integral $\frac{1}{2\hbar}\int \Delta E(t)dt$ remains in the phase parts.

6.2.2. *Diabatically avoided crossing case: $\beta \leq 0$ (see Fig. 6.1(b))*

As was explained above, with use of σ_0 and δ defined by the complex integral Eq. (6.32) and the Stokes constant U_1 given by Eqs. (5.108) and (5.109), all the necessary quantities to evaluate the transition matrix can be obtained (see the beginning of Sec. 5.2.1). As in the previous section, we can avoid the complex calculus and derive the compact explicit expressios for σ_0 and δ. From Eq. (6.32) we have

$$\sigma_0 + i\delta = \frac{1}{2\sqrt{\alpha}}\int_{|\beta|}^i \left(\frac{1+\zeta^2}{\zeta - |\beta|}\right)^{1/2} d\zeta.$$

(6.38)

The integral range can be divided into $(|\beta|, 0)$ and $(0, i)$, and we have

$$\sigma_0 + i\delta = \frac{i}{2\sqrt{\alpha}} \int_0^{|\beta|} \left(\frac{1+\zeta^2}{|\beta| - \zeta} \right)^{1/2} d\zeta + \frac{1}{2\sqrt{\alpha}} \int_0^i \left(\frac{1+\zeta^2}{\zeta - |\beta|} \right)^{1/2} d\zeta, \quad (6.39)$$

namely,

$$\sigma_0 + i\delta_0 = \frac{1}{2\sqrt{\alpha}} \int_0^i \left(\frac{1+\zeta^2}{\zeta - |\beta|} \right)^{1/2} d\zeta \quad (6.40)$$

and

$$\delta = \frac{1}{2\sqrt{\alpha}} \int_0^{|\beta|} \left(\frac{1+\zeta^2}{|\beta| - \zeta} \right)^{1/2} d\zeta + \delta_0 . \quad (6.41)$$

The complex integral over $(0, i)$ can be replaced by the simple quantities as before; namely, σ_0 and δ_0 can be replaced by Eqs. (5.100) and (5.101), respectively, where α and β should be used instead of a^2 and b^2.

The parameters α and β are obtained by fitting the portion of the minimum of $(\Delta E)^2$ by a quartic polynomial as

$$\alpha = \frac{\hbar^2 XY}{(XZ - Y^2)^{3/2}}, \quad (6.42)$$

$$\beta = -\frac{Y}{XZ - Y^2}, \quad (6.43)$$

where

$$X = \frac{1}{24} \left[\frac{d^4}{dt^4} (\Delta E(t))^2 \right]_{t=0}, \quad (6.44)$$

$$Y = \left[\frac{d^2}{dt^2} (\Delta E(t))^2 \right]_{t=0}, \quad (6.45)$$

$$Z = (\Delta E(t))^2_{t=0}. \quad (6.46)$$

Below we give some numerical examples taken from Ref. [71]. In Ref. [71] the expressions of σ_0 and δ which are different from and less accurate than those given above were used (see Eqs. (3.14)–(3.16) in [71]). In the parameter ranges in these calculations, however, both expressions give essentially the same results. It should also be noted that two terms are missing in Eq. (3.15) of [71] (see Eqs. (3.2)–(3.4) of [62]). The first numerical example is a simple two-state model given by

$$
\begin{pmatrix}
A\cosh(\omega t) + B & V_0 \\
V_0 & -(A\cosh(\omega t) + B)
\end{pmatrix}.
\tag{6.47}
$$

The complex crossing point t_* can be easily obtained as

$$
t_* = \frac{1}{\omega}\left\{\ln\left[\frac{-B \pm iV_0}{A} \pm \sqrt{\left(\frac{-B \pm iV_0}{A}\right)^2 - 1}\right] \pm 2\pi i\right\}.
\tag{6.48}
$$

The quantities σ_0 and δ can be calculated easily from the complex integral with use of the complex zero closest to the real axis. The numerical results are shown in Figs. 6.2(a) and 6.2(b), which correspond to $A/V_0 = 1.0$ and $A/V_0 = 5.0$, respectively. The semiclassical theory works well even in the region $B/V_0 > -A/V_0$ where the diabatic curves do not cross.

The second example is a LASIN (laser assisted surface ion neutralization) process, in which the neutralization of an ion by a collision with solid surface is enhanced by a laser with frequency η. This process is described by the following differential equations [72],

$$
i\hbar\frac{d}{dt}\begin{pmatrix} c_1(t) \\ c_2(t) \end{pmatrix} = \begin{pmatrix} E_2(t) & g(t) \\ g(t) & E_1(t) + \hbar\eta \end{pmatrix}\begin{pmatrix} c_1(t) \\ c_2(t) \end{pmatrix}
\tag{6.49}
$$

with

$$
E_1(t) = \frac{1}{2}[\epsilon_1 + \epsilon_2 - \sqrt{\omega^2 + 4V^2(t)}],
\tag{6.50}
$$

$$
E_2(t) = \frac{1}{2}[\epsilon_1 + \epsilon_2 + \sqrt{\omega^2 + 4V^2(t)}],
\tag{6.51}
$$

$$
g(t) = \frac{W_0\omega f(\lambda t)}{2\sqrt{\omega^2 + 4V^2(t)}},
\tag{6.52}
$$

$$
V(t) = V_0 f(\lambda t) = V_0 \cosh^{-1}(\lambda t),
\tag{6.53}
$$

where $\omega = \epsilon_2 - \epsilon_1 > 0$ is the energy defect between the ionic state of the surface and the atomic state, $V(t)$ and $W_0 f(\lambda t)$ represent the interaction between the two states ϵ_1 and ϵ_2, and the interaction with the laser field, respectively.

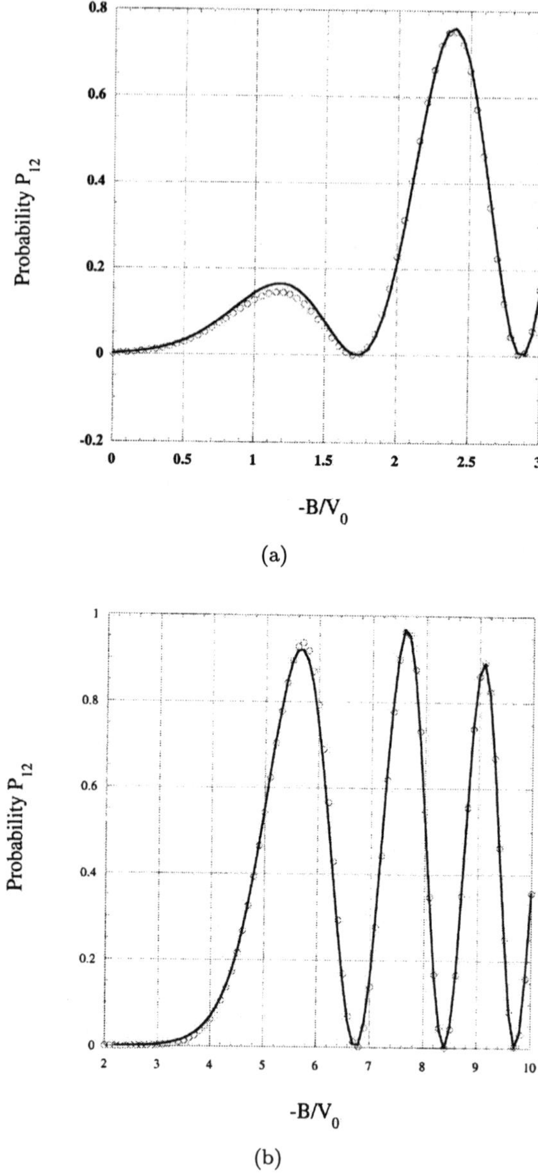

Fig. 6.2. Transition probability in the case of hyperbolic cosine potential $[\Delta\epsilon(\tau) = 2A/V_0 \cosh\tau + 2B/V_0]$ against B/V_0. —: exact numerical result, open circle: semiclassical theory. (a) $A/V_0 = 1.0$, (b) $A/V_0 = 5.0$. (Taken from Ref. [71] with permission.)

The adiabatic state $E_1(t)$ without laser field is dressed with one photon absorption to $E_1(t) + \hbar\eta$. The adiabatic states we finally need are obtained by further diagonalizing the matrix in Eq. (6.49). Introducing the dimensionless parameters,

$$X = \lambda t, \qquad r = \frac{\hbar\eta}{2V_0}, \qquad q = \frac{\hbar\omega}{2V_0}, \qquad v = \frac{2V_0}{\lambda}, \qquad E = \frac{W_0}{2V_0}, \qquad (6.54)$$

we obtain the complex crossing point as

$$X_* = \ln\left[\pm\frac{1}{\sqrt{x-q}} \pm \sqrt{\frac{1}{x-q^2} - 1}\right] \pm 2n\pi i \qquad (n = 0, 1, 2, \ldots), \qquad (6.55)$$

where x is a solution of

$$x^4 + (2E^2q^2 - 2r^2)x^3 + (E^4q^4 + 2E^2q^2r^2 + r^4 - 2E^2q^4)x^2$$

$$- (E^2q^6 + 2E^2 + q^4r^2)x + E^4q^8 = 0. \qquad (6.56)$$

By appropriately choosing the solution X_* closest to the real axis and evaluating σ_0 and δ, the LASIN probability can be calculated as a function of r which represents the laser frequency. Although in Ref. [72] only the weak laser intensity, $\alpha \simeq 10^{-4}$, is considered, here we consider a strong intensity, $\alpha \simeq 0.7$, in order to demonstrate the validity of the present semiclassical theory. The results are shown in comparison with the exact numerical results in Fig. 6.3 for the parameters $E = 0.25$, $q = 0.2923$, and $v = 23.585$ (cf. Ref. [72]). As is seen, the theory works well.

Let us next discuss about the neutrino conversion in matter. The MSW effect was proposed by Mikheyev, Smirnov and Wolfenstein in order to solve the solar neutrino problem that the intensity of the solar neutrino does not agree with that of the standard solar model [73, 74]. When electron neutrinos pass through an inhomogeneous matter, they interact with electrons to get extra potential and thus their "diabatic" energy changes as a function of electron density. The other neutrinos, for instance muon neutrino, are not affected by electrons. Thus we have a curve crossing picture as is shown in Fig. 2.3 and electron neutrinos produced in the center of the sun can be converted to muon neutrinos when they pass through the crossing region which is called the region of resonance electron density. If this kind of conversion really occurs, this guarantees that neutrino has a finite mass and gives a big influence on the standard theory of elementary particles. This conversion problem has

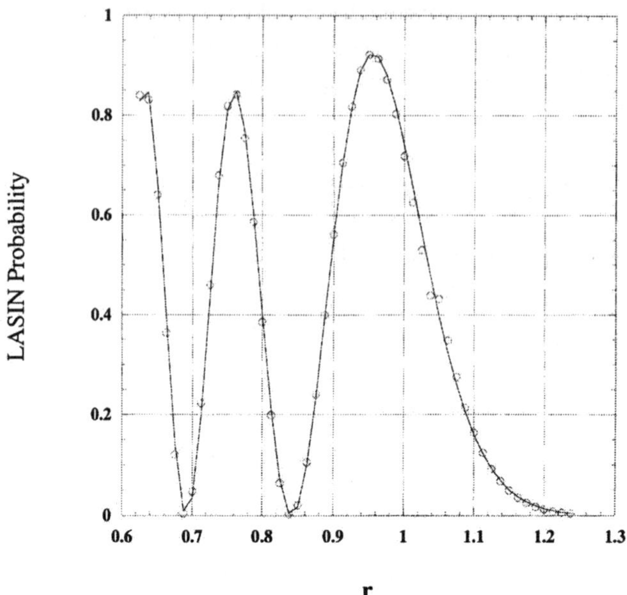

Fig. 6.3.   LASIN probability versus dimensionless photon energy r. $v = 23.585$, $q = 0.29225$, and $E = 0.25$. $r > 1.04183$ corresponds to the diabatically avoided crossing region. —: numerical exact result, o: similassical theory. (Taken from Thesis of Y. Teranishi with permission.)

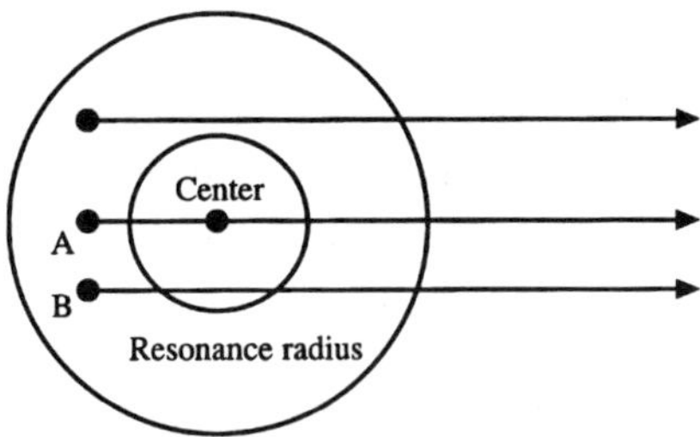

Fig. 6.4.   Schematic diagram for the production of neutrinos.

been discussed in terms of the nonadiabatic transition due to curve crossing basically with use of the Landau–Zener formula (see for instance [75]). They treated the case that the neutrino produced at A in Fig. 6.4 passes through the resonance radius twice just like in the ordinary head-on collision. In this case it may be all right to use the Landau–Zener fomula, but obviously it is not good when the neutrino is produced at the position B or C. In the case of B the neutrino experiences the closely lying two crossings and in the case of C it experiences the diabatically avoided crossing. The theory presented in this chapter can be directly applied to these problems.

The coupled differential equations we have to solve are [76]

$$
i\hbar \frac{d}{dt} \begin{pmatrix} \psi_e(t) \\ \psi_\mu(t) \end{pmatrix} = \Delta E \begin{pmatrix} \zeta(t) - \cos(2\theta) & \sin(2\theta) \\ \sin(2\theta) & -\zeta(t) + \cos(2\theta) \end{pmatrix} \begin{pmatrix} \psi_e(t) \\ \psi_\mu(t) \end{pmatrix}, \quad (6.57)
$$

where θ is the mixing angle in vacuum and $\zeta(t)$ is defined by

$$
\zeta(t) = \frac{2\sqrt{2}G_{\mathrm{F}} N_{\mathrm{e}}(t)}{\Delta E}. \quad (6.58)
$$

Here G_{F} is the Fermi coupling constant, $N_{\mathrm{e}}(t)$ is the electron density at time t. If we assume that the neutrino follows a straight line trajectory with the velocity v, then we have

$$
N_{\mathrm{e}}(t) = N_0 \exp\left(-\frac{\sqrt{(vt)^2 + b^2}}{r_{\mathrm{res}}} \right), \quad (6.59)
$$

where b is the impact parameter and r_{res} is the resonance radius. Numerical results are shown in Fig. 6.5, where the following parameter values are used: $\sin^2(2\theta) = 0.01$ and $\Delta E \equiv \delta m^2 / E = 10^{-9}$ eV^2/MeV, where δm^2 is the vacuum mass-squared splitting. The solid line is the result of the Landau–Zener formula and naturally does not work well at $b/r_{\mathrm{res}} \gtrsim 1$. On the other hand, the present theory (dash line) works very well in good agreement with the exact numeical results (dotts).

6.3. Other Exactly Solvable Models

As was discussed already in Chap. 3, the Landau–Zener theory presents an exact solution of the linear potential model, i.e., two linear potentials coupled by a constant interaction in the time-dependent framework (see the discussions below Eq. (3.7)), and the Rosen–Zener theory is an exact solution for the time-dependent problem of two constant potentials coupled by a sec-hyperbolic function (see the discussions in Sec. 3.2). The semiclassical solutions in high

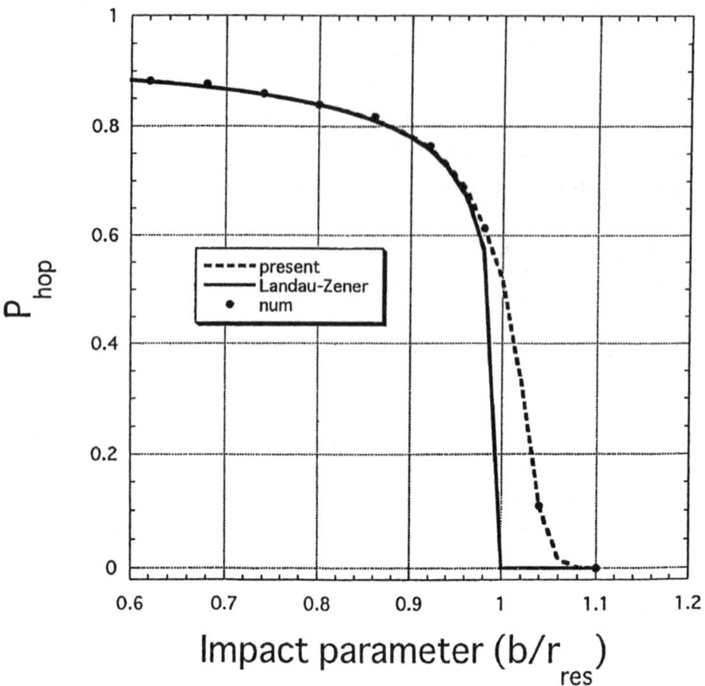

Fig. 6.5.   Neutrino production probability versus impact parameter. (Taken from Thesis of Y. Teranishi with permission.)

energy approximation for various types of exponential potential models discussed in Chap. 5 can also present exact solutions of the corresponding time-dependent problems. In addition to these there have been found some other time-dependent models which can be solved exactly. One is a model of exponential potentials coupled by a constant interaction [77] and the second is the two types of Demkov–Kunike model in which tanhyp potentials are coupled by a constant or a sechyp interaction [78, 79]. Furthermore, Hioe and Carroll obtained exact analytical solutions for a class of models which can be reduced to the hypergeometric differential equation [81, 82].

In the first case of exponential potentials coupled by a constant interaction, we solve the following coupled differential equations:

$$i\frac{d}{dt}\begin{pmatrix} c_1 \\ c_2 \end{pmatrix} = \begin{pmatrix} \alpha e^{-\beta t} + \delta & \gamma \\ \gamma & -\alpha e^{-\beta t} - \delta \end{pmatrix}\begin{pmatrix} c_1 \\ c_2 \end{pmatrix}. \tag{6.60}$$

Expressing c_1 in terms of c_2 from the second equation and inserting that into the first equation, we obtain

$$\frac{d^2}{dt^2}c_2 + [\gamma^2 + i\alpha\beta e^{-\beta t} + (\alpha e^{-\beta t} + \delta)^2]c_2 = 0 \,. \tag{6.61}$$

Introducing the new variables by $z = 2i(\alpha/\beta)e^{-\beta t}$ and $u(z) = z^{1/2}c_2$, we can transform the above equation into

$$\frac{d^2}{dt^2}u(z) + \left\{-\frac{1}{4} + \frac{\kappa}{z} - \frac{(\mu^2 - 1/4)}{z^2}\right\}u(z) = 0 \tag{6.62}$$

with

$$\kappa = \frac{1}{2} - i\frac{\delta}{\beta} \tag{6.63}$$

and

$$\mu = i\frac{\sqrt{\gamma^2 + \delta^2}}{\beta} \,. \tag{6.64}$$

The differential equation Eq. (6.62) is nothing but the Whittaker equation and the two independent solutions are given by $W_{\kappa,\mu}(z)$ and $W_{-\kappa,\mu}(-z)$ [80].

There are two types of Demkov–Kunike models. The first one is represented by the Hamiltonian matrix,

$$H(t) = \begin{pmatrix} \alpha(t) & V(t) \\ V(t) & -\alpha(t) \end{pmatrix} \tag{6.65}$$

with

$$\alpha(t) = a + b\tanh\left(\frac{t}{T}\right), \qquad V(t) = c\,\mathrm{sech}\left(\frac{t}{T}\right), \tag{6.66}$$

where a, b, T, and c are arbitrary constants. Inserting this Hamiltonian into Eq. (6.1) with $\hat{\epsilon}_{1(2)}(t) = +(-)\alpha(t)$ and $V_0 = V(t)$ and taking into account that the diabatic coupling $V(t)$ is time-dependent, we obtain

$$\frac{d^2 c_1}{dt^2} + \frac{1}{T}\tanh\left(\frac{t}{T}\right)\frac{dc_1}{dt}$$

$$+ \left\{c^2\mathrm{sech}^2\left(\frac{t}{T}\right) + \left[a + b\tanh\left(\frac{t}{T}\right)\right]^2 + \frac{i}{T}\left[a\tanh\left(\frac{t}{T}\right) + b\right]\right\}c_1$$

$$= 0 \,. \tag{6.67}$$

Using the transformations

$$z = \frac{1}{2}\left(\tanh\left(\frac{t}{T}\right) + 1\right) \tag{6.68}$$

and

$$c_1 = z^{-iT(a-b)/2}(z-1)^{iT(a+b)/2}u(z), \tag{6.69}$$

it can be easily shown that $u(z)$ satisfies the hypergeometric equation. The desired solution which satisfies the initial condition,

$$c_1(t = -\infty) = 1, \qquad c_2(t = -\infty) = 0, \tag{6.70}$$

is given by

$$c_1(z) = \pm z^{-iT(a-b)/2}(1-z)^{iT(a+b)/2}$$

$$\times F\left(iT[b+\sqrt{b^2-c^2}], iT[b-\sqrt{b^2-c^2}]; \frac{1}{2} - iT(a-b); z\right), \tag{6.71}$$

where $F(a, b; c; z)$ is the hypergeometric function. Finally, the transition probability P for the transition $1 \to 2$ is expressed as

$$P = \begin{cases} \dfrac{\sinh \pi T(b+\sqrt{b^2-c^2})\sinh \pi T(b-\sqrt{b^2-c^2})}{\cosh \pi T(a+b)\cosh \pi T(a-b)} & (a \geq b\text{: non-crossing case}) \\[3mm] \dfrac{\cosh \pi T(a+\sqrt{b^2-c^2})\cosh \pi T(a-\sqrt{b^2-c^2})}{\cosh \pi T(a+b)\cosh \pi T(a-b)} & (a \leq b\text{: crossing case}) \end{cases} \tag{6.72}$$

In the second case of the Demkov–Kunike model, the Hamiltonian of Eq. (6.65) is given by

$$\alpha(t) = a + b\tanh\left(\frac{t}{T}\right) \tag{6.73}$$

and

$$V(t) = c, \tag{6.74}$$

where c is a constant. Now we use the substitution of Eq. (6.68) and

$$c_1 = z^{iE_aT/2}(z-1)^{iE_cT/2}v(z) \tag{6.75}$$

with

$$E_a = \sqrt{(a-b)^2 + c^2}, \qquad E_c = \sqrt{(a+b)^2 + c^2}, \tag{6.76}$$

then $v(z)$ satisfies the hypergeometric equation. In this model the adiabatic potentials are different from the diabatic ones at $t \rightarrow \pm\infty$ and the initial condition has to be set in the adiabatic representation in which c_1 and c_2 are mixed. The coefficient c_2 is obtained from c_1 by

$$c_2 = \frac{i}{c}\frac{dc_1}{dt} - \frac{(a + b\tanh(t/T)}{c}c_1. \tag{6.77}$$

Finally, the nonadiabatic transition probability P is given by

$$P = \frac{\sinh[\pi T(E_c - E_a + 2b)/2]\sinh[\pi T(E_a - E_c + 2b)/2]}{\sinh(\pi T E_a)\sinh(\pi T E_c)}. \tag{6.78}$$

As was mentioned before, the theory of time-dependent nonadiabatic transitions has gained renewed significance because of its importance in optics activated by a remakable progress in recent laser technology. In the basic two-state coupled equations given by Eq. (6.1), the diagonal terms, $\hat{\epsilon}_{1(2)}(t)$ are as usual the unperturbed energy levels and the off-diagonal term $V_0(t)$, which is generally time-dependent, represents the laser-system interaction and is given in the semiclassical representation of the light field by

$$V_0(t) = dE(t)\cos\left(\int_{-\infty}^{t} \omega(t)dt\right), \tag{6.79}$$

where d, $E(t)$, and $\omega(t)$ are, respectively, the dipole moment, the square root of the field intensity, and the laser frequency. Making the following transformations:

$$\begin{aligned}
c_1(t) &= b_1(t)\exp\left[\frac{i}{2}\int_{-\infty}^{t}\omega(t)dt\right], \\
c_2(t) &= b_2(t)\exp\left[-\frac{i}{2}\int_{-\infty}^{t}\omega(t)dt\right],
\end{aligned} \tag{6.80}$$

and applying the so called rotating-wave approximation [26] in which terms containing the rapidly oscillating factors are neglected, we have

$$i\hbar\frac{d}{dt}\begin{pmatrix} b_1(t) \\ b_2(t) \end{pmatrix} = \frac{\hbar}{2}\begin{pmatrix} -\Delta & -\Omega \\ -\Omega & \Delta \end{pmatrix}\begin{pmatrix} b_1 \\ b_2 \end{pmatrix} \tag{6.81}$$

with

$$\Omega = \frac{dE(t)}{\hbar} \tag{6.82}$$

and

$$\Delta = \omega_0 - \omega, \tag{6.83}$$

where ω_0 is the transition frequency between the two unperturved states, and Ω and Δ are referred to as the Rabi frequency and the detuning. Further transforming Eq. (6.81) by

$$b_{1(2)}(t) = a_{1(2)}(t) \exp\left[+(-)\frac{i}{2} \int_{-\infty}^{t} \Delta dt\right], \tag{6.84}$$

we obtain finally

$$i\frac{d}{dt}\begin{pmatrix} a_1(t) \\ a_2(t) \end{pmatrix} = \begin{pmatrix} 0 & -\dot{A}e^{-iB}/2 \\ -\dot{A}e^{iB}/2 & 0 \end{pmatrix}\begin{pmatrix} a_1 \\ a_2 \end{pmatrix}, \tag{6.85}$$

where $\dot{A}$ and $\dot{B}$ represent the Rabi frequency Ω and the detuning Δ, respectively. Elimination of $a_2(t)$ from the above equation leads to the second order differential equation,

$$\frac{d^2}{dt^2}a_1 + \left(i\dot{B} - \frac{\ddot{A}}{\dot{A}}\right)\frac{d}{dt}a_1 + \left(\frac{\dot{A}}{2}\right)^2 a_1 = 0. \tag{6.86}$$

Then we introduce the following variable transformation from t to z:

$$z = z(t) \geq 0 \tag{6.87}$$

subject to the restrictions,

$$\dot{z} \geq 0, \quad z(-\infty) = 0, \quad z(\infty) = 1. \tag{6.88}$$

This transformation changes Eq. (6.86) to

$$\frac{d^2}{dz^2}a_1 + \frac{1}{\dot{z}}\left[\frac{d}{dt}\ln\left(\frac{\dot{z}}{\dot{A}}\right) + i\dot{B}\right]\frac{d}{dz}a_1 + + \left(\frac{\dot{A}}{2\dot{z}}\right)^2 a_1 = 0. \tag{6.89}$$

This equation can be compared with the hypergeometric differential equation,

$$\frac{d^2}{dz^2}a_1 + \frac{c - (a+b+1)z}{z(1-z)}\frac{d}{dz}a_1 - \frac{ab}{z(1-z)}a_1 = 0, \tag{6.90}$$

where a, b, and c are arbitrary constants. Equation (6.89) can be cast into the hypergeometric equation, if the following relations are satisfied:

$$\frac{d}{dt} \ln \frac{\dot{z}}{\dot{A}} + i\dot{B} = \dot{z} \frac{c - (a+b+1)z}{z(1-z)} \tag{6.91}$$

and

$$\left(\frac{\dot{A}}{\dot{z}} \right)^2 = -\frac{4ab}{z(1-z)} . \tag{6.92}$$

If we take

$$\dot{z} = \frac{z(1-z)}{h(z)} , \tag{6.93}$$

$$\dot{A} = \frac{\alpha}{\pi} \frac{[z(1-z)]^{1/2}}{h(z)} , \tag{6.94}$$

and

$$\dot{B} = \frac{1}{\pi} \frac{\beta z + \gamma}{h(z)} , \tag{6.95}$$

then the above conditions are satisfied and we can obtain the final analytical solution under the initail conditions $|a_1(-\infty)| = 1$ and $a_2(-\infty) = 0$ as

$$a_1(t) = F(a^*, b^*, c^*, z) \tag{6.96}$$

and

$$a_2(t) = \frac{(-ab)^{1/2}}{|1-c|} z^{1-c} F(a-c+1, b-c+1, 2-c, z) , \tag{6.97}$$

where $F(a, b, c, z)$ is the hypergeometric function and the constants a, b, and c are given in terms of the parameters α, β, and γ by

$$a = \frac{1}{2\pi} [(\alpha^2 - \beta^2)^{1/2} - i\beta] , \tag{6.98}$$

$$b = \frac{1}{2\pi} [-(\alpha^2 - \beta^2)^{1/2} - i\beta] \tag{6.99}$$

and

$$c = \frac{1}{2} + i\frac{\gamma}{\pi} . \tag{6.100}$$

Finally, the transition probabilities at $t = \infty$ are given by

$$|a_1(\infty)|^2 = \text{sech}\,\gamma\,\text{sech}(\beta + \gamma)$$

$$\times \left\{ \sinh^2\left(\frac{1}{2}\beta + \gamma\right) + \cos^2\left[\frac{1}{2}(\alpha^2 - \beta^2)^{1/2}\right] \right\} \qquad (6.101)$$

and

$$|a_2(\infty)|^2 = \text{sech}\,\gamma\,\text{sech}(\beta + \gamma)$$

$$\times \left\{ \sinh^2\left(\frac{\beta}{2}\right) + \sin^2\left[\frac{1}{2}(\alpha^2 - \beta^2)^{1/2}\right] \right\}. \qquad (6.102)$$

It should be noted that the constant α is equal to the total area of the pulse amplitude,

$$A \equiv \int_{-\infty}^{\infty} \dot{A}\,dt$$

$$= \frac{\alpha}{\pi} \int_{-\infty}^{\infty} \frac{[z(1 - z)]^{1/2}}{h(z)}\,dt$$

$$= \frac{\alpha}{\pi} \int_{0}^{1} z^{-1/2}(1 - z)^{-1/2}\,dz = \alpha\,. \qquad (6.103)$$

The function $h(z)$ is an arbitrary function and this solution can cover quite a wide class of models. For $h(z) = (\pi\omega)^{-1}(\beta z + \gamma)$, $\omega = constant$, the present model reduces to the one solved by Bambini and Berman [83], which includes the Rosen–Zener case as a special case with $\beta = 0$. For $h(z) = $ constant, the model covers the case considered by Hioe [84].

Hioe and Carroll [82] further considered the case that the function $h(z)$ and the variable z are given by

$$z = \frac{1}{2}[1 + f(g(t))] \qquad (6.104)$$

and

$$h(z) = [\dot{g}(t)]^{-1}[z(1 - z)]^{\delta} = [\dot{g}(t)]^{-1}\left[\frac{1 - f^2}{4}\right]^{\delta}, \qquad (6.105)$$

where $g(t)$ is an arbitrary function, δ is an arbitrary constant, and the $f(g)$ is a function to be determined. From Eqs. (6.93)–(6.95), we obtain

$$\frac{1}{2}\frac{df}{dg} = \left[\frac{1}{4}(1 - f^2)\right]^{1-\delta},\tag{6.106}$$

$$\dot{A} = \frac{\alpha}{\pi}\dot{g}(t)\left\{\frac{1}{4}[1 - f^2(g(t))]\right\}^{1/2-\delta}\tag{6.107}$$

and

$$\dot{B} = \frac{1}{2\pi}\dot{g}(t)\frac{(\beta + 2\gamma) + \beta f(g(t))}{(1/4[1 - f^2(g(t))])^\delta}.\tag{6.108}$$

It is interesting to see some simple special cases of δ.

(i) Case I: $\delta = 0$.

We have

$$f = \tanh\left[\frac{1}{2}g(t)\right],\tag{6.109}$$

$$\dot{A} = \frac{\alpha}{2\pi}\dot{g}(t)\text{sech}\left[\frac{1}{2}g(t)\right]\tag{6.110}$$

and

$$\dot{B} = \frac{1}{2\pi}\dot{g}(t)\left\{\beta + 2\gamma + \beta\tanh\left[\frac{1}{2}g(t)\right]\right\},\tag{6.111}$$

where the arbitrary function $g(t)$ should satify $g(\pm\infty) = \pm\infty$ and is chosen to be 0 at $t = 0$. This case covers the models mentined above in the previous paragraph.

(ii) Case II: $\delta = -1/2$.

For the arbitrary function $g(t)$ which satisfies $g(\pm\infty) = \pm\infty$, we have

$$f = \frac{g}{(16 + g^2)^{1/2}},\tag{6.112}$$

$$\dot{A} = \frac{4\alpha}{\pi}\dot{g}(t)\frac{1}{16 + g^2(t)}\tag{6.113}$$

and

$$\dot{B} = \frac{1}{\pi} \frac{\dot{g}(t)}{(16 + g^2(t))^{1/2}} \left[\beta + 2\gamma + \frac{\beta g(t)}{(16 + g^2(t))^{1/2}} \right]. \qquad (6.114)$$

If we take $g(t) \propto t$, then $\dot{A}$ gives the Lorentzian pulse.

(iii) Case III: $\delta = 1/2$.

In this case, we have

$$f = \sin[g(t)], \qquad (6.115)$$

$$\dot{A} = \frac{\alpha}{\pi} \dot{g}(t) \qquad (6.116)$$

and

$$\dot{B} = \frac{\dot{g}(t)}{\pi} \frac{\beta + 2\gamma + \beta \sin[g(t)]}{\cos[g(t)]}. \qquad (6.117)$$

The function $g(t)$ should satisfy $g(\pm\infty) = \pm\pi/2$. This case can represent a square pulse, a Gaussian pulse, and an exponential pulse [82].

Chapter 7
Two-State Problems

In Chap. 5 the new theories for both curve-crossing and non-curve-crossing types of nonadiabatic transitions are presented. These theories describe the most basic parts of variety of dynamic processes involving nonadiabatic transitions, namely they present nonadiabatic transition amplitude matrices at the transtion points; thus by combining with the other basic processes such as adiabatic wave propagation, wave reflection at turning point, and quantum mechanical tunneling, they can describe various physical processes such as inelastic scattering, elastic scattering with resonances, predissociation, and perturbed bound states. The recipe how to deal with these processes are presented in this chapter.

7.1. Diagrammatic Technique

As was frequently mentioned before, the various dynamic processes can be decomposed into sequential basic events such as wave propagation and nonadiabatic transition. The conceivable basic elements (or events) are (1) wave propagation along adiabatic potential without any transition, (2) wave reflection at turning point, (3) single potential barrier penetration and reflection, (4) nonadiabatic transition at crossing point, (5) nonadiabatic tunneling at the energy lower than the bottom of the upper adiabatic potential, and (6) noncrossing type of nonadiabatic transition. These can be expressed by the diagrams depicted in Figs. 7.1–7.5.

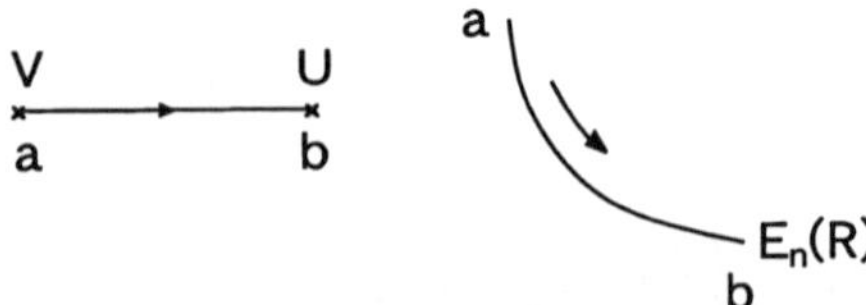

Fig. 7.1.   Diagram for wave propagation from a to b along the potential $E_n(R)$. (Taken from Ref. [9] with permission.)

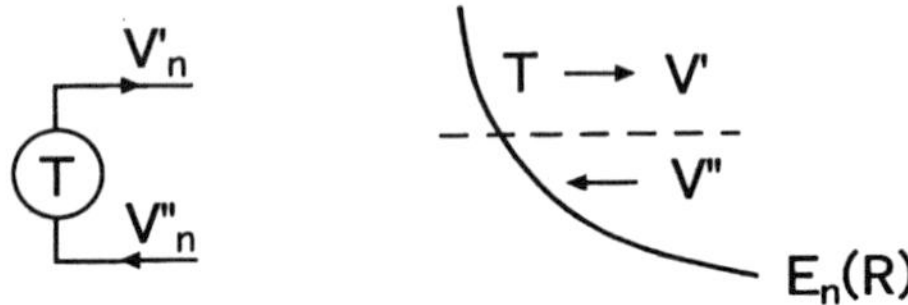

Fig. 7.2.   Diagram for wave reflection at the turning point T. (Taken from Ref. [9] with permission.)

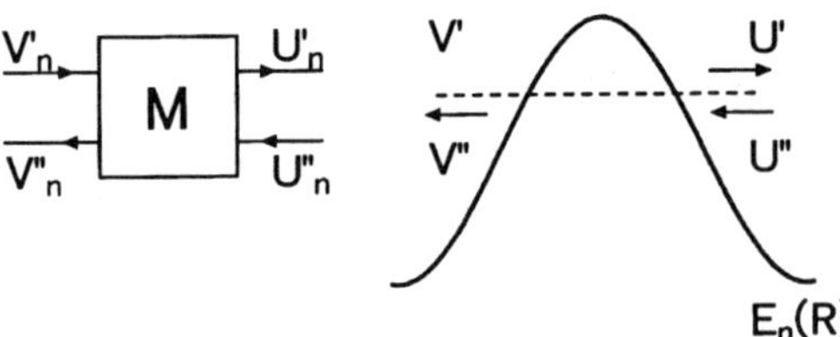

Fig. 7.3.   Diagram for single-potential barrier penetration and reflection. (Taken from Ref. [9] with permission.)

When we write the semiclassical wave function on the adiabatic potential $E_n(R)$ as

$$\psi_n(R) \simeq -V'_n \varphi_n^{(+)}(R:a) + V''_n \varphi_n^{(-)}(R:a) \tag{7.1}$$

with

$$\varphi_n^{(\pm)}(R:a) = \sqrt{\frac{\mu}{2\pi k_n(R)}} \exp\left[\pm i \int_a^R k_n(R)dR\right], \tag{7.2}$$

the connections of the coefficients are given as follows for each basic element.

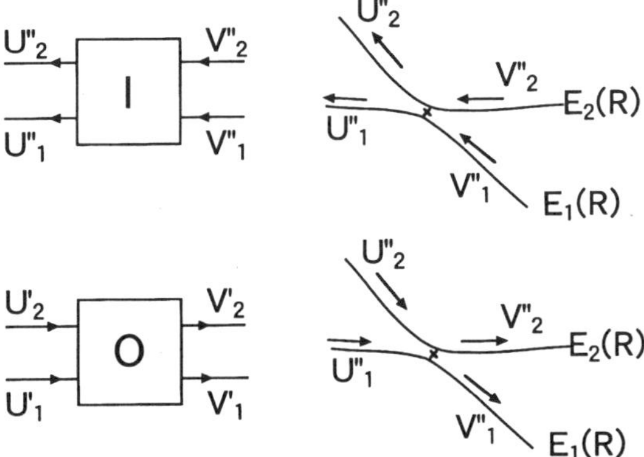

Fig. 7.4. Diagram for nonadiabatic transition. (Taken from Ref. [9] with permission.)

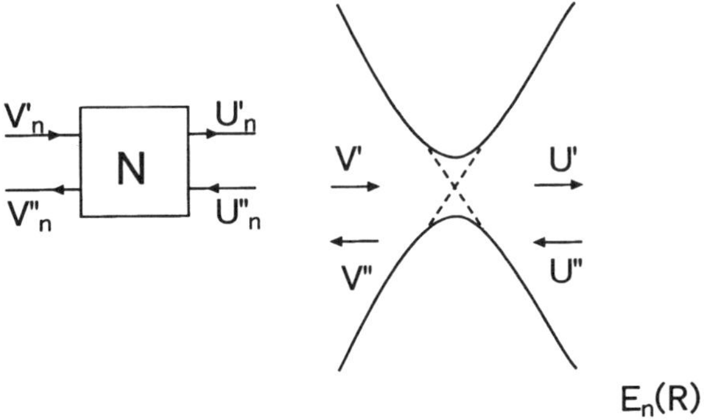

Fig. 7.5. Diagram for nonadiabatic tunneling. (Taken from Ref. [9] with permission.)

(1) Wave propagation (see Fig. 7.1):

$$\mathcal{U} = \begin{pmatrix} e^{i\gamma_n(a,b)} & 0 \\ 0 & e^{-i\gamma_n(a,b)} \end{pmatrix} \mathcal{V} \equiv P_{ab}\mathcal{V}, \qquad (a < b), \qquad (7.3)$$

where $\gamma_n(a,b)$ is defined by Eq. (3.33), $\mathcal{V}$ is a column vector with components V' and V'', and $\mathcal{U}$ is the similar vector which represents the wave function with b as a reference point. This gives the shift of reference point from a to b.

(2) Wave reflection at turning point (Fig. 7.2):

$$\mathcal{V} = \begin{pmatrix} e^{\pi i/4} & 0 \\ 0 & e^{-\pi i/4} \end{pmatrix} \begin{pmatrix} A \\ A \end{pmatrix} \equiv Q \begin{pmatrix} A \\ A \end{pmatrix}, \tag{7.4}$$

where A simply represents the wave amplitude. When the turning point is on the right side, the signs of $\pi i/4$ should be interchanged. If we eliminate the amplitude A, then we have $V' = \exp(i\pi/2)V''$.

(3) Single potential barrier penetration and reflection (Fig. 7.3):

$$\mathcal{U} = M\mathcal{V} \equiv Q \begin{pmatrix} \sqrt{1+e^{2\pi\epsilon}}e^{-i\phi} & ie^{\pi\epsilon}e^{i\theta} \\ -ie^{\pi\epsilon}e^{-i\theta} & \sqrt{1+e^{2\pi\epsilon}}e^{i\phi} \end{pmatrix} Q^*\mathcal{V}, \tag{7.5}$$

where

$$\epsilon = \begin{cases} \pi \displaystyle\int_a^b |k(R)|dR & \text{for } E \leq E_t \\[3mm] \dfrac{1}{\pi i} \displaystyle\int_{R_+}^{R_-} k(R)dR & \text{for } E \geq E_t \,, \end{cases} \tag{7.6}$$

$$\theta = \begin{cases} 0 & \text{for } E \leq E_t \\[3mm] \displaystyle\int_{R_+}^{R_t} k(R)dR + \int_{R_-}^{R_t} k(R)dR & \text{for } E \geq E_t \,, \end{cases}$$

$$\phi = \arg \Gamma \left(\frac{1}{2} + i\epsilon \right) - \epsilon \ln|\epsilon| + \epsilon \,, \tag{7.7}$$

$R_\pm$ are the complex turning points and E_t represents the energy of the barrier top. When $E > E_t$, Q should be replaced by unit matrix. This matrix connects the waves on the right side of the barrier to those on the left side.

(4) Nonadiabatic transition at crossing point (Fig. 7.4):

The matrices I_X and $O_X = I_X^t$ (transpose of I_X) introduced before take care of this transition by

$$\mathcal{U}'' = I_X\mathcal{V}'' \quad \text{and} \quad \mathcal{V}' = O_X\mathcal{U}' \,, \tag{7.8}$$

where $\mathcal{V}'$ is a column vector with components V_1' and V_2', and the other vectors have the similar meanings. At $E \leq E_X$ in the LZ-case we have

$$\mathcal{V}' = QS^{\mathrm{R}(a)}\mathcal{V}'', \tag{7.9}$$

where $S^{\mathrm{R}(a)}$ is the reduced scattering matrix in the adiabatic representation given by Eqs. (5.85)–(5.87).

(5) Nonadiabatic tunneling at $E \leq E_b$ (Fig. 7.5):

The coefficient vector of wave function is given by

$$\mathcal{U} = QNQ^*\mathcal{V}, \tag{7.10}$$

where N is a transfer matrix defined by Eq. (5.128) from the reduced scattering matrix $S^{\mathrm{R}(a)}$ which is defined by Eq. (5.120) with the various quantities given in Sec. 5.2.2.

(6) Rosen–Zener–Demkov and exponential potential models (Fig. 7.4):

We need I_X and O_X matrices which are given in Secs. 3.2 and 5.4. The reference point R_X is taken to be $\mathrm{Re}\,R_*$.

Various kinds of dynamic processes involving curve crossings can now be diagrammatically expressed by assembling the basic elements given above and can be described by combining the appropriate matrices [7, 9, 12, 85]. Examples are given below.

It should further be noted that the adiabatic state representation is much better than the diabatic representation and can be used uniformly whatever the coupling strength is. The diabatic representation is usually easier to use, but works well only when the diabatic coupling is weak. In the examples below, basic physical quantities such as scattering matrices, scattering phase shifts, resonance widths, and eigenvalues are explicitly expressed in terms of the nonadiabatic transition matrices derived before without relying on the wavefunctions directly. When we need wavefunctions, we can use Eq. (7.1) with the coefficients V_n' and V_n'' obtained appropriately from the diagarammatic technique. For the quantitative applications, however, it should be noted that the wavefunctions $\psi_n^{\pm}$ of Eq. (7.2) are not accurate at turning points where they actually diverge and thus they should be replaced by more accurate ones such as those obtained in the uniform semiclassical approximation [5, 12]. All the effects of nonadiabatic transitions can be properly included in the wavefunctions through the coefficients V_n. This treatment would also be useful for analyzing the transient time-evolution of a system under a certain non-stationary initial condition.

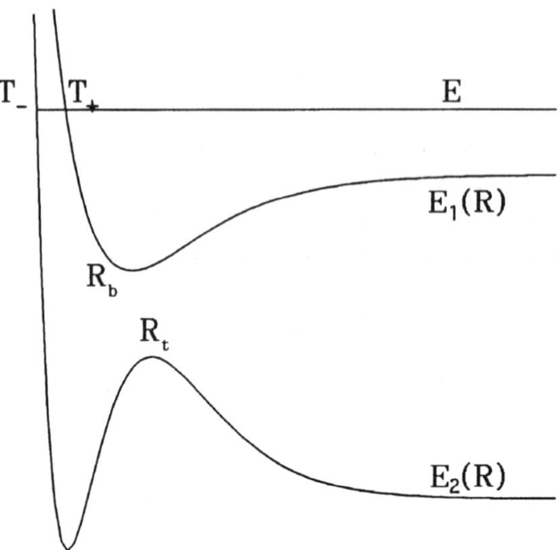

Fig. 7.6.   Schematic adiabatic potentials for inelastic scattering.

7.2.   Inelastic Scattering

As was already discussed in Chap. 3, the inelastic scattering processes depicted
in Fig. 2.1(a) for LZ-type and in Fig. 7.6 for NT-type can be directly described
in terms of the nonadiabatic transition matrices I and $O = I^t$ or by the reduced
scattering matrices S^R. The two potentials $E_1(R)$ and $E_2(R)$ are the functions
of the radial scattering coordinate R and naturally, the energy E is assumed
to be higher than the asymptotic value of the upper potential $E_2\,(R = \infty)$.

In the Landau–Zener case, the total scattering matrix is given by

$$S_{mn} = (S_{\mathrm{LZ}}^{\mathrm{R}(a)})_{mn} \exp[i(\eta_1^{(a)} + \eta_2^{(a)})], \qquad n,m = 1,2, \qquad (7.11)$$

where the reduced scattering matrix is given by Eqs. (5.85)–(5.87) and the
elastic scattering phases $\eta_j^{(a)}$ $(j = 1, 2)$ are defined by Eqs. (5.83) and (5.84).
When the energy E is higher than the crossing point E_X, then the scattering
matrix can be rewritten as (see Eq. (3.16))

$$S = P_{\infty X} O_X P_{XTX} I_X P_{X\infty}, \qquad (7.12)$$

where the matrices P represent the adiabatic wave propagation and are defined
by Eqs. (3.17) and (3.18) with $R_X = R_0$ (minimum energy separation) and the

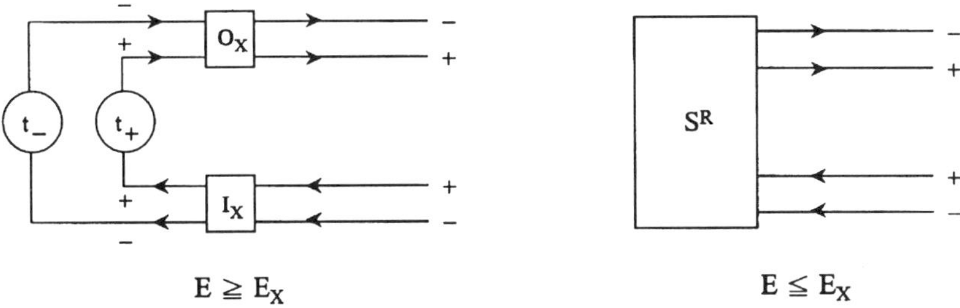

Fig. 7.7. Diagrams corresponding to Fig. 7.6. In the LZ-case at $E \leq E_X$, the reduced scattering matrix S^{R} is used. In the NT-case this is applicable only to $E > E_2(\infty)$. (Taken from Ref. [62] with permission.)

matrix I_X is given by Eq. (5.105). In the NT-case, the total scattering matrix is given by the same equation (7.11) with $X = R_t$ and the various basic quantities, $p_{\mathrm{ZN}}, \phi_{\mathrm{S}}, \sigma_0$, given by those defined in Sec. 5.2.2.3. The diagrams shown in Fig. 7.7 are helpful to understand the physical meaning of the expression Eq. (7.12), as was explained in the previous section.

It should be noted again that the above equations are defined in the adiabatic state representation which is generally better than the diabatic state representation, as was pointed out before.

In the case of the Rosen–Zener–Demkov type non-curve-crossing and the exponential potential models, I_X-matrix should be simply replaced by the one given by Eq. (3.36) or by Eq. (5.259), respectively.

7.3. Elastic Scattering with Resonances and Predissociation

The potential schemes and the corresponding diagrams are depicted in Figs. 7.8 and 7.9.

In the LZ-case the elastic scattering phase shift η is obtained with the help of the diagram as

$$\eta = \eta^{(0)} + \frac{1}{2}\arg\left\{\frac{\tilde{S}_{11} + \det(\tilde{S}(P_{XT_0})_{22}^2}{1 + \tilde{S}_{22}(P_{XT_0})_{22}^2}\right\}, \tag{7.13}$$

where

$$\eta^{(0)} = \int_X^{\infty}[k_1(R) - k_1(\infty)]dR - k_1(\infty)X + \frac{\pi}{4}, \tag{7.14}$$

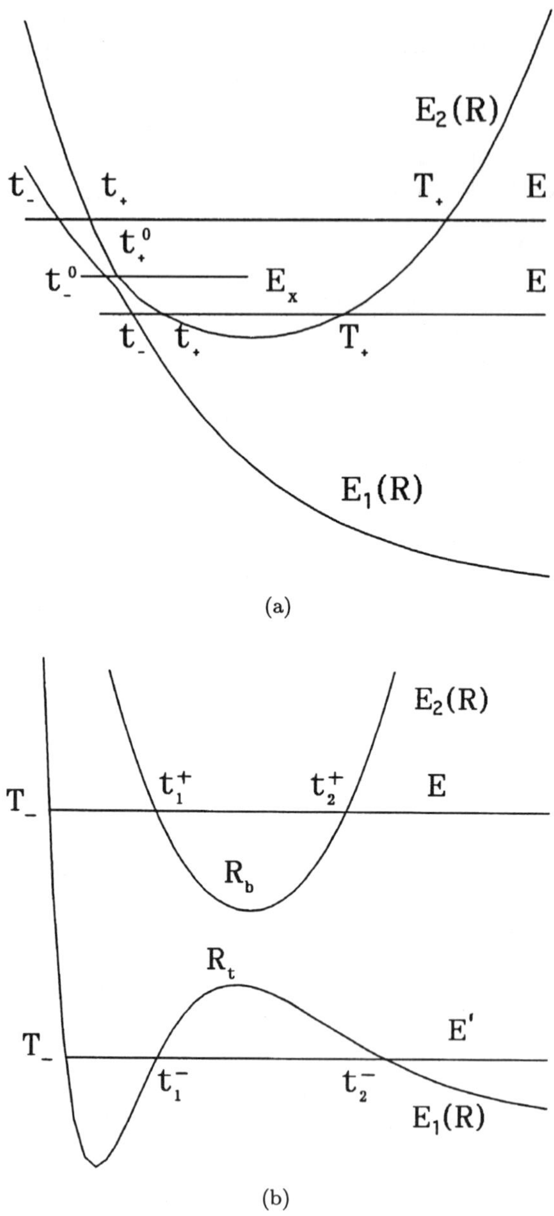

Fig. 7.8. Schematic adiabatic potentials for elastic scattering with resonance. (a) LZ-case and (b) NT-case.

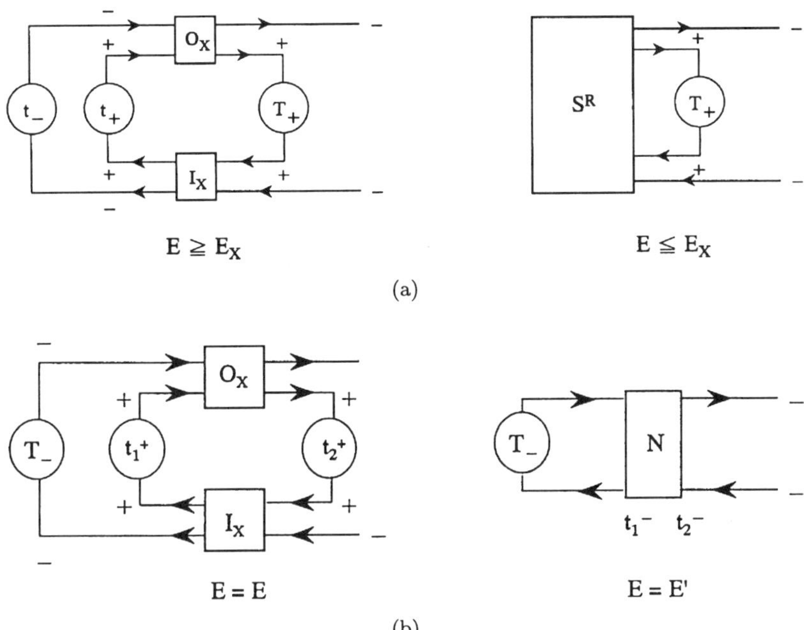

Fig. 7.9. Diagrams corresponding to Fig. 7.8. (Taken from Refs. [61] and [62] with permission.)

$$\tilde{S} = \begin{cases} P_{TX} S^{\mathrm{R}(a)} P_{TX} = -i O_X P_{XTX} I_X & E \geq E_X, \\ S^{\mathrm{R}(a)} & E \leq E_X \end{cases}, \tag{7.15}$$

$$X = \begin{cases} R_0 & E \geq E_X, \\ t_{1,2} & E \leq E_X \end{cases}, \tag{7.16}$$

$$T = t_{1,2}, \tag{7.17}$$

and

$$T_0 = T_+, \tag{7.18}$$

where T_+ is the right turning point on the upper potential (see Fig. 7.8(a)). The reduced scattering matrix $S^{\mathrm{R}(a)}$ is the same as before.

In the NT-case (see Figs. 7.8(b) and 7.9(b)), we have

$$\eta = \eta^{(0)} + \arg S_{12}^{\mathrm{R}(a)} - \frac{1}{2} \arg S_{22}^{\mathrm{R}(a)}$$

$$+ \arctan \left\{ \frac{1 - |S_{22}^{\mathrm{R}(a)}|}{1 + |S_{22}^{\mathrm{R}(a)}|} \tan \left[\gamma_1(t_1, t_1^l) + \frac{1}{2} \arg S_{22}^{\mathrm{R}(a)} + \frac{\pi}{4} \right] \right\}, \qquad (7.19)$$

where $t_1^l = R_t$ when $E \geq E_t$. The reduced scattering matrix $S^{\mathrm{R}(a)}$ in the adiabatic representation is given by Eq. (5.120). It should be noted that the index 1 (2) of this reduced scattering matrix does not correspond to the state number but to the left (right) side of the barrier of the lower adiabatic potential. At $E \geq E_b$, Eq. (7.19) can be rewritten as

$$\eta = \eta^{(0)} - \frac{\pi}{4} + \frac{1}{2} \arg \left[\frac{\tilde{S}_{11} - |\tilde{S}|(P_{XT_0X})_{22}}{1 - \tilde{S}_{22}(P_{XT_0X})_{22}} \right], \qquad (7.20)$$

where $T_0 = t_2^+$ (right turning point on the upper potential), $X = R_b$, and

$$\tilde{S} = O_X P_{XTX} I_X \qquad (7.21)$$

with $T_2 = t_1^+$ (see Fig. 7.8(b)).

As is seen above, when a resonance exists, the elastic phase shift can be generally expressed as follows:

$$\eta = \eta_{\mathrm{bg}} + \arctan[P \tan Q], \qquad (7.22)$$

where η_{bg} is the background phase which is a slowly varying function of energy and the second arctan term represents the resonance. The quantities P and Q in this term are appropriate functions of energy E. In the vicinity of resonance, the phase shift can be expressed as

$$\eta \simeq \eta_{\mathrm{bg}} + \arctan \left[\frac{\Gamma/2}{E_r - E} \right], \qquad (7.23)$$

where Γ and E_r represent the resonance width and the resonance position, respectively. Comparing the two equations, Eqs. (7.22) and (7.23), we can easily obtain the equation for E_r as

$$Q(E_r) = \left(n + \frac{1}{2} \right) \pi \quad \text{for } n = 0, 1, 2, \ldots \qquad (7.24)$$

Expanding $Q(E)$ around $E = E_r$ as

$$Q(E) = Q(E_r) + \left[\frac{dQ(E)}{dE}\right]_{E_r} (E - E_r) + \cdots , \qquad (7.25)$$

we obtain

$$\Gamma = 2P \left[\frac{dQ(E)}{dE}\right]_{E_r}^{-1} . \qquad (7.26)$$

Corresponding to the various cases discussed above, i.e. the LZ and NT-cases, the appropriate expressions should be inserted into P and Q.

7.4. Perturbed Bound States

When the two potentials are energetically closed at $R \to \infty$ (see Figs. 7.10 and 7.11), we have bound state problems whose eigenvalues are perturbed by the nonadiabatic coupling.

In the LZ-case, using the above diagrammatic technique, we can obtain the following secular equation,

$$\det[i\tilde{S}P_{XT_0X} - 1] = 0 , \qquad (7.27)$$

where

$$\tilde{S} = \begin{cases} P_{TX}S^{R(a)}P_{TX} = -iO_X P_{XTX} I_X & E \geq E_X \\ S^{R(a)} & E \leq E_X \end{cases} , \qquad (7.28)$$

$$X = \begin{cases} R_0 & E \geq E_X \\ t_\pm & E \leq E_X \end{cases} , \qquad (7.29)$$

$$T = t_\pm , \qquad (7.30)$$

$$T_0 = T_\pm , \qquad (7.31)$$

and $S^{R(a)}$ is the reduced scattering matrix defined by Eqs. (5.85)–(5.87).

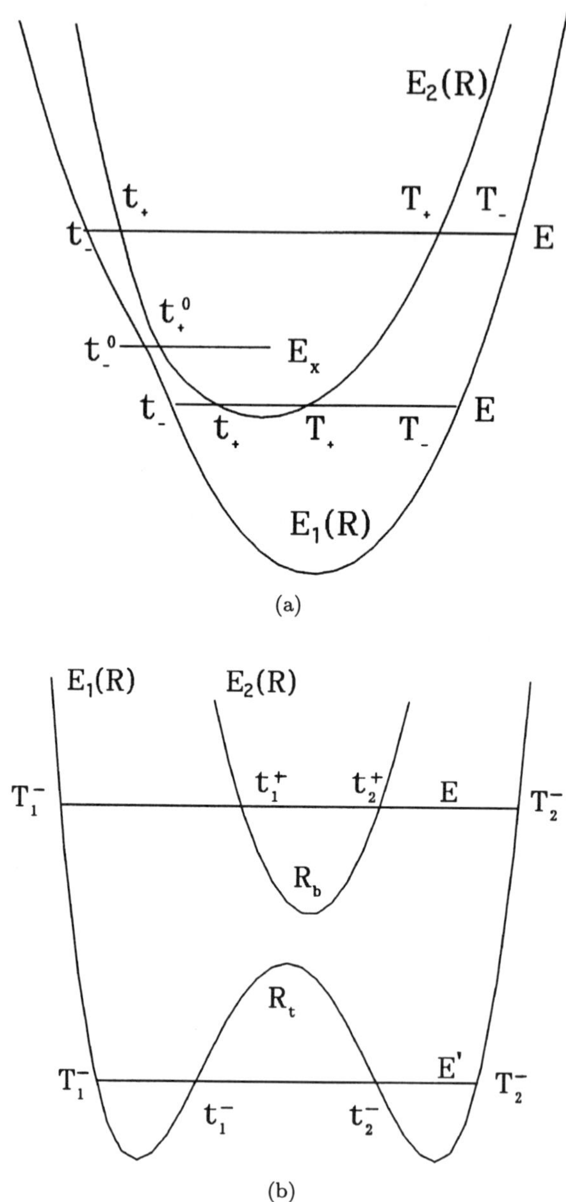

Fig. 7.10.   Schematic adiabatic potentials for perturbed bound states.   (a) LZ-case and (b) NT-case.

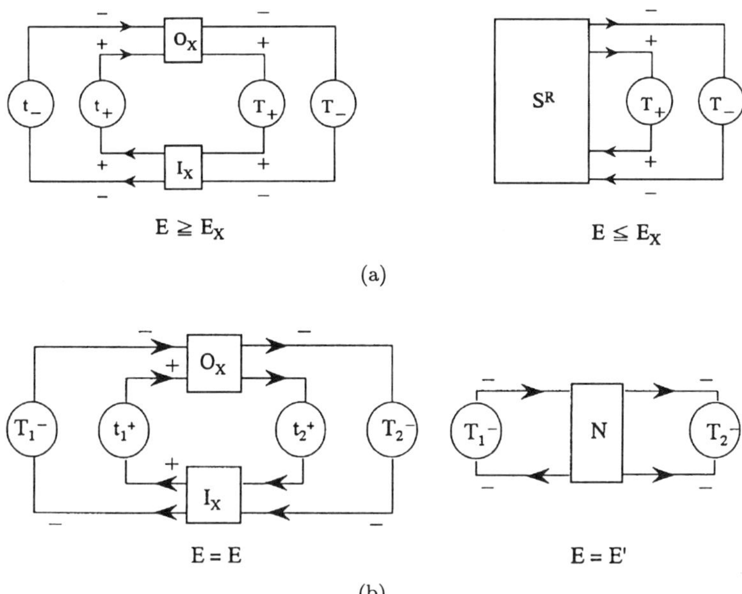

Fig. 7.11. Diagrams corresponding to Fig. 7.10. (Taken from Refs. [61] and [62] with permission.)

In the NT-case, we obtain the following secular equation,

$$\sum_{j=1}^{2}(J_{1j} - J_{2j}) = 0 \tag{7.32}$$

with

$$J = L(t_2^-, T_2^-)NL(T_1^-, t_1^-), \tag{7.33}$$

where $t_1^- = t_2^- = R_t$ when $E \geq E_t$, and

$$L_{nm}(a,b) = \delta_{nm} \exp\left\{i(1 - 2\delta_{n,2})\left[\gamma_1(a,b) + \frac{\pi}{4}\right]\right\}. \tag{7.34}$$

When $E \geq E_b$, the secular equation can be rewritten as

$$|O_X P_{XT_lX} I_X P_{XT_rX} - 1| = 0 \tag{7.35}$$

with $T_l = (T_1^-, t_1^+)$ and $T_r = (T_2^-, t_2^+)$ (see Figs. 7.11(a)).

7.5. Time-Dependent Periodic Crossing Problems

When two diabatic states cross many times as a function of time t, the transition probability naturally changes periodically. This kind of process can be treated without difficulty with use of the two-state theory presented in Sec. 6.2. For instance, when crossing points are well separated from each other as in Fig. 7.12(a), the overall transition matrix T is given by

$$T = O_X X I_X X O_X X \cdots, \tag{7.36}$$

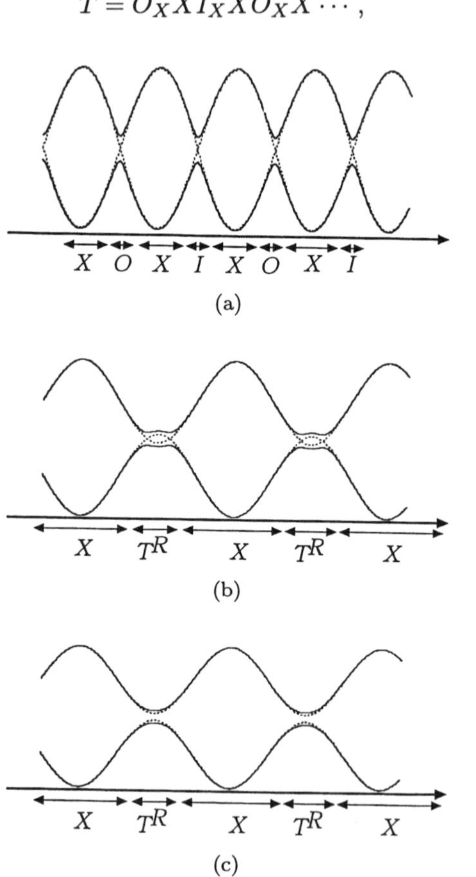

(a)

(b)

(c)

Fig. 7.12. (a) Schematic time-dependent periodic potentials in the case that avoided crossings are well separated. ———: adiabatic, - - -: diabaitc. (Taken from Thesis of Y. Teranishi with permission.) (b) The same as Fig. 7.12(a) in the case of closely lying avoided crossings. (c) The same as Fig. 7.12(a) in the case of diabatic avoided crossings.

where I_X and $O_X = I_X^t$ are defined by Eq. (6.21) and X is the same as P-matrix defined by Eq. (6.22).

When the two crossings are located close to each other or the two diabatic potentials are diabatically avoided as in Fig. 7.12(b) or 7.12(c), the overall effect of the nonadiabatic transitions there can be described by the transition matrix T^R which is equivalent to the reduced scattering matrix in the time-independent scheme and the overall transition matrix T is given by

$$T = T^{\mathrm{R}} X T^{\mathrm{R}} X \cdots , \tag{7.37}$$

where X denotes the adiabatic propagation matrix as in Eq. (6.22) from the right-side avoided crossing point (real part of the complex crossing point) to the left-side of the next transition region. The Stokes constant and other necessary quantities to define T^{R} are given in Sec. 6.2.2. Numerical examples are presented in Figs. 7.13, where the diabatic potential energy difference is given by

$$\Delta\epsilon(t) = \frac{2A}{V_0 \sin(\omega t)} + \frac{2B}{V_0} , \tag{7.38}$$

and V_0 is the constant diabatic coupling. As is seen in this figure, the semi-classical theory works very well even in the non-crossing region $B/V_0 > A/V_0$, where the Landau–Zener theory does not work at all. This kind of periodic crossing case appears frequently in the problems of periodic external field such as the Zener tunneling [29, 86–88].

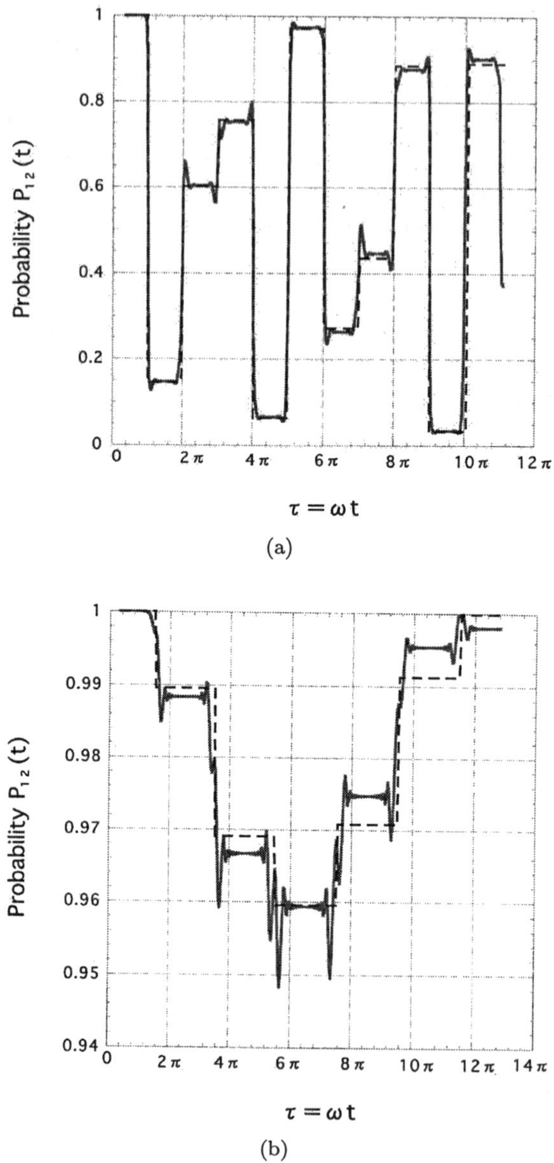

(a)

(b)

Fig. 7.13.   Transition probability as a function of time in the case of $\Delta\epsilon(\tau) = 2A/V_0 \sin(\tau) + 2B/V_0$ with $A/V_0 = 15.0$. The region $B/V_0 > A/V_0$ corresponds to the non-crossing regime. ———: exact numerical result, - - -: semiclassical theory. (a) $B/V_0 = 0.0$, (b) $B/V_0 = 18.0$. (Taken from Ref. [71] with permission.)

Chapter 8

Effects of Dissipation and Fluctuation

As was mentioned in Introduction, nonadiabatic transitions due to curve-crossings play important roles in condensed matter also. In this case, effects of dissipation and fluctuation should be taken into account. Theoretical studies have been carried out only for the time-dependent Landau–Zener model. Here analytical as well as numerical studies done by Kayanuma and his collaborators are summarized [89–93]. The Hamiltonian to describe the problem is written as [93]

$$H(t) = H_{\mathrm{el}}(t) + H_{\mathrm{ph}} + H_{\mathrm{I}}, \qquad (8.1)$$

with

$$H_{\mathrm{el}}(t) = \frac{1}{2}vt(|1\rangle\langle1| - |2\rangle\langle2|) + J(|1\rangle\langle2| + |2\rangle\langle1|), \qquad (8.2)$$

$$H_{\mathrm{ph}}(t) = \sum_k \omega_k b_k^\dagger b_k \qquad (8.3)$$

and

$$H_{\mathrm{I}}(t) = \frac{1}{2}\sum_k \alpha_k \omega_k (b_k + b_k^\dagger)(|1\rangle\langle1| - |2\rangle\langle2|), \qquad (8.4)$$

where v is the rate of the change of energy, α_k represents the coupling to the kth phonon mode of frequency ω_k, and $b_k^\dagger$ and b_k are the creation and annihilation operators of the kth phonon. The Hamiltonian $H_{\mathrm{el}}(t)$ represents the electronic system composed of the two states $|1\rangle$ and $|2\rangle$ which depend linearly on time t and couple to each other by J, H_{ph} is the Hamiltonian of the phonon environment, and H_I represents the interaction between the electronic system and the phonons. Here we are interested in the transition probability P which represents a transition from the initial diabatic electronic state $|1\rangle$ at $t = -\infty$ to the final diabatic state $|2\rangle$ at $t = \infty$. Some important parameters to describe the system are introduced. Because of the Gaussian character of the linear electron-phonon interaction, the dynamics of this quantum dissipation system can be specified by the spectral density function $\phi(\omega)$ defined by

$$\phi(\omega) = \sum_k \alpha_k^2 \omega_k^2 [(n_k + 1)\delta(\omega - \omega_k) + n_k \delta(\omega + \omega_k)], \tag{8.5}$$

where

$$n_k = \frac{1}{\{e^{\omega_k/k_B T} - 1\}}. \tag{8.6}$$

The relaxation energy ΔE and the amplitude of energy fluctuation D are defined by

$$\Delta E = \int_{-\infty}^{\infty} \phi(\omega)\omega^{-1} d\omega \tag{8.7}$$

and

$$D^2 = \int_{-\infty}^{\infty} \phi(\omega) d\omega. \tag{8.8}$$

The dimensionless coupling constant S and the representative phonon energy $\bar{\omega}$ are introduced as,

$$S = \sum_k \alpha_k^2, \tag{8.9}$$

and

$$\Delta E = S\bar{\omega}. \tag{8.10}$$

The whole transition process can be characterized by the degree of coherence which is measued by the ratio of $\tau_{\rm ph}$ and $\tau_{\rm tr}$, where $\tau_{\rm ph} \simeq D^{-1}$ represents the fluctuation time scale and $\tau_{\rm tr} \simeq J/|v|$ represents the nonadiabatic transition time. When $\tau_{\rm ph} \ll 1/\bar{\omega}$, the relative phase memory is completely lost within the short time scale $\sim \tau_{\rm ph}$. When $\tau_{\rm ph} \gtrsim 1/\bar{\omega}$, the phase relaxation is incomplete. Kayanuma and Nakayama used the formal perturbation expansion with repsect to the diabatic coupling strength J^2 and obtained analytical formulas in the following limiting cases. Only the final results are summarized.

(1) The case of rapid passage ($\tau_{\mathrm{ph}} \gg \tau_{\mathrm{tr}}$):
This corresponds to the large velocity $|v|$ and the coherence is preserved. Thus the Landau–Zener formula is recovered,

$$P = P_{\mathrm{LZ}} \equiv 1 - e^{-2\pi J^2/|v|} . \tag{8.11}$$

(2) The case of slow passage ($\tau_{\mathrm{tr}} \gg \tau_{\mathrm{ph}}$):
In this case the coherence is always interrupted in the time-interval τ_{tr} and the system makes transitions between $|1\rangle$ and $|2\rangle$, while the relaxation toward the equilibrium is occuring within the respective electronic state.

First, we consider the case that the energy dissipation is negligible, $\Delta E \to 0$. In the case of high temperature limit with small coupling to phonon, the following result is obtained:

$$P = P_{\mathrm{SD}} \equiv \frac{1}{2}\{1 - e^{-4\pi J^2/|v|}\} . \tag{8.12}$$

It should be noted that in the limit $J^2/|v| \to \infty$,

$$P_{\mathrm{SD}} \to \frac{1}{2} . \tag{8.13}$$

This indicates that the strong decoherence reduces the whole transition process to diffusion-like so that the electronic states are equally populated after the slow passage.

Next, let us consider the effect of the finite energy dissipation. In the case of weak coupling and high temperature, we have

$$P = \frac{1}{2}\{1 - e^{-4\pi J^2/|v|}\} , \tag{8.14}$$

which recovers the formula Eq. (8.12). In the limit of strong coupling at low temperature, on the other hand, the value of P strongly depends on the sign of v. For positive v, we have

$$P_+ = 1 - e^{-2\pi J^2/|v|} = P_{\mathrm{LZ}} . \tag{8.15}$$

This gives an interesting result that in both limits of rapid and slow passages the Landau–Zener formula is recovered (see Eqs. (8.11) and

(8.12)). When the velocity v is negative, namely the initial state is the upper one, then the probability P becomes

$$P_- = \{1 - e^{-2\pi J^2/|v|}\}e^{-2\pi J^2/|v|} = P_{\mathrm{LZ}}(1 - P_{\mathrm{LZ}}). \qquad (8.16)$$

Interestingly, this means that the transition occurs twice at $t = \pm\Delta E/v$.

(3) The case of energy fluctuation alone:

The Gaussian stochastic fluctuation of the energy levels with the amplitude D is treated as an environmental perturbation. The probability P generally lies in between the two limits, namely, $P_{\mathrm{SD}} \le P \le P_{\mathrm{LZ}}$ [90, 91]. In the limit of large amplitude fluctuation, P agrees with P_{SD} as it should.

Some numerical studies in comparison with the above analytical formulas are shown below [93]. The numerical calculations were carried out with use of the interaction mode model by separating out the phonon modes into system mode and reservoir modes [94]. The following dimensionless parameters are introduced: $\tilde{J} \equiv J/\bar{\omega}, \tilde{v} \equiv v/\bar{\omega}^2, \tilde{\kappa} \equiv \kappa/\bar{\omega}, \tilde{T} \equiv k_{\mathrm{B}}T/\bar{\omega}$, and $\tilde{D} \equiv D/\bar{\omega}$. Figure 8.1 shows P_+(circle) and P_-(diamond) against $1/|v|$ for a fixed value of

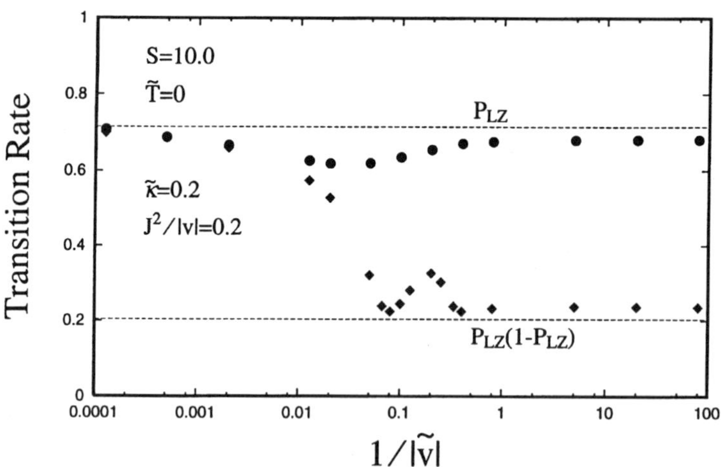

Fig. 8.1.   The transition rate P_+(circle) and P_-(diamond) as a function of $|\tilde{v}|^{-1}$ for a fixed value of $J^2/|v|$ in the zero temperature, strong coupling case. The predicted values by the formulas, P_{LZ}, and $P_{\mathrm{LZ}}(1 - P_{\mathrm{LZ}})$ are shown by dashed lines. (Taken from Ref. [93] with permission.)

$J^2/|v| = 0.2$ at $T = 0$ with $S = 10.0$ and $\tilde{\kappa} = 0.2$. The values of P_{LZ} and $P_{\mathrm{LZ}}(1 - P_{\mathrm{LZ}})$ are also shown. In the rapid passage limit, $P_+ = P_- = P_{\mathrm{LZ}}$ is attained as discussed above. As the speed of passage decreases, both P_+ and P_- deviate from P_{LZ} to lower values by the same amount. As $|v|$ decreases further, P_+ takes a minimum value at an intermediate value of $1/|v|$ and then increases again to approach to P_{LZ} in the slow passage limit, consistently with the formula Eq. (8.15). On the other hand, P_- decreases dramatically from P_{LZ} to $P_{\mathrm{LZ}}(1 - P_{\mathrm{LZ}})$ as $|v|$ decreases.

The dependence on S of P_+ and P_- at zero temperature is shown in Fig. 8.2 for a fixed value of $\tilde{J} = 0.5$ in the slow passage, $|v| = 1.25$. It is remarkable that P_+ is essentially independent of S and is given by P_{LZ}, but P_- is reduced strongly by the coupling with phonons even in the weak coupling region.

Figure 8.3 shows the time variation of $P_+(t)$ and $P_-(t)$ in the high temperature weak coupling limit for $\tilde{T} = 10.0, \tilde{D} = 1.0, \tilde{J} = 0.5, \tilde{\kappa} = 0.2$, and $|\tilde{v}| = 1.0$. The transition rate is almost independent of the sign of v and is strongly reduced from P_{LZ}. The dependence of P_+ and P_- on $J^2|v|$ is shown in Fig. 8.4 for $\tilde{J} = 0.5$ and $\tilde{\kappa} = 0.2$. The coupling constant S and the temperature $\tilde{T}$ are chosen so that the fluctuation dominance condition, $\Delta E \to 0$ and $k_{\mathrm{B}}T \to \infty$ with $D = $ finite, is satisfied: $S = 0.0499, \tilde{T} = 10.0$ for $\tilde{D} = 1.0$ and

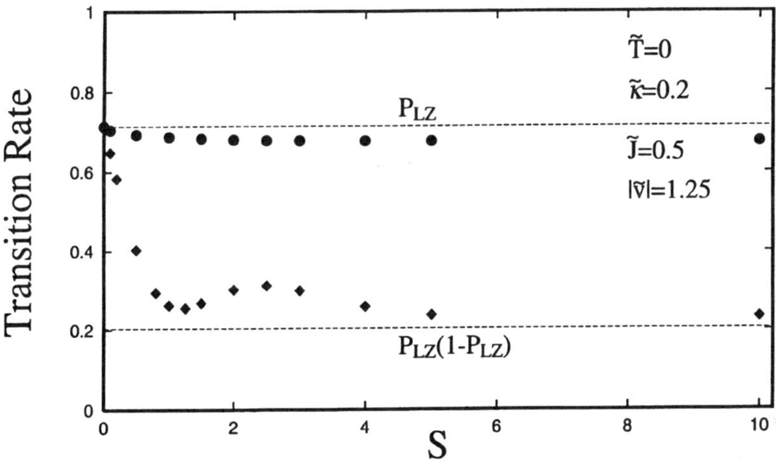

Fig. 8.2.   The transition rate P_+(circle) and P_-(diamond) as a function of S in the zero temperature, slow passage case for fixed values of $\tilde{J}, |v|$ and $\tilde{k}$. The values of P_{LZ} and $P_{\mathrm{LZ}}(1 - P_{\mathrm{LZ}})$ are given by dashed lines. (Taken from Ref. [93] with permission.)

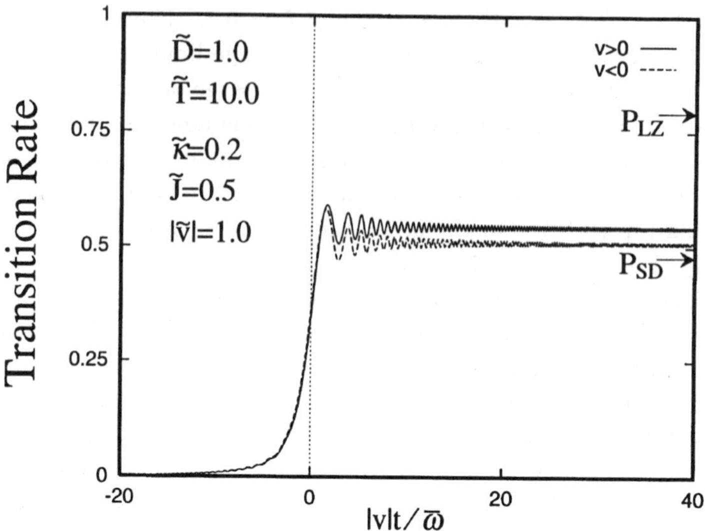

Fig. 8.3.   The time-dependent probability $p_+(t)$ (solid line) and $p_-(t)$ (dashed line) in the case of high temperature, weak coupling. (Taken from Ref. [93] with permission.)

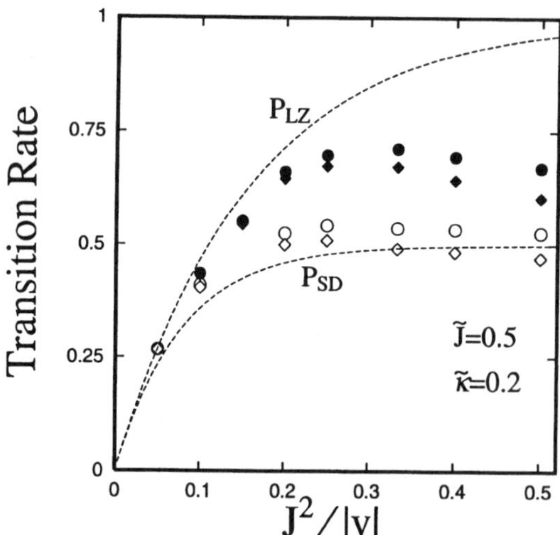

Fig. 8.4.   The transition rate P_+ (circle) and P_- (diamond) in the case of high temperature, weak coupling. The results for $\tilde{D} = 0.5$ and $\tilde{D} = 1.0$ are shown by solid and open symbols, respectively. (Taken from Ref. [93] with permission.)

$S = 0.0249$, $\tilde{T} = 5.0$ for $\tilde{D} = 0.5$. As noted before, P generally lies in between P_{SD} and P_{LZ} except for small deviations.

Kayanuma also considered the effects of fluctuation of the off-diagonal diabatic coupling J^2 [92]. He employed the following Hamiltonian,

$$H(t) = \frac{1}{2}vt(|1\rangle\langle 1| - |2\rangle\langle 2|) + g(t)(|1\rangle\langle 2| + |2\rangle\langle 2|)\,, \tag{8.17}$$

where the Markoffian–Gaussian fluctuation is assumed for $g(t)$,

$$\langle g(t)g(t')\rangle = J^2 e^{-\gamma|t-t'|}\,. \tag{8.18}$$

Here $\langle\cdots\rangle$ denotes the expectation value and γ represents the decay constant of the fluctuation correlation. In the slow fluctuation limit ($\gamma = 0$), the

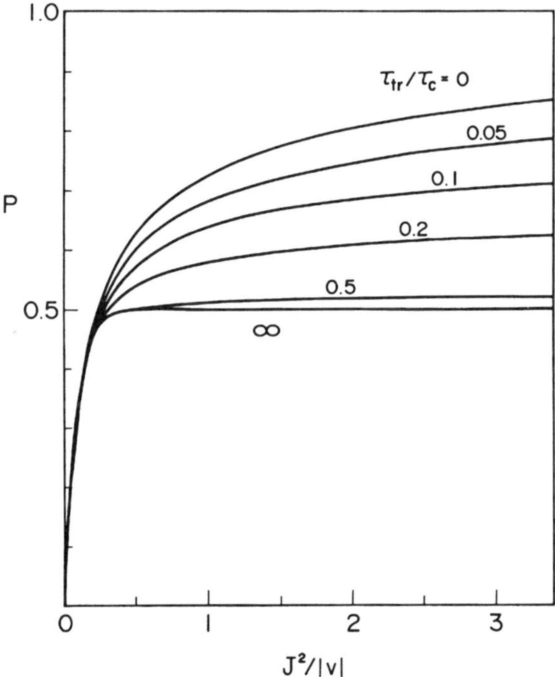

Fig. 8.5. Transition probability from $|1\rangle$ to $|2\rangle$ versus $J^2/|v|$ for various values of $\tau_{\mathrm{tr}}/\tau_{\mathrm{c}}$ ($= J\gamma/|v|$). The case of $\tau_{\mathrm{tr}}/\tau_{\mathrm{c}} = 0$ and ∞ correspond to $P = P_{\mathrm{SF}}$ and $P = P_{\mathrm{RF}}$, respectively. (Taken from Ref. [92] with permission.)

probability P is given by

$$P = P_{\mathrm{SF}} = 1 - \left\{ 1 + \left(\frac{4\pi J^2}{|v|} \right) \right\}^{-1/2} . \tag{8.19}$$

In the rapid fluctuation limit $(\gamma = \infty)$, on the other hand, we have

$$P = P_{\mathrm{RF}} = \frac{1}{2}\{ 1 - e^{-4\pi J^2/|v|} \} . \tag{8.20}$$

Figure 8.5 shows the numerical results of P against $J^2/|v|$ for various values of $\tau_{\mathrm{tr}}/\tau_{\mathrm{c}}(= J\gamma/|v|)$, where $\tau_{\mathrm{c}} = 1/\gamma$ represents the rapidity of the fluctuation. The probability P changes continuously from $P = P_{\mathrm{SF}}$ to $P = P_{\mathrm{RF}}$ as the parameter $\tau_{\mathrm{tr}}/\tau_{\mathrm{c}}$ increases.

Chapter 9

Multi-Channel Problems

So far we have discussed various types of two-state or two-channel problems and demonstrated that once the basic accurate two-state theory is prepared, three different types of physical problems, inelastic scattering, elastic scattering with resonances and perturbed bound states, can be uniformly formulated with use of the theory. However, physical and chemical processes in realty naturally require us to deal with various types of multi-dimensional and/or multi-channel problems. In this chapter, we will consider one-dimensional multi-channel problems which contain many curve crossings in both time-independent and time-dependent schemes. First, some exactly solvable models will be introduced briefly. Since they are very limited, however, main efforts will be paid to explain how the basic two-state theories presented in previous chapters can be employed to analyze multi-channel problems.

9.1. Exactly Solvable Models

9.1.1. *Time-independent case*

As can be easily understood, it is very difficult to solve analytically exactly any multi-channel problems. This is especially true for time-independent case The only potential curve system exactly solved so far is the so called Demkov–Osherov model [95]. In this model, one slanted level with the slope α crosses many horizontal levels n $(n = 1, 2, 3, \ldots, N)$ with the level dependent constant coupling h_n (see Fig. 9.1). The subspace composed of the horizontal levels is assumed to be diagonalized. Namely, the corresponding Schrödinger equation

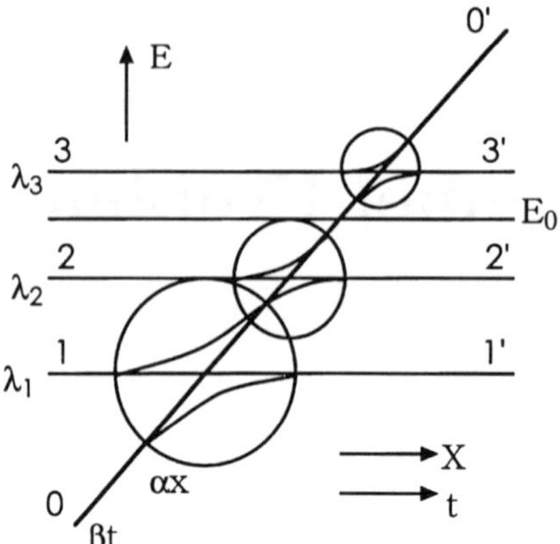

Fig. 9.1.   Crossings of many horizontal states with a slanted state. (Taken from Ref. [95] with permission.)

is

$$\left(-\frac{1}{2\mu}\frac{d^2}{dx^2} + V\right)\Psi = 0 \tag{9.1}$$

with

$$\Psi = \begin{pmatrix} \psi_0 \\ \psi_1 \\ \psi_2 \\ \cdot \\ \cdot \\ \psi_N \end{pmatrix} \tag{9.2}$$

and

$$V = \begin{pmatrix} \alpha x & h_1 & h_2 & \cdots & \cdots & \cdots \\ h_1 & \lambda_1 & 0 & \cdots & \cdots & \cdots \\ h_2 & 0 & \lambda_2, & \cdots & \cdots & \cdots \\ \cdots & \cdots & \cdots & \cdots & \cdots & \cdots \end{pmatrix}. \tag{9.3}$$

Using the same complex contour integral method as that for the time-dependent case described in the next subsection, Demkov and Osherov solved this problem exactly and obtained the following results for the overall transition probabilities P_{ij} for the transition $i \to j$:

$$P_{nm'} = 0 \quad \text{for } n > m', \tag{9.4}$$

$$P_{n'm} = 0 \quad \text{for } m > n', \tag{9.5}$$

$$P_{n'm'} = 0, \tag{9.6}$$

$$P_{00'} = 0, \tag{9.7}$$

$$P_{nm} = (1 - p_n)p_{n+1} \cdots p_{n_0}p_{n_0-1} \cdots p_{m+1}(1 - p_m) \tag{9.8}$$

$$P_{00} = p_1 \cdots p_{n_0}p_{n_0} \cdots p_1, \tag{9.9}$$

where $n, m \geq 1$, p_n represents the Landau–Zener transition probability between the slant level and the nth horizontal level,

$$p_n = e^{-2\pi h_n^2/\alpha}, \tag{9.10}$$

and the totel energy E is assumed to be in between λ_{n_0} and λ_{n_0+1}. The primed state n' indicates the state n on the right side of the crossing region (see Fig. 9.1). The above results can be easily understood in terms of the semiclassical idea of nonadiabatic transition at avoided crossing and adiabatic wave propagation along adiabatic potential. It is interesting to note that there appear no phase terms in the above results. This is easily understood, because for any one of the above transitions there is only one possible path connecting initial and final states and no phase interference occurs. It should be noted, however, that the above result is quantum mechanically exact for any number N of the horizontal states and independent of the spacing among the latter.

9.1.2. *Time-dependent case*

The time-dependent version of the above model, namely,

$$(H_0 + |\psi_0\rangle\beta t\langle\psi_0|)|\Psi\rangle = i\frac{\partial}{\partial t}|\Psi\rangle \qquad (\beta > 0), \tag{9.11}$$

was solved by Demkov and Osherov with use of the contour integral [95],

$$|\Psi\rangle = \int_C G(E)|\psi_0\rangle F(E)e^{-iEt}dE, \tag{9.12}$$

where

$$H_0 = \begin{pmatrix} 0 & h_1 & h_2 & \cdots & \cdots & \cdots \\ h_1 & \lambda_1 & 0 & \cdots & \cdots & \cdots \\ h_2 & 0 & \lambda_2 & \cdots & \cdots & \cdots \\ \cdots & \cdots & \cdots & \cdots & \cdots & \cdots \end{pmatrix} \tag{9.13}$$

and

$$G(E) = (H_0 - E)^{-1}. \tag{9.14}$$

With use of the integration by parts, we obtain

$$F(E) = i\beta \frac{d}{dE}[F(E)\langle\psi_0|G(E)|\psi_0\rangle], \tag{9.15}$$

$$|\Psi\rangle = N \int_C \frac{G(E)|\psi_0\rangle}{\langle\psi_0|G(E)|\psi_0\rangle} \exp\left[-\frac{i}{\beta}\int^E \frac{dE'}{\langle\psi-0|G(E')|\psi_0\rangle} - iEt\right] dE \tag{9.16}$$

and

$$\exp\left[-\frac{i}{\beta}\int^E \frac{dE'}{\langle\psi_0|G(E')|\psi_0\rangle} - iEt\right]\Bigg|_C = 0. \tag{9.17}$$

By evaluating the resolvent operator $G(E)$ from Eq. (9.14), we can have

$$\langle\psi_0|\Psi\rangle = N \int_C L(E)dE \tag{9.18}$$

and

$$\langle\psi_n|\Psi\rangle = N h_n \int_C (\lambda_n - E)^{-1} L(E)dE \tag{9.19}$$

with

$$L(E) = \prod_m (\lambda_m - E)^{-ih_m^2/\beta} \exp\left(i\frac{E^2}{2\beta} - iEt\right). \tag{9.20}$$

By appropriately choosing the contour C to satisfy the initail condition at $t \to -\infty$ and Eq. (9.17), we can obtain the normalization constant N and finally explicit expressions for the various transition probabilities P_{ij} for the transition $i \rightarrow j$ as follows:

$$P_{nm} = 0 \quad \text{for } 0 < m < n, \tag{9.21}$$

$$P_{nn} = 0, \tag{9.22}$$

$$P_{nm} = (1 - p_n)p_{n+1} \cdots p_{m-1}(1 - p_m) \quad \text{for } m > n \tag{9.23}$$

$$P_{n0} = (1 - p_n)p_{n+1} \cdots p_N. \tag{9.24}$$

The physical picture of the above result is much clearer than the time-independent case.

The above Demkov–Osherov model was extended to the case that two bands composed of an infinite number of equidistant states cross with each other [96, 97]. They analyzed the behaviour of adiabatic states against the change of the diabatic coupling strength, and found a new kind of pseudocrossing named as "anti-diabatic states" which show saw-like behaviour and lie halfway between the original adjacent diabatic states. Some important features of time evolution of the system was also analyzed.

The other model which has been analytically exactly solved is the so called bow-tie model, in which an arbitrary number N of linear time-dependent diabatic potential curves cross at one point and only a particular horizontal curve among them has interactions with the others. Carroll and Hioe first motivated the study by the applications to quantum optics and derived exact solutions for the three-level case [98–100]. Later, Brundobler and Elser noted that the method used by Carroll and Hioe can be generalized to write the solutions in the form of complex contour integrals for arbitrary number N of states [101]. The explicit expressions of transition probabilities could not be obtained, however. Recently, Ostrovsky and Nakamura have succeeded to derive the exact compact analytical solutions for arbitrary N [102]. The Hamiltonian matrix in the model is given by

$$H_{00} = 0, \quad H_{jj} = \beta_j t, \quad H_{j0} = H_{0j} = V_j \quad (j \neq 0),$$
$$H_{jk} = 0 \quad \text{for } j \neq 0 \quad \text{and} \quad k \neq 0, \tag{9.25}$$

where the states with positive slope β_j are labelled with positive subscripts j in order of increasing β_j, and the states with negative β_j are labelled by negative indices j, the larger j corresponding to the larger $|\beta_j|$, i.e.,

$$\cdots \beta_{-3} < \beta_{-2} < \beta_{-1} < \beta_0 < \beta_1 < \beta_2 \cdots . \tag{9.26}$$

The expansion coefficients c_j of the total wave function in the above diabatic representation satisfy the following coupled equations as usual:

$$i\frac{dc_0}{dt} = \sum_n V_n c_n, \qquad i\frac{c_j}{dt} = \beta_j t c_j + V_j c_0, \qquad (j \neq 0). \tag{9.27}$$

As in the Demkov–Osherov model, the Laplace transformation reduces this system of equations to a single first order differential equation which is easily solved and we can obtain the following contour-integral representations:

$$c_0(t) = Qt \int_C \frac{du}{\sqrt{-u}} e^{-iut^2/2} \prod_n \left(\frac{-u+\beta_n}{-u}\right)^{ih_n} \tag{9.28}$$

and

$$c_j(t) = -QV_j \int_C \frac{du}{\sqrt{-u}} e^{-iut^2/2} \frac{1}{-u+\beta_j} \prod_n \left(\frac{-u+\beta_n}{-u}\right)^{ih_n}, \tag{9.29}$$

where

$$h_j = \frac{V_j^2}{(2\beta_j)} \tag{9.30}$$

and Q is a certain normalization constant. By properly taking the contour C in the complex u-plane, we can finally obtain the expression of Q and the transition probabilities P_{ij} for the transition $i \to j$ as

$$P_{00} = \left[1 - \prod_{n>0} p_n - \prod_{n<0} p_n\right]^2, \tag{9.31}$$

$$P_{0j} = P_{j0} = \left(\prod_{n>j} p_n\right)(1-p_j)\left(\prod_{n>0} p_n + \prod_{n<0} p_n\right) \qquad (j > 0), \tag{9.32}$$

$$P_{0j} = P_{j0} = \left(\prod_{n<j} p_n\right)(1-p_j)\left(\prod_{n>0} p_n + \prod_{n<0} p_n\right) \qquad (j < 0), \tag{9.33}$$

$$P_{jj} = \left[1 + \left(\prod_{n>j} p_n\right) p_j - \prod_{n>j} p_n\right]^2 \qquad (j > 0), \tag{9.34}$$

$$P_{jk} = \left(\prod_{n>j} p_n\right)\left(\prod_{n>k} p_n\right)(1-p_j)(1-p_k) \qquad (j > 0, k > 0), \tag{9.35}$$

$$P_{jk} = \left(\prod_{n>j} p_n\right)\left(\prod_{n<k} p_n\right)(1-p_j)(1-p_k) \qquad (j > 0, k < 0), \tag{9.36}$$

$$P_{jj} = \left[1 + \left(\prod_{n<j} p_n\right) p_j - \prod_{n<j} p_n\right]^2 \qquad (j < 0), \tag{9.37}$$

$$P_{jk} = \left(\prod_{n<j} p_n\right)\left(\prod_{n<k} p_n\right)(1-p_j)(1-p_k) \qquad (j < 0, k > 0) \tag{9.38}$$

and

$$P_{jk} = \left(\prod_{n<j} p_n\right)\left(\prod_{n<k} p_n\right)(1-p_j)(1-p_k) \qquad (j < 0, k < 0), \tag{9.39}$$

where p_j is the Landau–Zener transition probability,

$$p_j = e^{-2\pi|h_j|} = \exp\left(-\frac{\pi V_j^2}{|\beta_j|}\right). \tag{9.40}$$

The physical interpretation of the above result has also been succesfully obtained in terms of the multipath pairwise successive two-state transitions [103]. Brundobler and Elser [101] put forward the hypothesis that for the most general multi-level linear model, in which the Hamiltonian matrix is given by

$$H_{ij} = \beta_j t \delta_{ij} + A_{ij} \tag{9.41}$$

with arbirary constants β_j and A_{ij}, the probability that the system remains in the initial jth diabatic state is given by a simple formlua,

$$P_{jj} = \prod_k \exp\left(-\frac{2\pi|V_{jk}|^2}{|\beta_j - \beta_k|}\right), \tag{9.42}$$

provided that this state corresponds to the extremum (maximum or minimum) slope β_j. This hypothesis is confirmed within the present model. Indeed, if the 0-state corresponds to the maximum or minimum slope (namely, if all $\beta_j < 0$ or all $\beta_j > 0$), then we have from Eq. (9.31)

$$P_{00} = \prod_n p_n^2 \,. \tag{9.43}$$

If a certain jth state corresponds to the maximum or minimum slope, then we have from Eq. (9.34) or Eq. (9.37)

$$P_{jj} = p_j^2 \,. \tag{9.44}$$

The above hypothesis is obvious, of course, from the view point of the semi-classical idea.

9.2.   Semiclassical Theory of Time-Independent Multi-Channel Problems

Since the exactly solvable models are very much limited, as is clear from the previous section, it is inevitable to use some kind of approximate theories to deal with general multi-channel problems. In this sense, the newly completed two-state semiclassical theory presented in Sec. 5.2 can present not only physically meaningful interpretation of multi-channel problems, but also a nice computational tool to calculate physical quantities such as scattering matrix, resonance width and bound state energies. When avoided crossings are well separated from each other, nonadiabatic transitions well localized at each avoided crossing can be treated as in a pure two-state problem. Even when some of avoided crossings come close together and their nonadiabatic couplings overlap with each other, the present semiclassical theory can still work surprisingly well. This is because the present semiclassical theory can take multi-state coupling effects into consideration by using adiabatic potentials and also the underlying analytical sturcture of the two-state problem is most properly taken into account. Actually, the important two parameters a^2 and b^2 which are defined in terms of the two adiabatic potential curves at one avoided crossing include all the interaction information coming from the other neighbouring avoided crossings; namely when two avoided crossings come close together, the corresponding a^2 and b^2 change from the corresponding values when they are far apart. This is simply because the adiabatic potentials are obtained by diagonalizing the whole multi-channel electronic Hamiltonian matrix. In the diabatic representation the interactions among other states are completely neglected, so that the theory cannot work better than in the adiabatic representation [106, 107].

Since in the special Demkov–Osherov model [95] the overall transition probability can be expressed as a simple product of the Landau–Zener formula, many people thought that phases may not be so important in general. This is not correct, of course. For example, ocillation and resonance structure of overall nonadiabatic transition probabilities depend on various phases strongly. The I-matrix propagation method developed by Nakamura [8, 108] made an important step to properly take phases into consideration. The present version of I-matrix propagation scheme [104–107] enables us to deal with multi-channel curve crossing problems more conveniently by absorbing all adiabatic phases between avoided crossings into the redefined I-matrix. Generally speaking, the better the two-state theory is, the more accurate results we can obtain for multi-channel curve crossing problems. Even the effects of closed channels can be incorporated. Firstly here, the basic idea is explained by taking an example shown in Fig. 9.2, which is a simple three-channel problem including both open and closed channels. The diagrammatic technique explained before in Sec. 7.1 is sometimes useful. The corresponding diagram is depicted in Fig. 9.2(b) [8]. When closed channels are included, it is convenient to construct the following χ-matrix given below, in terms of which the S-matrix is expressed as,

$$S = P_{\infty R_0}\{\chi_{\mathrm{oo}} - \chi_{\mathrm{oc}}(\chi_{\mathrm{cc}} - P^*_{R_0 T^{\mathrm{R}} R_0})^{-1}\chi_{\mathrm{co}}\}P_{R_0 \infty}, \qquad (9.45)$$

where the suffix o(c) means open (closed), and χ_{oc} represents the open channel-closed channel block of the whole χ-matrix, and so on.

The suffix T^{R} represents the right-side, i.e. outer, turning point T^{R}_n, and R_0 is an appropriately chosen position larger than the right-most basic element (diagram) and smaller than T^{R} if the latter exists. R_0 can be taken to be either dependent on or independent of the channel number. The matrices $P_{\infty R_0} = P_{R_0 \infty}$ and $P_{R_0 T^{\mathrm{R}} R_0}$ are the same as those defined by Eqs. (3.17)–(3.18) except that the integration limits in $P_{R_0 T^{\mathrm{R}} R_0}$ are interchanged since $R_0 < T_0$. Now, the matrix χ in the case of Fig. 9.2 can be simply expressed as

$$\chi = P_{C_1 R_0} O_{C_1} P'_{C_1 C_2} O_{C_2} P'_{C_2 T C_2} I_{C_2} P'_{C_1 C_2} I_{C_1} P_{C_1 R_0}, \qquad (9.46)$$

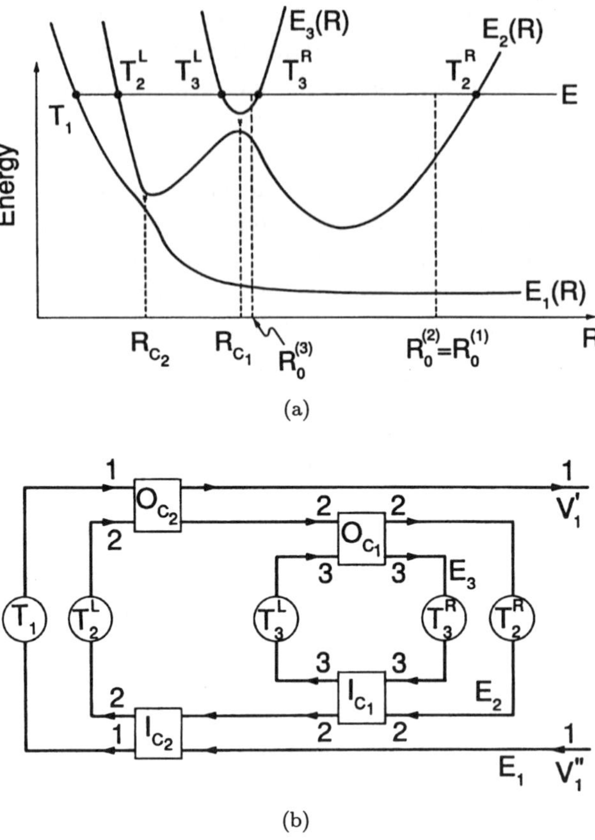

Fig. 9.2.   (a) Model three-state potential curve diagram.   (b) Diagram corresponding to Fig. 9.2(a). (Taken from Ref. [9] with permission.)

where the matrices P_{AB}, I_X and O_X ($= I_X^t$: transpose of I_X) are the same as before except that I_X and I_X^t are now 3×3 matrices having off-diagonal elements only between the channels coupled by curve crossing. The matrices $P'_{C_1 C_2}$ and $P'_{C_2 T C_2}$ are defined as

$$
[P'_{C_1 C_2}]_{nm} = \begin{cases} \delta_{nm} \exp\left[i \int_{R_{C_1}}^{R_{C_2}} k_n(R)dR \right] & \text{for } n,m = 1,2 \\[4mm] \delta_{nm} \exp\left[i \int_{R_{C_1}}^{T_3^L} k_3(R)dR \right] & \text{otherwise} \end{cases} \tag{9.47}
$$

and

$$[P'_{C_2TC_2}]_{nm} = \begin{cases} \delta_{nm} \exp\left[2i \int_{T_n^L}^{R_{C_2}} k_n(R)dR + \frac{\pi i}{2}\right] & \text{for } n, m = 1, \\ \delta_{nm} & \text{otherwise}, \end{cases} \qquad (9.48)$$

where the $(3,3)$-element of $P'_{C_1C_2}$ can either be put in $P'_{C_2TC_2}$ with the exponent doubled.

In this formulation the effects of closed channels can be straightforwardly incorporated into the S-matrix without difficulty. Resonances in the scattering clearly correspond to the complex solutions of

$$\det[\chi_{cc} - P^*_{R_0}T^R R_0] = 0. \qquad (9.49)$$

The above method can handle the general cases which involve tunneling and nonadiabatic tunneling, since the corresponding basic matrices are available now. Below, the general framework will be presented in more detail, in which the transitions considered can be either LZ- or NT-type. Some numerical examples will be presented to demonstrate the accuracy of the present theory. This general framework can be used for any type of transitions other than LZ and NT as far as the corresponding I-matrix is known. The Rosen–Zener type non-curve-crossing and the exponential potential models are such examples.

9.2.1. *General framework*

The multi-channel WKB wave function can be defined almost anywhere for both incoming and outgoing branches, and the internal reduced scattering matrix can be defined at a certain finite distance $R = R_0$, where all channels are energetically open. This is a connection matrix which connects the coefficients associated with incoming WKB solutions to the coefficients associated with outgoing WKB solutions. Then, we further propagate the solutions to asymptotic region where the final S-matrix is defined. As is well known, the exact quantum mechanical close-coupling calculations have to be carried out far into the asymptotic region to obtain converged solutions. In the semiclassical propagation method, however, we can terminate the propagation at the position just beyond the outmost avoided crossing.

Let us assume that we consider a general multi-channel system which contains totally $n + m$ states, in which n represents the number of asymptotically open channels and m is the number of closed channels (see Fig. 9.3). Of course, n and m vary as the collision energy changes.

9.2.1.1. *Case of no closed channel $(m = 0)$*

In this case, all channels (n) are open, and the avoided crossings are assumed to be distributed in the order,

$$R_N < R_{N-1} < \cdots < R_1 , \tag{9.50}$$

where R_N represents the innermost avoided crossing and R_1 is the outermost avoided crossing. Each avoided crossing at R_i $(i = 1, 2, \ldots, N)$ is identified by the channel indices α and β $(\alpha < \beta = 1, 2, \ldots, n)$, and the $n \times n$ I_{R_i}-matrix is given by

$$(I_{R_i})_{\alpha\alpha} = I_{11}, \qquad (I_{R_i})_{\alpha\beta} = I_{12} , \tag{9.51}$$

$$(I_{R_i})_{\beta\alpha} = I_{21}, \quad \text{and} \quad (I_{R_i})_{\beta\beta} = I_{22} \tag{9.52}$$

with the other elements

$$(I_{R_i})_{\nu\gamma} = \delta_{\nu\gamma}, \quad (\nu, \gamma) \neq (\alpha\alpha), (\beta\beta), (\alpha\beta), (\beta\alpha) , \tag{9.53}$$

where I_{11}, I_{12}, I_{21} and I_{22} are the matrix elements of the two-by-two I-matrix defined in Eq. (5.107) for LZ-case and Eq. (5.157) for NT-case. Final reduced scattering matrix can be expressed as

$$S^{\mathrm{R}} = (I_{R_N} I_{R_{N-1}} \cdots I_{R_1})^t (I_{R_N} I_{R_{N-1}} \cdots I_{R_1}) . \tag{9.54}$$

When there are closed channels, i.e., $m \neq 0$, this matrix gives the internal reduced scattering matrix χ. We will always denote it as χ in order to distinguish from the final reduced scattering matrix S^{R}.

9.2.1.2. *Case of $m \neq 0$ at energies higher than the bottom of the highest adiabatic potential*

In this case m channels are closed, as is shown in Fig. 9.3(a). Although in Fig. 9.3(a), we have chosen the NT-type avoided crossing at $R = R_j$ for the highest one, it can be LZ-type or any other type as far as we know its I-matrix. We first find χ-matrix at $R = R_0$ where all $n + m$ channels are open. Actually, χ can be otained exactly in the same way as in Eq. (9.46), but **now**

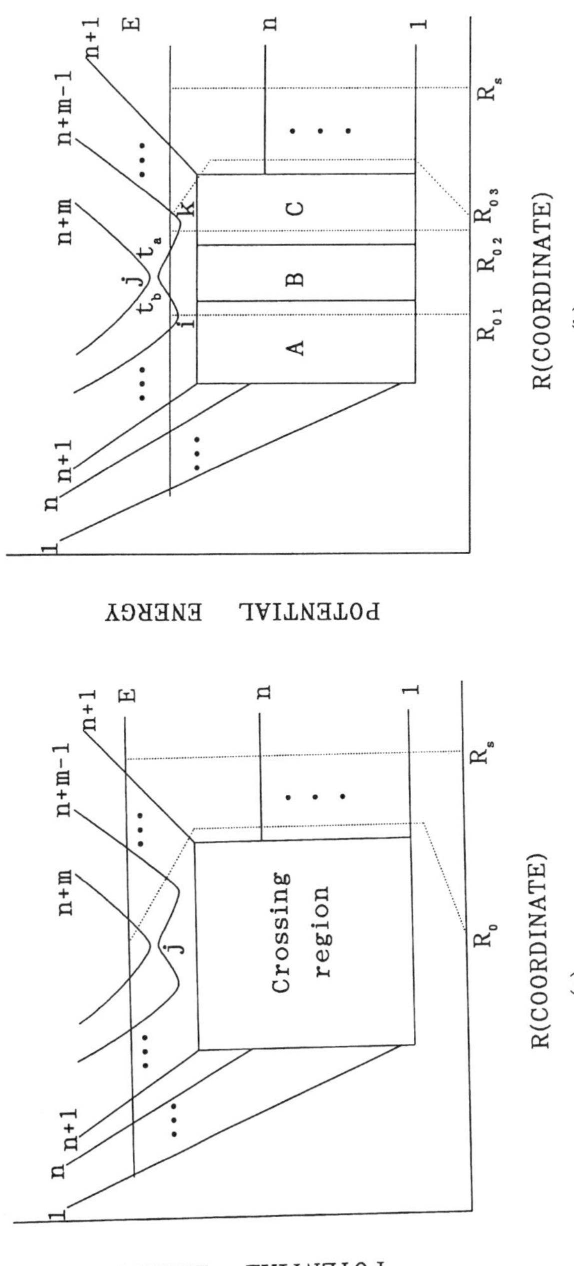

Fig. 9.3. Schematic $(n+m)$-state potential curves with n open and m closed channels. The turning points T_α ($\alpha = 1, 2, \ldots, n+m$) and t_β ($\beta = n+1, n+2, \ldots, n+m$) mentioned in the text (omitted in this figure) are, respectively, the leftmost and rightmost turning points on the αth and βth adiabatic potentials. $1, \ldots, n, n+1, \ldots, n+m-1, n+m$ correspond to the adiabatic potentials $E_1(R) \cdots E_n(R), E_{n+1}(R) \cdots E_{n+m-1}(R), E_{n+m}(R)$. (a) In the case of energy lower than the bottom of $E_{n+m}(R)$. (b) In the case of energy higher than the bottom of $E_{n+m}(R)$.

it becomes $(n+m) \times (n+m)$. It should be noted that this χ-matrix may now also include contributions from the other avoided crossings, if any, lying in the region $R > R_0$. This means that R_i does not have to be a definite single value, but can designate different positions depending on channels.

Let us write χ-matrix as

$$\chi = \left[\begin{array}{c|c} \chi_{\mathrm{oo}}(n \times n) & \chi_{\mathrm{oc}}(n \times m) \\ \hline \chi_{\mathrm{co}}(m \times n) & \chi_{\mathrm{cc}}(m \times m) \end{array}\right], \tag{9.55}$$

where o(c) means open (closed). Then, the final $n \times n$ reduced S^{R}-matrix can be found as

$$S^{\mathrm{R}} = \chi_{\mathrm{oo}} - \chi_{\mathrm{oc}}D^{-1}\chi_{\mathrm{co}} \tag{9.56}$$

with

$$D_{\alpha\beta} = \delta_{\alpha\beta}e^{-i2\theta_\alpha} + [\chi_{\mathrm{cc}}]_{\alpha\beta} \qquad (\alpha, \beta = 1, 2, \ldots, m), \tag{9.57}$$

where the additional adiabatic phase θ_α in Eq. (9.55) represents the WKB phase integral on the αth adiabatic potential,

$$\theta_\alpha = \int_{T_{n+\alpha}}^{t_{n+\alpha}} K_{n+\alpha}(R)dR, \qquad \alpha = 1, 2, \ldots, m, \tag{9.58}$$

which is an ordinary phase integral on a single adiabatic potential well. The derivation of S^{R}-matrix is essentially the same as that of the multi-channel quantum defect theory [7, 109–111], and actually has been used in the heavy particle scattering theory [57, 112]. But now we have compact analytical expressions for all elements.

Equation (9.56) indicates that resonance information is totally included in the D-matrix given by Eq. (9.57), and actually the complex zeros of $\det(D)$ provide positions and widths of resonances. It can be easily checked that D-matrix goes to the correct adiabatic limit when all avoided crossings turn to be adiabatic, $a^2 \longrightarrow 0$.

9.2.1.3. *Case of $m \neq 0$ at energies lower than the bottom of the highest adiabatic potential*

Now we turn to the situation shown in Fig. 9.3(b) in which the highest avoided crossing j is assumed to be NT-type and energy E is lower than the bottom of the highest adiabatic potential. The NT- and LZ-cases require different approaches. In the case of LZ, we can still use the I-matrix, even if energies are lower than the corresponding avoided crossing, and thus the whole procedure is the same as in the previous subsection. But in the case of NT, I-matrix no longer exsits at energies lower than the corresponding avoided crossing, and we must use the transfer matrix N^R obtained from S^R in Eq. (5.120) by Eq. (5.128) for the reason that it represents the local transmission phenomenon. This N^R-matrix represents nonadiabatic tunneling through the avoided crossing j in Fig. 9.3(b), since the highest state $E_{n+m}(R)$ is closed everywhere. Now the propagation scheme becomes a little bit complicated because of this nonadiabatic tunneling process. We have to classify avoided crossings into three regions: A, B and C in Fig. 9.3(b);

$$\text{Avoided crossings in } A\colon R_\alpha < R_i\,,$$

$$\text{Avoided crossings in } B\colon R_i < R_\alpha < R_k\,,$$

and

$$\text{Avoided crossings in } C\colon R_k < R_\alpha\,,$$

where α is a running index that covers all avoided crossing in the corresponding region. Thus, we can define the I-matrices for these three regions as

$$I_A = I_{R_N} I_{R_{N-1}} \cdots I_{R_i}\,, \tag{9.59}$$

$$I_B = I_{R_{i-1}} \cdots I_{R_{j-1}} I_{R_{j+1}} \cdots I_{R_{k+1}} \tag{9.60}$$

and

$$I_C = I_{R_k} I_{R_{k-1}} \cdots I_{R_1}\,, \tag{9.61}$$

where $R_N < R_{N-1} \cdots < R_i < R_{i-1} < R_{j-1} < R_{j+1} \cdots < R_{k+1} < R_k \cdots < R_1$, and each I_{R_α} $(\alpha \neq j)$ can be calculated in the same way as shown in Eqs. (9.51) and (9.52).

Let us define the internal $(n + m - 1) \times (n + m - 1)$ χ-matrix at $R \lesssim R_{01}$ as

$$\chi^{[1]} = (I_A I_B)^t (I_A I_B)\,, \tag{9.62}$$

where $R_{01} > R_i$ $(R_{02} < R_k)$ is a certain position in the left (right) well of $E_{n+m-1}(R)$ (see Fig. 9.3(b)). Here we have combined I_B with I_A, since the tunneling through j can commute with nonadiabatic transitions in region B. Next we consider the WKB wave function $\psi_{n+m-1}(R)$ in the region of tunneling through the highest barrier (crossing j) from R_{01} to R_{02}, and then we have the internal $(n+m-1) \times (n+m-1)$ χ-matrix, $\chi^{[2]}$:

$$\chi_{\alpha\beta}^{[2]} = \chi_{\alpha\beta}^{[1]} - \frac{\chi_{\alpha(n+m-1)}^{[1]} N_{22}^{\mathrm{R}} \chi_{(n+m-1)\beta}^{[1]}}{D_b}, \tag{9.63}$$

$$\chi_{\alpha(n+m-1)}^{[2]} = \frac{i N_{21}^{\mathrm{R}} e^{-i\theta_b}}{D_b} \chi_{\alpha(n+m-1)}^{[1]}, \tag{9.64}$$

$$\chi_{(n+m-1)\beta}^{[2]} = \chi_{\beta(n+m-1)}^{[2]} \tag{9.65}$$

$$\chi_{(n+m-1)(n+m-1)}^{[2]} = N_{11}^{\mathrm{R}} - \frac{\chi_{(n+m-1)(n+m-1)}^{[1]} (N_{12}^{\mathrm{R}})^2}{D_b}, \tag{9.66}$$

where $\alpha, \beta = 1, 2, \ldots (n+m-2)$, and D_b is given by

$$D_b = e^{-i2\theta_b} + \chi_{(n+m-1)(n+m-1)}^{[1]} N_{22}^{\mathrm{R}} \tag{9.67}$$

with

$$\theta_b = \int_{T_{n+m-1}}^{t_b} K_{n+m-1}(R)\,dR, \tag{9.68}$$

which is the phase integral along the left well of $E_{n+m-1}(R)$ (see Fig. 9.3(b)). N_{11}^{R}, N_{22}^{R} and N_{12}^{R} are evaluated from Eqs. (5.120) and (5.128). The third step is to propagate the WKB wave functions to $R \sim R_{03}$ where all avoided crossings in region C contribute to give

$$\chi^{[3]} = I_C^t \chi^{[2]} I_C \equiv \left[\begin{array}{c|c} \chi_{oo}(n \times n) & \chi_{oc}(n \times (m-1)) \\ \hline \chi_{co}((m-1) \times n) & \chi_{cc}((m-1) \times (m-1)) \end{array} \right]. \tag{9.69}$$

The final step is to propagate the WKB wave functions from R_{03} to R_s where the asymptotic region is reached, and we finally obtain S^{R}-matrix ($n \times n$) as

$$S^{\mathrm{R}} = \chi_{oo} - \chi_{oc} D^{-1} \chi_{co} \tag{9.70}$$

with

$$D_{\alpha\beta} = \delta_{\alpha\beta} e^{-i2\theta_\alpha} + [\chi_{cc}]_{\alpha\beta}, \qquad (\alpha, \beta = 1, 2, \ldots, m-1), \tag{9.71}$$

where the phase integrals along adiabatic potentials are defined by

$$\theta_\alpha = \int_{T_{n+\alpha}}^{t_{n+\alpha}} K_{n+\alpha}(R)dR, \qquad (\alpha = 1, 2, \ldots, m-1).$$ (9.72)

Note that

$$\theta_{m-1} \equiv \theta_a = \int_{t_a}^{t_{n+(m-1)}} K_{n+(m-1)}(R)dR,$$ (9.73)

which is the phase integral on the right well along $E_{n+m-1}(R)$ in Fig. 9.3(b). It should be noted that the turning points t_a and t_b in Fig. 9.3(b) must be replaced by R_{t_j} when the energy is located in the gap of the top avoided crossing j. Now, resonances come from the following two parts: one is from the complex zeros of $D_b = 0$ in Eqs. (9.63)–(9.66) and the other from the complex zeros of $\det D = 0$ in Eq. (9.70). The great advantage of the present semiclassical method is that the resonance part can be completely separated from the other transition processes and thus can be easily analyzed. If there is no open channel, the χ-matrix contains only closed channel portion, χ_{cc}, and $\det D = 0$ becomes the secular determinant whose roots give the solutions of the perturbed bound states.

When the energy further goes down, lower than the bottom of the adiabatic potential $E_{n+m-1}(R)$ in Fig. 9.3(b), the present semiclassical theory cannot take the highest, i.e. jth, crossing contribution into consideration. However, when the energy is low enough, the nonadiabatic tunneling at j almost coincides with single potential barrier tunneling. In that case we can neglect adiabatic potential $E_{n+m}(R)$, or we can treat the effect of the highest avoided crossing perturbatively. Then, the present I-matrix propagation method can still be formulated in a similar way as above. In the way presented here, the present semiclassical theory can deal with multi-channel curve crossing problems without any restriction for energy and number of channels, as far as all avoided crossings are relatively well separated from each other.

9.2.2. *Numerical example*

Here, two examples are presented to demonstrate the accuracy of the present semiclassical theory. First one is a model system of two Morse potentials crossed with two repulsive exponential potentials [106, 107]. The model potentials in diabatic representation are given by

$$V_1(R) = 0.037e^{-1.3(R-3.25)} - 0.034 \,,$$

$$V_2(R) = 0.037e^{-1.3(R-3.25)} - 0.012 \,,$$

$$V_3(R) = 0.4057[1 - e^{-0.344(R-3)}]^2 - 0.03 \,,$$

$$V_4(R) = 0.4057[1 - e^{-0.344(R-3)}]^2 \,. \tag{9.74}$$

Coupling terms are given as

$$V_{13}(R) = V_{14}(R) = V_{23}(R) = V_{24}(R) = \frac{2V_0}{1 + e^{R-3}}$$

$$V_{12}(R) = V_{34}(R) = 0 \,. \tag{9.75}$$

All quantities are in atomic units, and the reduced mass of the system is chosen to be that of an oxygen molecule ($m = 29377.3$). This model system was taken from some states of O_2 and the coupling represents the spin-orbit interaction among vibrational states of the oxygen molecule [113]. We have chosen $V_0 = 0.002$ at which any perturbation theory does not work at all, as it is realized as the strong coupling regime [113].

Figure 9.4 shows adiabatic potential diagram with $V_0 = 0.002$ from which we estimate all effective coupling constants a^2 for four nonzero coupling terms from Eq. (5.123) as:

$$V_0 = 0.002 \Rightarrow (a_1^2, a_2^2, a_3^2, a_4^2) = (7.88, 4.5, 3.61, 1.8) \,. \tag{9.76}$$

It should be noted that although the diabatic coupling constant V_0 is the same, their corresponding effective coupling parameters a^2 are very different from crossing to crossing. Actually that $0.1 \gtrsim a^2 \gtrsim 10$ corresponds to the most significant region for nonadiabatic transition, and in this region any attempt to use perturbation method will fail. Figure 9.5 show an excellent agreement between exact quantum results (Fig. 9.5(a)) and the present semiclassical results (Fig. 9.5(b)) for a wide range of energy. Even very detailed resonance sturctures are very well reproduced by the present semiclassical theory.

The second model is composed of one Morse potential crossing with five repulsive exponential potentials [107]. In the diabatic representation, this is

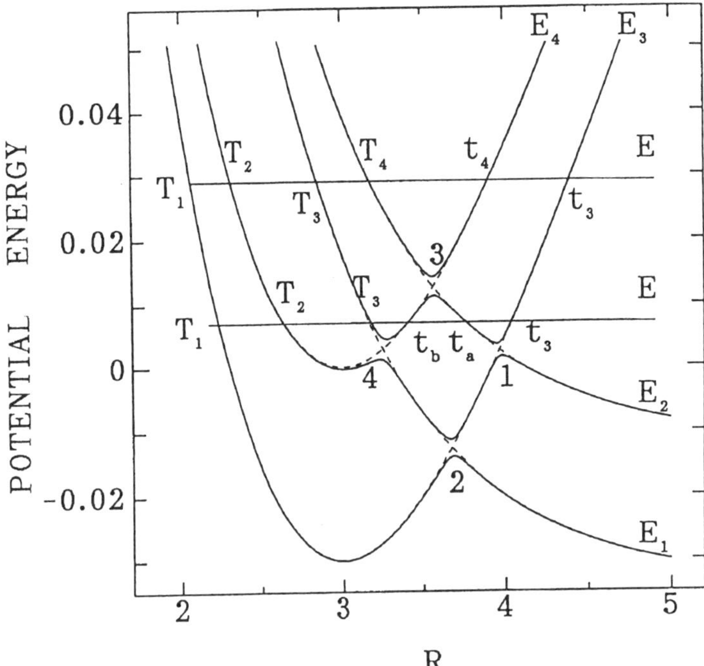

Fig. 9.4. Four-state potential diagram of Eqs. (9.74) and (9.75) with $V_0 = 0.002$. Solid lines for adiabatic potentials $E_i(R)$ and dashed lines for diabatic potentials $V_i(R)$ ($i = 1 - 4$). (Taken from Ref. [181] with permission.)

given by

$$V_j(R) = 0.037e^{-1.3(R-3.25)} + V_j^{(0)}, \qquad j = 1, 5,$$

$$V_6(R) = 0.4057[1 - e^{-0.344(R-3)}]$$

$$V_{j6} = \frac{2V_0}{1 + e^{R-3}}, \qquad j = 1, 5,$$

(9.77)

where

$$V_{j0} = -0.034, -0.012, +0.010, +0.032, +0.060 \quad \text{for } j = 1 - 5. \tag{9.78}$$

Everything is expressed in atomic units and the reduced mass is taken to be $m = 29373.3$. Figure 9.6 shows the potential system for $V_0 = 0.006$. The numerical results for the $i \rightarrow j$ transition probability P_{ji} are shown in Figs. 9.7 and 9.8 for $V_0 = 0.002$ and $V_0 = 0.006$, respectively. The corresponding

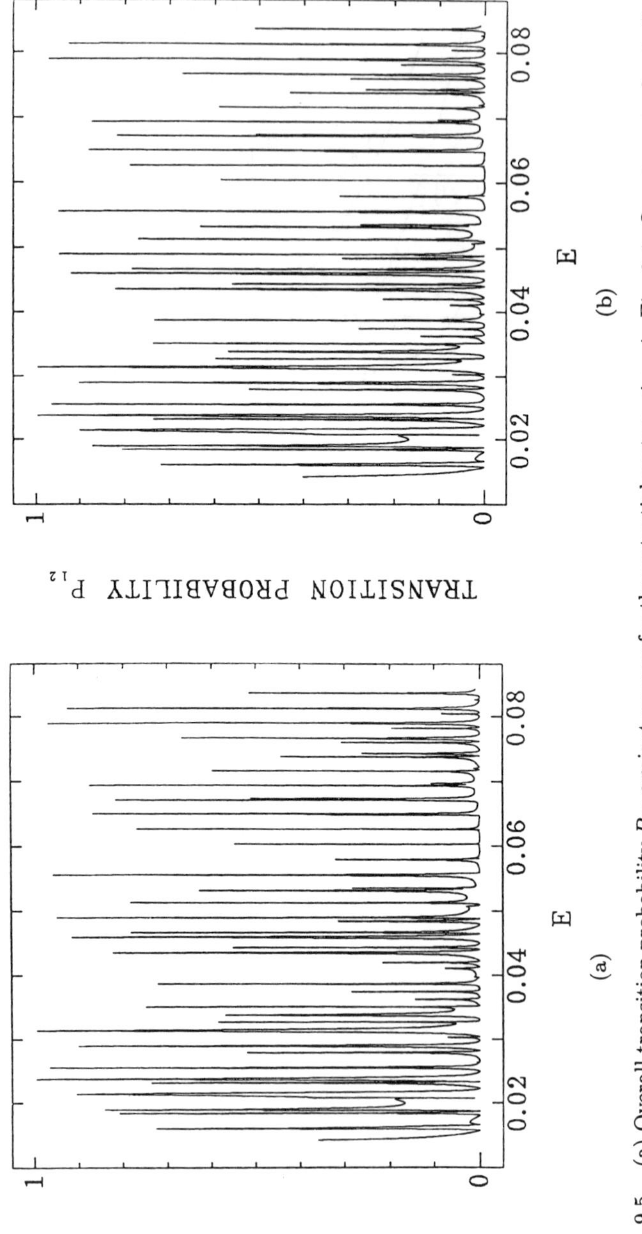

Fig. 9.5. (a) Overall transition probability P_{12} against energy for the potential system given in Fig. 9.4. Quantum mechanically exact numerical solution of coupled equations. (b) The same as Fig. 9.5(a) for the semiclassical theory in the adiabatic representation. The starting energy and the energy step used in the calculations are exactly the same as those in Fig. 9.5(a). (Taken from Ref. [181] with permission.)

effective coupling strengths are

$$V_0 = 0.002 \Rightarrow (a_1^2, a_2^2, a_3^2, a_4^2, a_5^2) = (21.8, 11.8, 6.48, 3.66, 1.78),$$
$$V_0 = 0.006 \Rightarrow (a_1^2, a_2^2, a_3^2, a_4^2, a_5^2) = (0.79, 0.40, 0.24, 0.14, 0.07),$$
$$(9.79)$$

where a_j^2 represents the parameter a^2 at the crossing between $V_j(R)$ $(j = 1-5)$ with $V_6(R)$. As is clearly seen from these figures, the semiclassical theory (solid line) reproduces the exact quantum mechanical numerical results (dotted line) very nicely.

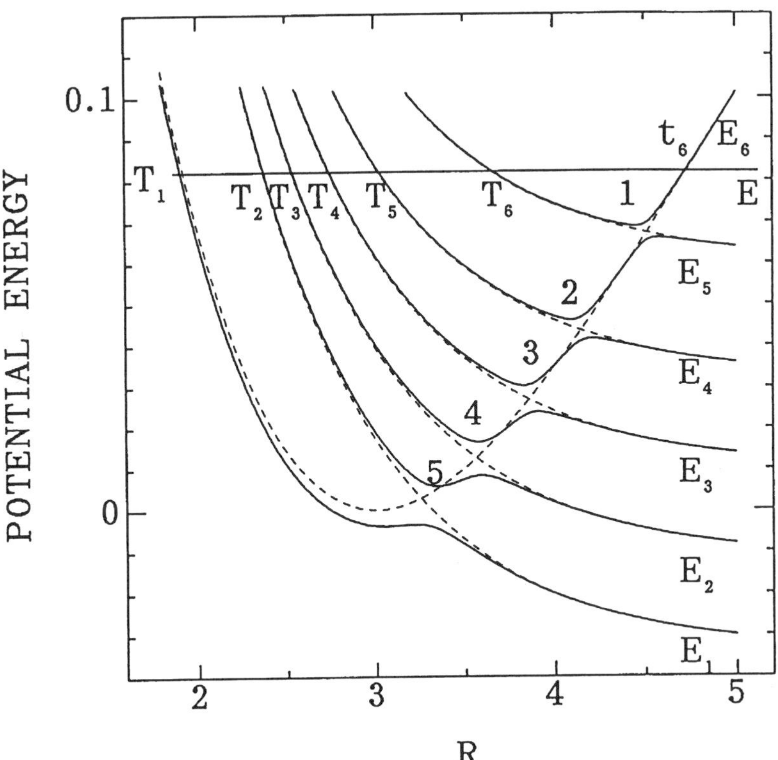

Fig. 9.6. Six-state potential diagram of Eqs. (9.77) and (9.78) with $V_0 = 0.006$. Solid lines for adiabatic potentials $E_i(R)$ and dashed lines for diabatic potentials $V_i(R)$ $(i = 1 - 6)$. (Taken from Ref. [107] with permission.)

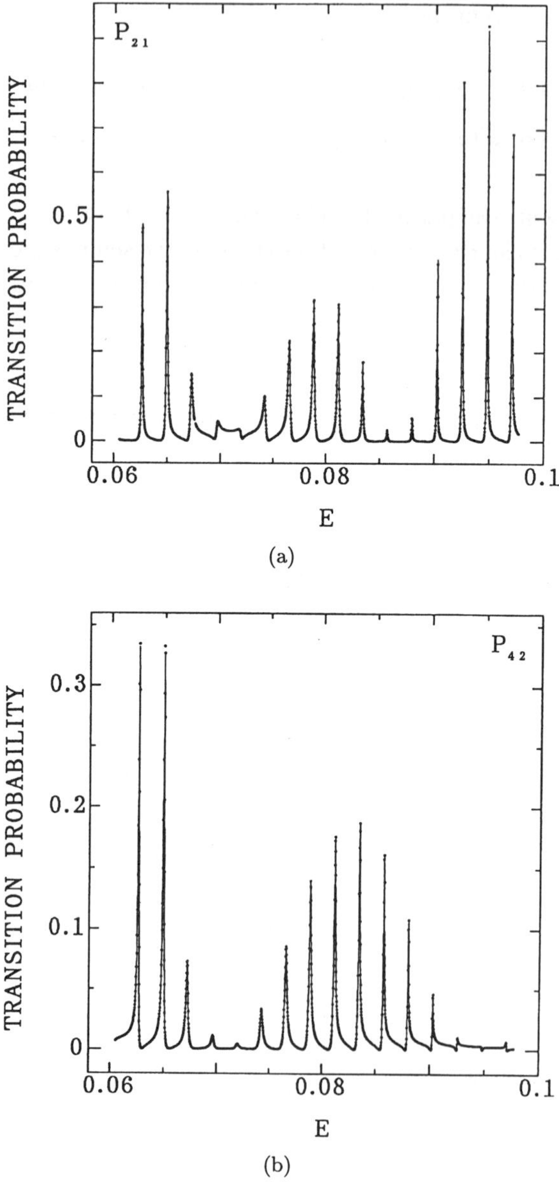

Fig. 9.7. Overall transition probabilities against energy for the potential system given in Eqs. (9.77) and (9.78) with $V_0 = 0.002$. $\cdots$: exact quantum, —: semiclassical theory in the adiabatic representation. (a) P_{21}, (b) P_{42}, (c) P_{54}. (Taken from Ref. [107] with permission.)

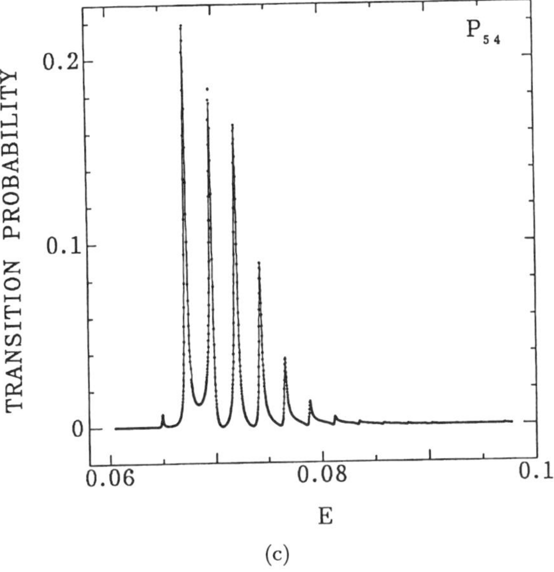

(c)

Fig. 9.7. (*Continued*)

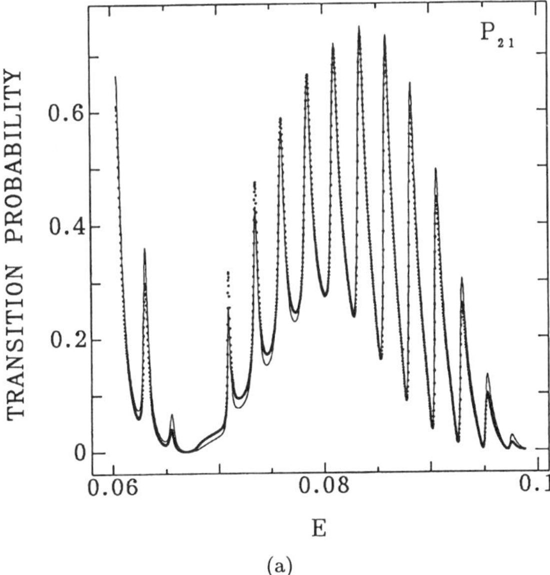

(a)

Fig. 9.8.   The same as Fig. 9.7 except for $V_0=0.006$. (Taken from Ref. [107] with permission.)

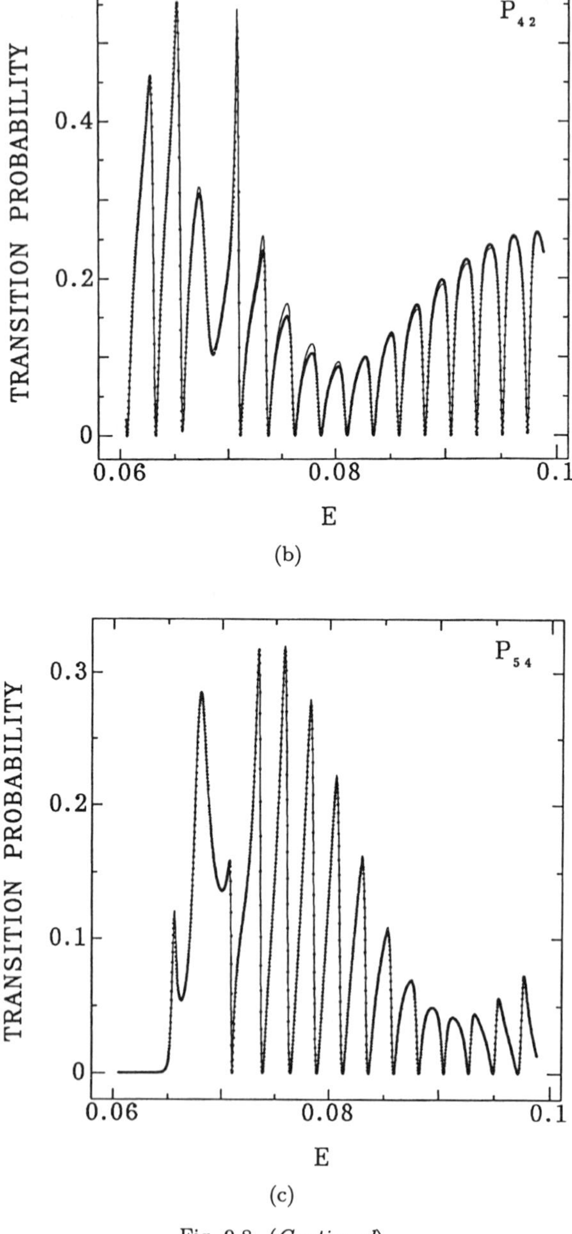

(b)

(c)

Fig. 9.8. (*Continued*)

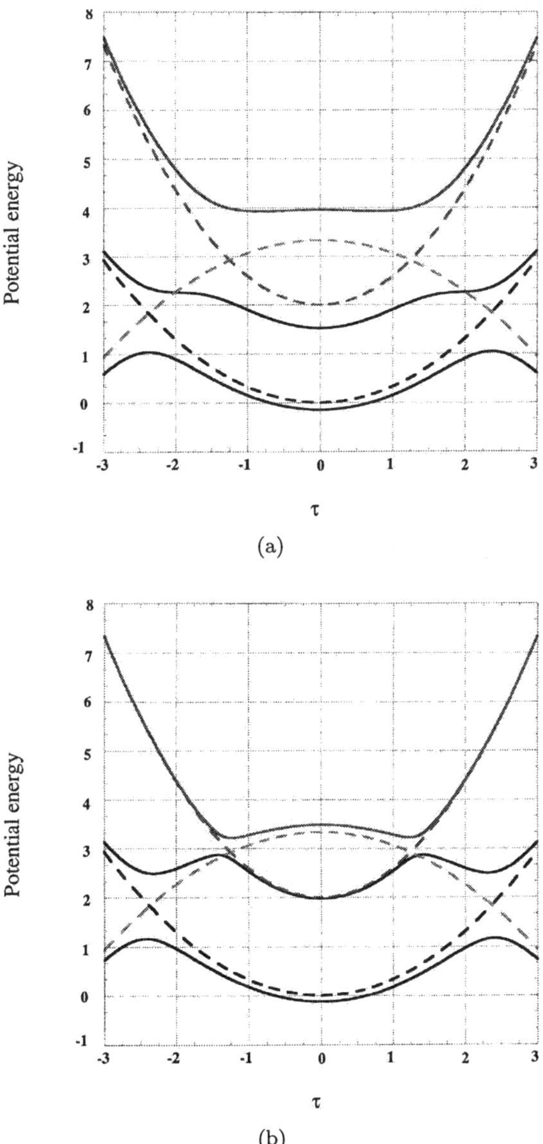

(a)

(b)

Fig. 9.9. Diabatic (dashed line) and full adiabatic (solid line) potentials in the case of three-level problems of Eq. (9.80). The ordinate and abscissa are the scaled energy, $\alpha_j \tau^2 + \beta_j$, and the scaled time, $\tau = 2V_{12}t/\hbar$, respectively. (a) $V_{13}/V_{12} = 1.0$, (b) $V_{13}/V_{12} = 0.2$. (Taken from Ref. [71] with permission.)

9.3.   Time-Dependent Problems

The matrix multiplication method explained in the previous section can be easily extended to time-dependent multi-channel crossing problems. The matrices I_X and P_{BA} in Sec. 6.2, and T^R and X in Sec. 7.5 become $N \times N$-matrices, where N is the number of channels. From the analysis of the time-independent multi-channel curve crossing problems [104–107], we can safely expect that this kind of two-by-two approximation works well, because our basic two-state theory is very accurate. The matrix I_X (or T^R) contains a 2×2 submatrix given by Eq. (6.21), representing a transition at the relevant crossing; otherwise this matrix is diagonal. For the evaluation of the basic parameters σ_0 and δ, the following three methods are possible. The first and the most accurate one is to diagonalize the whole $N \times N$ potential matrix to obtain fully adiabatic potentials and then evaluate the complex contour integral given by Eq. (6.32) for each relevant avoided crossing. This is quite difficult, unfortunately, because it is practically very hard to find complex crossing points accurately, especially when the number of states N exceeds three. The simplest, yet still not bad method is the two-by-two diabatic approach, in which only the relevant two states are considered at each crossing. In the two-state case, it is, of course, not difficult to evaluate the integral in Eq. (6.32). If the couplings at other crossings are strong and affect the relevant crossing, however, this method naturally breaks down. In this case, the adiabatic potentials obtained from the relevant two diabatic states are different from those obtained by the full diagonalization. The third method is to evaluate the parameters σ_0 and δ from fully adiabatic potentials on the real axis only (Eqs. (5.100)–(5.104)). See also the description below Eq. (6.32). This method can avoid the annoying complex calculus and is quite convenient, being more accurate than the two-by-two diabatic approach.

Here we take the following three-level problem as an example:

$$
H = \begin{bmatrix} -a_1 t^2 + b_1 & V_{12} & V_{13} \\ V_{12} & a_2 t^2 & 0 \\ V_{13} & 0 & a_3 t^2 + b_3 \end{bmatrix},
\tag{9.80}
$$

where V_{ij} are constant couplings and $a_1, a_2, a_3, b_1, b_3 > 0$. The dimensionless parameters defined as $\alpha_j \equiv \hbar^2 a_j / 8 V_{12}^3$, $\beta_j \equiv b_j / 2 V_{12}$ are taken to be $\alpha_1 = 0.05$, $\alpha_2 = 0.061$, $\alpha_3 = 0.111$, $\beta_1 = 1.66$ and $\beta_3 = 1.0$. Figure 9.9 depict the diabatic and adiabatic potentials for (a) $V_{13}/V_{12} = 1.0$ and $V_{13}/V_{12} = 0.2$. Figures 9.10(a)–9.10(c) show the results of the three methods mentioned above in comparison with the exact results as a function of V_{13}/V_{12}. The method based on the complex crossing points in the full adiabatic representation works very well even in a strong coupling region (see Fig. 9.10(a)). The two-by-two diabatic method, on the other hand, does not work well in a strong coupling region (see Fig. 9.10(b)). The third method, which employs the parameters given by Eqs. (5.100)–(5.104) with Eqs. (6.23)–(6.31) evaluated from the full adiabatic potentials on the real axis, gives much better results than the two-by-two diabatic method, even though the required computational effort is not much at all. This method works well until the full adiabatic potentials become flat (like in Fig. 9.9(a)) and the parameters cannot be estimated. Since the

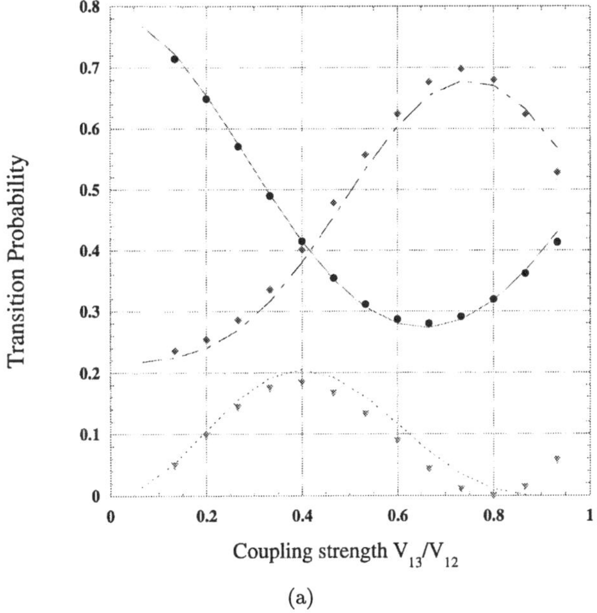

Fig. 9.10. Transition probabilities against the coupling strength V_{13}/V_{12} in the case of three-level problem given in Eq. (9.80). Solid line: $P_{1\to1}$ (exact), dash-dot line: $P_{1\to2}$ (exact), dotted line: $P_{1\to3}$ (exact), solid circle: $P_{1\to1}$ (semiclassical), solid rhomb: $P_{1\to2}$ (semiclassical), solid triangle: $P_{1\to3}$ (semiclassical). (a) The semiclassical theory is based on the complex crossing points in the full adiabatic representation. (b) The semiclassical theory is based on the two-by-two diabatic approach. (c) The semiclassical theory is based on the parameters defined by Eqs. (6.23)–(6.26). (Taken from Ref. [71] with permission.)

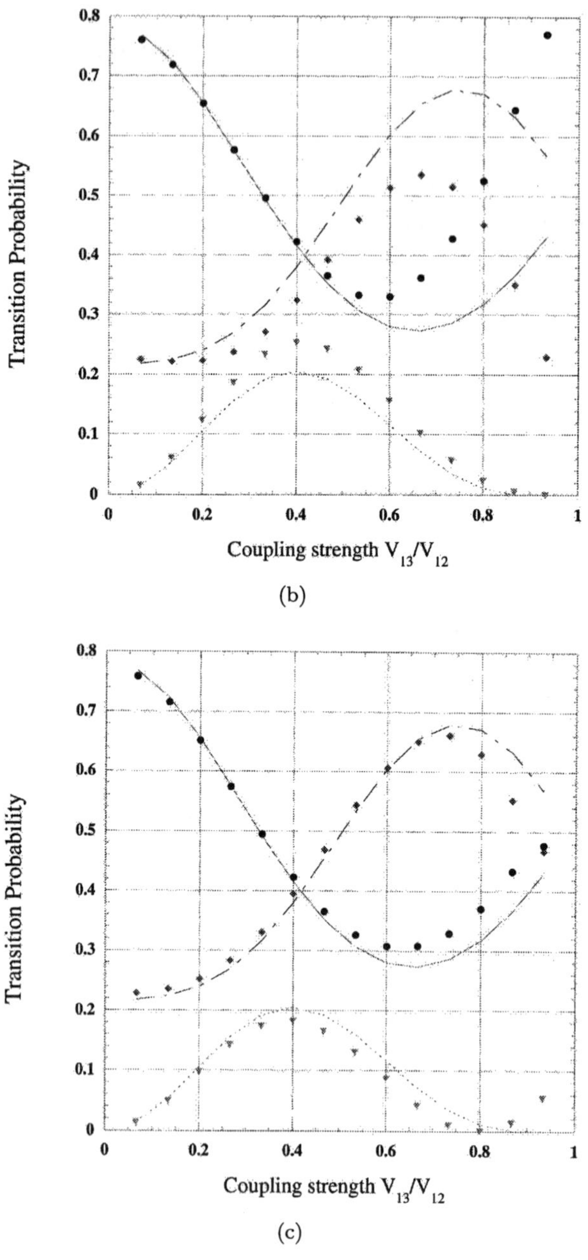

Fig. 9.10. (*Continued*)

search of complex crossing points becomes extremely difficult when the number of states exceeds three, the third method based on Eqs. (5.100)–(5.104) with Eqs. (6.23)–(6.31) is recommended.

Chapter 10

Multi-Dimensional Problems

Since most of physical, chemical, and biological dynamic processes in reality proceed in multi-dimensional configuration space, it is naturally very important to develop a useful methodology to deal with multi-dimensional processes [119]. Unfortunately, however, it is almost impossible to develop any intrinsically multi-dimensional analytical theory and thus it is necessary to figure out some useful approximate methods with the help of accurate one-dimensional theories. There are two ways to do that: one is to reduce any multi-dimensional problem to a one-dimensional multi-channel problem by expanding the total wave function in terms of appropriate internal states, as usually done in the quantum mechanical numerical solutions of Schrödinger equation. The second, which is more approximate but may be practically more useful, is to use classical trajectories which define curvi-linear one-dimensional space, and try to incorporate quantum mechainical effects as much as possible. The ordinary multi-dimensional semiclassical mechanics belong to this category [115–119]. The one-dimensional semiclassical theories of nonadiabatic transitions discussed so far can be incorporated in both methods, in principle.

Nonadiabatic transitions in multi-dimensional space are typically exemplified in terms of crossings or avoided crossings among multi-dimensional potential energy surfaces in real molecular systems [35, 38, 40, 120, 121]. As in the one-dimensional case, dynamic processes occur most effectively when these crossings or avoided crossings exist. When the system of our interest is described by N-independent variables, i.e. an N-dimensional problem, the corresponding potential energy surfaces can have real crossings of $(N-2)$-dimension. This is called Neuman–Wigner non-crossing rule and can be easily understood from Eq. (2.1). Suppose R is an N-dimensional vector. In order to have a real crossing, both $V_1(R) - V_2(R)$ and $V(R)$ should be zero at the same time, unless the electronic symmetry of the two states are different and $V(R) = 0$. Since the two conditions should be satisfied at the same time, the potential surface crossing, if it exists, should be of $(N-2)$- or lower-dimension. In the case of one-dimension $N = 1$, no crossing is possible, and in the case of $N = 2$, real crossing only at a point (conical intersection) is possible. In the following Sec. 10.1, various schemes of crossing in multi-dimensional space are classified and explained. In Secs. 10.2 and 10.3, the above mentioned two ways of semiclassical treatments are discussed.

10.1.   Classification of Surface Crossing

10.1.1.   *Crossing seam*

This is the direct extension of one-dimensional avoided crossing. A typical example of two-dimensional potential energy surfaces is shown in Fig. 10.1. The diabatic surfaces cross with each other along a line called "seam", and the adiabatic surfaces never cross and have the shape of hyperbolic cylinder oriented along the seam. In the direction perpendicular to the seam, we have the same situation as the one-dimensional avoided crossing case and the nonadiabatic coupling has a sharp peak at the avoided crossing. In the direction along

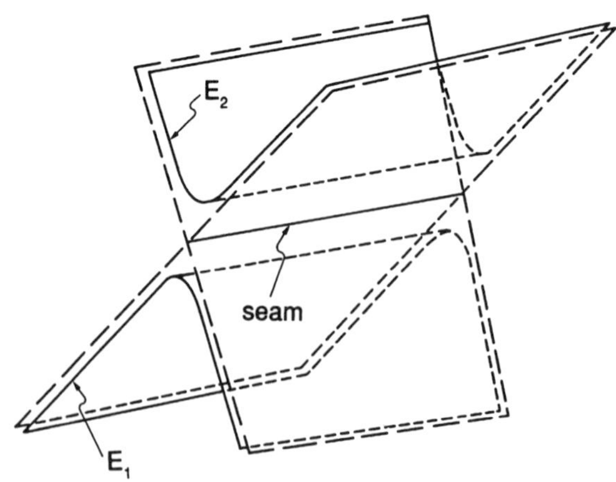

Fig. 10.1.   Schematic view of an avoided crossing of the two-dimensional potential energy surfaces. (Taken from Ref. [8] with permission.)

the seam, the Rosen–Zener type transition occurs, but it is much less effective. Practical examples can be found, for instance, in Ref. [120]. The simplest basic model in the two-dimensional (x, y) space is given by the following electronic Hamiltonial matrix in the diabatic representation as a direct extension of the Landau–Zener model:

$$H^{\text{el}}(x, y) = \begin{pmatrix} A_1 x + B_1 y & V_0 \\ V_0 & A_2 x + B_2 y \end{pmatrix}, \tag{10.1}$$

where $A_1 \sim B_2$ and V_0 are certain constants. The seam line between the two diabatic surfaces is defined by the equation,

$$y = -\frac{A_2 - A_1}{B_2 - B_1} x. \tag{10.2}$$

10.1.2.   *Conical intersection*

The conical intersection is schematically shown in Fig. 10.2. The adiabatic surfaces are discontinuous at the crossing point (appex). The Jahn–Teller intersection is one of the typical examples [122–124]. Many practical examples

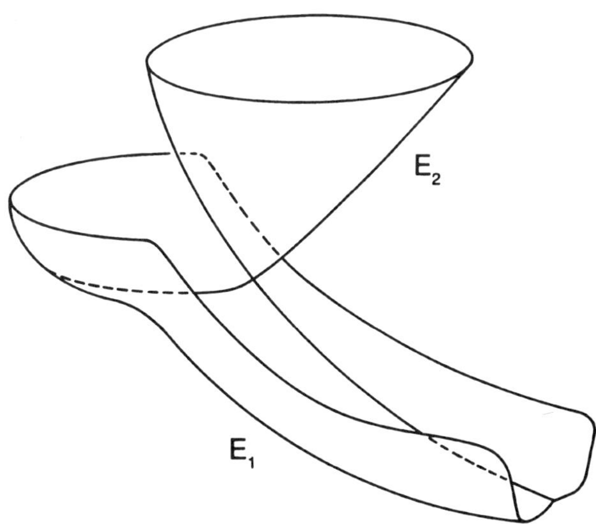

Fig. 10.2.   Schematic view of a conical intersection of the two-dimensional potential energy surfaces. (Taken from Ref. [8] with permission.)

can be found, for instance in Refs. [120, 121, 125]. In all directions passing through the appex, real curve crossing occurs and the nonadiabatic coupling has a delta function peak there. Along lines passing nearby the appex, the ordinary one-dimensional avoided crossing scheme holds. Because of the degeneracy at the appex, the electronic wave functions of adiabatic states is not single-valued, when the appex is encircled. Namely, the wave function does not come back to the original value, taking the different sign after one loop of the appex. In order to satisfy the sigle-valuedness of the total wave function, an additional phase should be introduced. This problem was first analyzed by Longuet–Higgins [126, 127], and has been investigated by many other authors [125, 128–133]. The additional phase due to conical intersection is called geometrical phase or Berry phase [125, 131, 136, 137]. Because of this additional effect, the nuclear wave functions in the adiabatic representation should change sign across some cut containing the locus of conical interesction, and accordingly satisfy an additional boundary condition [131]. The simplest basic model in the two-dimensional (x, y) space is given by

$$H^{\text{el}}(x, y) = \begin{pmatrix} Ax & Cy \\ Cy & Bx \end{pmatrix}, \tag{10.3}$$

where $A \sim C$ are certain constants. The nonadiabatic coupling along the direction x with coordinate y fixed at a constant y_c is given by (see Eq. (3.6))

$$T_{12}^{\text{rad}} = \left\langle \psi_2 \left| \frac{\partial}{\partial x} \right| \psi_1 \right\rangle = -\frac{(A'/4Cy_c)}{1 + (A'x/2Cy_c)^2}, \tag{10.4}$$

where $\psi_{1,2}$ are the adiabatic electronic wave functions and $A' = A - B$.

10.1.3.　*Renner–Teller effect*

In the case of doubly degenerate Π electronic states of a linear triatomic molecule belonging to the $D_{\infty h}$ symmetry, the two adiabatic potentials behave like in Fig. 10.3. They are continuous everywhere and tangentially touch at a reference nuclear geometry corresponding to the linear conformation. This is sometimes called "glancing intersection." The two-dimensional basic model in the diabatic representation is given by [120, 126]

$$H^{\text{el}}(x, y) = \begin{pmatrix} Ax^2 + By^2 & Cxy \\ Cxy & Bx^2 + Ay^2 \end{pmatrix}, \tag{10.5}$$

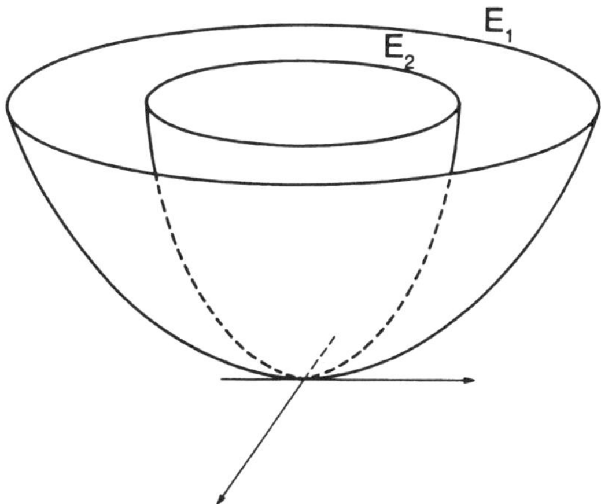

Fig. 10.3. Schematic view of a glancing intersection of the two-dimensional potential energy surfaces. (Taken from Ref. [8] with permission.)

where $A \sim C$ are certain constants. The corresponding adiabatic surfaces $E_1(x, y)$ and $E_2(x, y)$ are given by

$$E_1(x, y) = B(x^2 + y^2)$$
$$E_2(x, y) = A(x^2 + y^2). \tag{10.6}$$

In contrast to the Jahn–Teller case, there appears no discontinuity and no sign change of the adiabatic wave functions takes place. The nonadiabatic coupling is zero along any straight line going through the origin. Unfortunately, there is no basic analytical theory of nonadiabatic transition for this model, although there have been carried out some numerical calculations with use of the wave packet propagation method [134] based on the general analysis of the nonrelativistic spinless rovibronic Hamiltonian [135].

10.2. Reduction to One-Dimensional Multi-Channel Problem

10.2.1. *Linear Jahn–Teller problem*

The two-dimensional linear Jahn–Teller problem originally discussed by Longuet–Higgins [126] can be reduced to the following coupled equations after separating the angular motion which introduces the half-integer angular momentum quantum number m $(=1/2, 3/2, \ldots)$:

$$\frac{1}{2}\frac{d^2\psi_1}{d\rho^2} + \left(\epsilon - \frac{m^2}{2\rho^2} - \rho\right)\psi_1 = \frac{m}{2\rho^2}\psi_2,$$

$$\frac{1}{2}\frac{d^2\psi_2}{d\rho^2} + \left(\epsilon - \frac{m^2}{2\rho^2} + \rho\right)\psi_2 = \frac{m}{2\rho^2}\psi_1 \tag{10.7}$$

with

$$\epsilon = E\frac{\mu}{F^2\hbar^2},$$

$$\rho = \frac{F\mu}{\hbar}\sqrt{x^2 + y^2}, \tag{10.8}$$

where μ is the mass of the representative particle in the system, E is its energy, and the adiabatic potentials felt by that particle is $\pm F\sqrt{x^2 + y^2}$. The half-integer of the angular momentum quantum number m is the reflection of the geometrical phase of conical intersection. In the Renner–Teller problem the corresponding quantum number appears as an integer [126]. This simple model can represent a realistic triatomic system composed of three identical atoms in the ^2S state [126]. It should be noted that all parameters in the above equations are dimensionless.

As can be seen from Eq. (10.7), the nonadiabatic coupling is proportional to ρ^{-2} in the similar way to Coriolis coupling. Thus, as was explained in Sec. 3.4, it is better to introduce the dynamical state (or generalized adiabatic) representation, in which the dynamical states are given by

$$W_+(\rho) = \frac{m^2}{2\rho^2} + \sqrt{\rho^2 + \frac{m^2}{4\rho^4}}$$

$$W_-(\rho) = \frac{m^2}{2\rho^2} - \sqrt{\rho^2 + \frac{m^2}{4\rho^4}}. \tag{10.9}$$

In this representation the new nonadiabatic coupling is localized in coordinate space, and thus the semiclassical theory can be applied. The potentials W_+ and W_- for $m = 9/2$ and $m = 1/2$ are shown Figs. 10.4(a) and 10.4(b). As is clearly seen from these figures, resonance states are supported by the upper potential $W_+(\rho)$ and the decay process presents an interesting subject. Actually, this problem has attracted much attention so far [138–145].

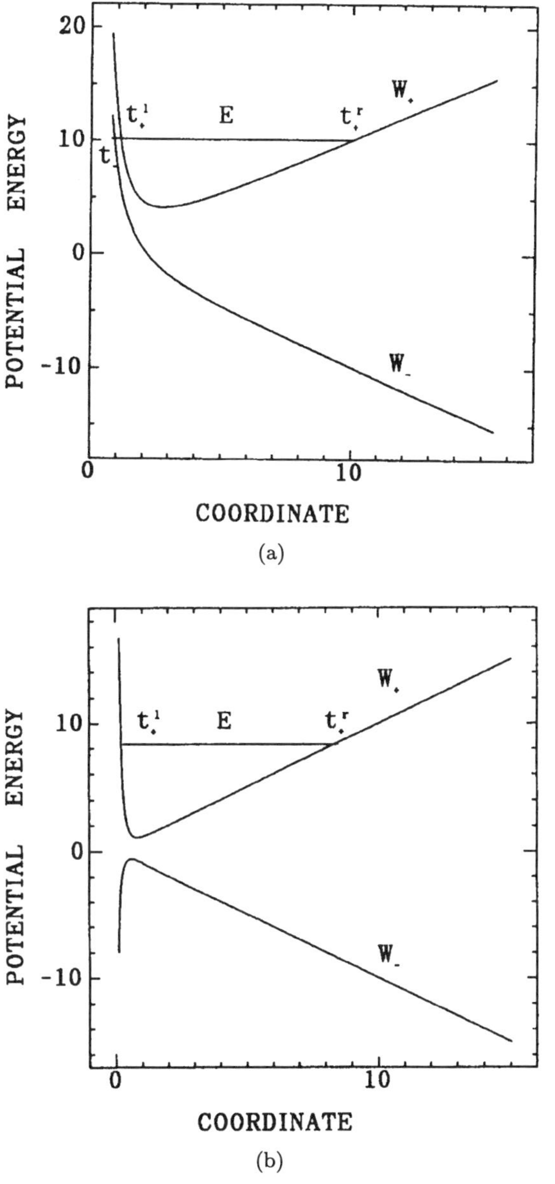

Fig. 10.4.   (a) The generalized adiabatic potentials $W_+(R)$ and $W_-(R)$ for $m = 9/2$. The potential energy and coordinate are dimensionless as defined by Eq. (10.8). (b) The same as Fig. 10.4(a) except that $m = 1/2$. (Taken from Ref. [145] with permission.)

Since the originally two-dimensional problem is reduced to the one-dimensional two-channel problem thanks to the symmetry of the problem, we can apply the semiclassical theory of nonadiabatic transitions to this problem [145]. The resonance positions ϵ_n and widths Γ_n for $n = 0, 1, 2, \ldots$ are given by the following expressions:

$$\theta(\epsilon_n) = \phi_L + \left(n + \frac{1}{2} \right) \pi \tag{10.10}$$

and

$$\Gamma_n = 2 \frac{1 - \sqrt{1 - P_L}}{1 + \sqrt{1 - P_L}} \left[\frac{d(\theta - \phi_l)}{d\epsilon} \right]^{-1}_{\epsilon - \epsilon_n} = 2 \frac{1 - \sqrt{1 - P_L}}{1 + \sqrt{1 - P_L}} \frac{1}{\pi} \frac{d\epsilon_n}{dn} \tag{10.11}$$

with

$$\theta(\epsilon) = \int_{t_+^l}^{t_+^r} K_+(\rho) d\rho \tag{10.12}$$

and

$$K_\pm(\rho) = \sqrt{2(\epsilon - W_\pm(\rho))} . \tag{10.13}$$

The overall nonadiabatic transition probability P_L and the phase ϕ_L are defined by the following nonadiabatic transition matrix N:

$$\begin{pmatrix} A_- \\ A_+ \end{pmatrix} = N \begin{pmatrix} B_- \\ B_+ \end{pmatrix} = \begin{pmatrix} \sqrt{1 - P_L}\, e^{2i\phi_L} & \sqrt{P_L}\, e^{i\sigma_L} \\ -\sqrt{P_L}\, e^{i\sigma_L} & \sqrt{1 - P_L}\, e^{-2i\phi_L} \end{pmatrix} \begin{pmatrix} B_- \\ B_+ \end{pmatrix}, \tag{10.14}$$

where A_- and B_- (A_+ and B_+) represent the coefficients of the adiabatic wave functions on the potential $W_-(\rho)$ for the outgoing and incoming portions, respectively (outgoing and incoming portions on $W_+(\rho)$). The nonadiabatic transitions are induced at $\rho \lesssim \rho_{\min}$, where $\rho_{\min}$ is the minimum point of the upper potential W_+. The complex crossing points of the potentials $W_\pm$ are

$$\rho_0^{(\pm)} = \pm i \left(\frac{m}{2} \right)^{1/3} ,$$

$$\rho_1^{(\pm)} = \left[\frac{\sqrt{3}}{2} \pm \frac{i}{2} \right] \left(\frac{m}{2} \right)^{1/3} , \tag{10.15}$$

$$\rho_2^{(\pm)} = \left[-\frac{\sqrt{3}}{2} \pm \frac{i}{2} \right] \left(\frac{m}{2} \right)^{1/3} .$$

The complex roots $\rho_2^{(\pm)}$ are neglected, since the real part is negative. The overall nonadiabatic transition matrix N defined by Eq. (10.14) is now given by

$$N = O_1 D_1 N_0 D_1 I_1 , \tag{10.16}$$

where

$$I_1 = \begin{pmatrix} \sqrt{1 - p_1}\, e^{i\phi_1} & -\sqrt{p_1}\, e^{i\sigma_{10}} \\ \sqrt{p_1}\, e^{-i\sigma_{10}} & \sqrt{1 - p_1}\, e^{-i\phi_1} \end{pmatrix} , \tag{10.17}$$

$$O_1 = I_1^T \qquad \text{(transposed)} , \tag{10.18}$$

$$N_0 = \begin{pmatrix} \sqrt{1 - p_0}\, e^{i\phi_0} & -\sqrt{p_0}\, e^{i\sigma_0} \\ \sqrt{p_0}\, e^{-i\sigma_0} & \sqrt{1 - p_0}\, e^{-i\phi_0} \end{pmatrix} \tag{10.19}$$

and

$$D_1 = \begin{pmatrix} \exp\left(-i \int_{\mathrm{Re}\,\rho_1}^{t_-} K_-(\rho)d\rho + i\frac{\pi}{4} \right) & 0 \\ 0 & \exp\left(-i \int_{\mathrm{Re}\,\rho_1}^{t_+^l} K_+(\rho)d\rho + i\frac{\pi}{4} \right) \end{pmatrix} . \tag{10.20}$$

Here N_0 represents the overall nonadiabatic transition at ρ_0, $I_1(O_1)$ describes the transition at ρ_1 in the incoming (outgoing) segment, and D_1 corresponds to the adiabatic wave propagation between ρ_1 and t_+^l (or t_-). The overall nonadaiabtic transition probability P_L and the phases ϕ_L in Eq. (10.14) are given by

$$P_\mathrm{L} = p_0 + (1 - p_0)[4p_1(1 - p_1)\sin^2(\sigma_1 - \phi_1 - \phi_0)] \tag{10.21}$$

and

$$\tan(2\phi_\mathrm{L}) = -\frac{p_1 \sin(2\sigma_1 - \phi_0) + (1 - p_1)\sin(2\phi_1 + \phi_0)}{p_1 \cos(2\sigma_1 - \phi_0) + (1 - p_1)\cos(2\phi_1 + \phi_0)} , \tag{10.22}$$

where

$$\sigma_1 = \sigma_{10} + \int_{t_-}^{\mathrm{Re}\,\rho_1} K_-(\rho)d\rho - \int_{t_+}^{\mathrm{Re}\,\rho_1} K_+(\rho)d\rho . \tag{10.23}$$

Interestingly, the nonadiabatic transitions at ρ_0 and ρ_1 cannot be described as the pure Landau–Zener or Rosen–Zener type of transitions, but in terms of mixture. The probabilities p_0 and p_1 are treated by the Rosen–Zener type of theory, and the phases ϕ_0 and ϕ_1 should be treated by the Landau–Zener type of theory (see Chaps. 3 and 5) [145]. Namely,

$$p_j = \frac{1}{1 + e^{2\delta_j}} \qquad (j = 0, 1), \tag{10.24}$$

$$\phi_j = \frac{\delta_j}{\pi} \ln\left(\frac{\delta_j}{\pi}\right) - \frac{\delta_j}{\pi} - \arg \Gamma\left(i\frac{\delta_j}{\pi}\right) - \frac{\pi}{4} \qquad (j = 0, 1) \tag{10.25}$$

and

$$\delta_j = I_{\mathrm{m}} \left\{ \int_{\mathrm{Re}\,\rho_j}^{\rho_j^+} [K_-(\rho) - K_+(\rho)]d\rho \right\} \qquad (j = 0, 1). \tag{10.26}$$

The numerical comparison between the exact solutions of the coupled Schrödinger equations and the present semiclassical theory is presented in Table 10.1 for $m = 9/2$. It can be easily seen that the agreement between the two is very good, especially for energy levels, and that it becomes better as the angular momentum m increases, although it is not shown here. The level width shows an interesting behaviour as a function of energy, that is, it increases slowly up to a certain maximum and then decreases very slowly. The

Table 10.1.   The resonance energy and width for $m = \frac{9}{2}$. "Exact" means the results of exact numerical calculation. "Adiabatic" and "generalized adiabatic" represent the results in the semiclassical approximation.

n	Exact	Adiabatic	Generalized adiabatic
0	(4.64, 0.0062)	(4.62, 0.0088)	(4.65, 0.005)
1	(5.60, 0.0131)	(5.60, 0.0154)	(5.61, 0.0128)
2	(6.49, 0.0192)	(6.50, 0.021)	(6.50, 0.0202)
3	(7.33, 0.0242)	(7.34, 0.0257)	(7.33, 0.0267)
4	(8.12, 0.0287)	(8.14, 0.0298)	(8.12, 0.0323)
21	(18.51, 0.0577)	(18.53, 0.0556)	(18.47, 0.061)
174	(70.30, 0.0728)	(70.23, 0.0570)	(70.36, 0.0543)
258	(91.08, 0.0707)	(91.00, 0.0540)	(91.12, 0.0503)

semiclassical results reproduce this behaviour. The exact numerical solution
was obtained by solving the coupled equations by the R-matrix propagation
method [146]. The details are given in Ref. [145].

The case of $m = 1/2$ requires a special treatment, because the lower potential W_- diverges to negative infinity at $\rho \to 0$ (see Eq. (10.9)). Potential
curves in this case are shown Fig. 10.4(b) and exibit the NT (nonadiabatic tunneling) type. Thus we need the NT-type reduced scattering matrix $(S^R_{\mathrm{NT}})_{11}$
(see Chap. 5 and Apendix A) given by

$$(S^{\mathrm{R}}_{\mathrm{NT}})_{11} = \frac{p}{1 + (1-p)e^{2i(\sigma - \phi_S)}} e^{i\Phi} \tag{10.27}$$

with

$$p = e^{-2\delta}, \tag{10.28}$$

where $\phi_S(\delta)$ is the same as Eq. (10.25), $\sigma = \theta$ given in Eq. (10.12), and δ in
Eq. (10.26), respectively. The phase Φ in Eq. (10.27) can be, of course, given
explicitly, but it is not necessary here. By the same analysis as before we can
obtain the following formulas for the resonance energy and width:

$$\theta(\epsilon_n) = \phi_S + \left(n + \frac{1}{2}\right)\pi, \qquad n = 0, 1, 2, \ldots \tag{10.29}$$

and

$$\Gamma_n = 2\frac{p}{2-p}\left[\frac{d(\sigma - \phi_S)}{d\epsilon}\right]^{-1} = 2\frac{p}{2-p}\frac{1}{\pi}\frac{d\epsilon_n}{dn}. \tag{10.30}$$

Table 10.2 shows a comparison between the numerical solution of the coupled
equations and the present semiclassical theory. The agreement is satisfactorily
good.

Before concluding this subsection, the following two things are better noted:
(1) Considering the the fact that the combination of Landau–Zener and Rosen–
Zener type of theories are required, it would be interesting to investigate the
applicability of the theory of exponential potential model. (2) The basic
Eqs. (10.7) in the case of $m = 1/2$ should probably be checked again, because the potential $W_-(\rho)$ is negatively divergent at $\rho \to 0$. Although this is
the critical case and does not lead to unphysical "fall of a particle to center"
[45], a certain small term positively proportional to ρ^{-2} might be missing in
the potential.

Table 10.2.   The same as Table 10.1, but for $m = \frac{1}{2}$.

n	Exact[a]	Adiabatic	Generalized adiabatic
0	(2.10, 0.238)	(1.88, 0.228)	(2.20, 0.231)
1	(3.44, 0.241)	(3.27, 0.260)	(3.54, 0.213)
2	(4.55, 0.236)	(4.40, 0.240)	(4.64, 0.200)
3	(5.54, 0.228)	(5.40, 0.227)	(5.62, 0.189)
4	(6.45, 0.226)	(6.32, 0.216)	(6.52, 0.181)
8	(9.60, 0.215)	(9.48, 0.189)	(9.65, 0160)
301	$\cdots$	(100.23, 0.068)	(100.4, 0.063)

[a]The exact result for $n = 301$ could not be obtained by the present method because of a strong overlapping of resonances.

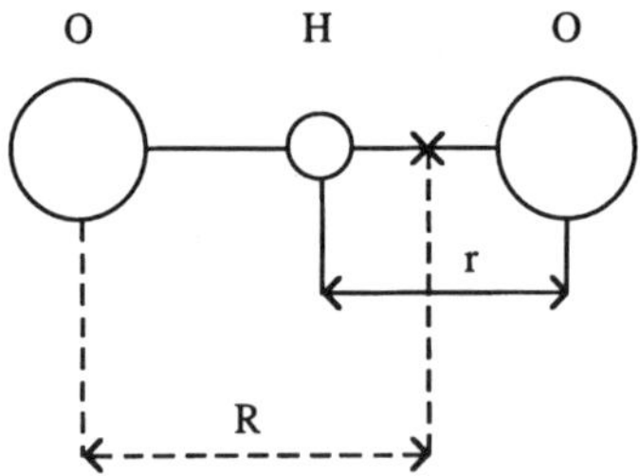

Fig. 10.5.   Collinear conformation of the triatomic system OHO.

10.2.2.   *Collinear chemical reaction*

Next example we consider is an electronically adiabatic chemical reaction, namely a reaction process on a single electronic potential energy surface. For simplicity, first we consider the hypothetical collinear chemical reaction of the symmetric triatomic system OHO. Three atoms are confined onto a physically one-dimensional space (see Fig. 10.5) and the Hydrogen atom H in the middle transfers between the two Oxygen atoms in this space. There are two independent Jacobi coordinates and this presents a mathematically two-dimensional problem. For convenience, we introduce the hyperspherical coordinates (ρ, ϕ) defined as,

$$r = \left[\frac{(m_O + m_H)^2}{(2m_O + m_H)m_H} \right]^{1/4} \rho \sin \phi \qquad (10.31)$$

and

$$R = \left[\frac{(2m_O + m_H)m_H}{(m_O + m_H)^2}\right]^{1/4} \rho \cos\phi \,, \tag{10.32}$$

where $m_O(m_H)$ is the mass of Oxygen (Hydrogen). The hyperangle ϕ ranges from 0 to $\phi_0 = \arctan([(2m_O + m_H)m_H]^{1/2}/m_O)$. In terms of these coordinates the Hamiltonian takes the form

$$H = K(\rho) + H_{\mathrm{ad}}(\phi; \rho) \,, \tag{10.33}$$

where $K(\rho)$ is the radial part of the total kinetic energy operator. As usual, the angular eigenvalue problem is first solved at fixed values of the hyperradius ρ,

$$H_{\mathrm{ad}}\Phi_n^{g,u}(\phi; \rho) = E_n^{g,u}(\rho)\Phi_n^{g,u}(\phi; \rho) \,, \tag{10.34}$$

where g and u indicate the gerade and ungerade symmetry of the state Φ_n. The solution of Eq. (10.34) gives adiabatic radial potentials $E_n^{g,u}(\rho)$, two lowest of which are shown in Fig. 10.6.

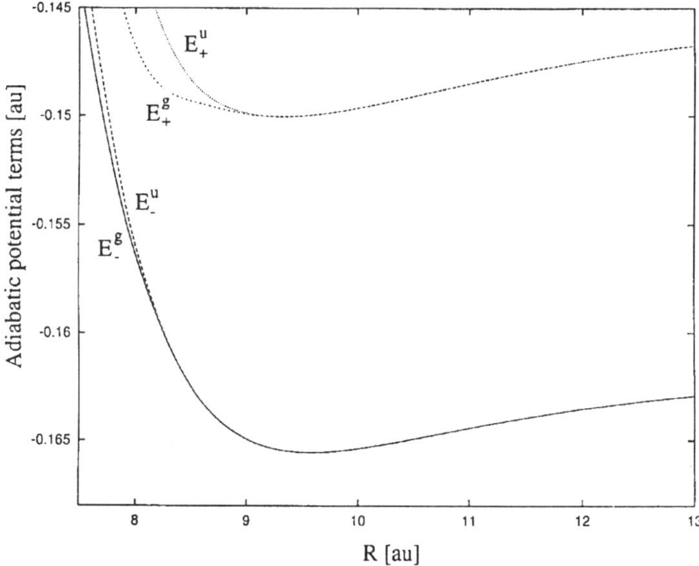

Fig. 10.6. Four lowest adiabatic potential curves as a function of hyperradius for the O+HO reaction system. $E_\pm^g$ are the gerade states, $E_\pm^u$ are the ungerade states. (Taken from Ref. [147] with permission.)

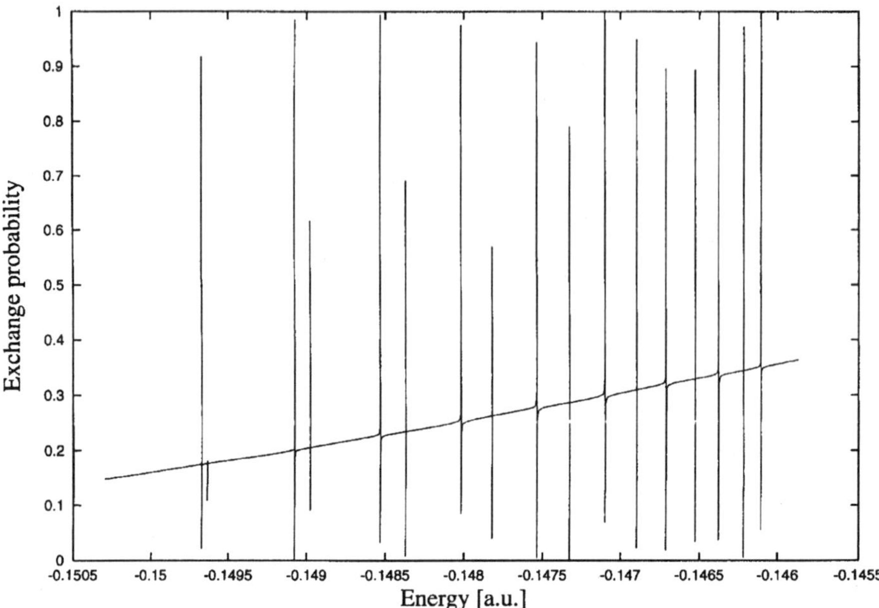

Fig. 10.7.   The exchange reaction probability versus energy. Sharp spikes represent Feshbach resonances. Eight sharp and short spikes correspond to those due to the excited ungerade state. The other nine spikes are due to the gerade state. See Tables 10.3 and 10.4. (Taken from Ref. [14] with permission.)

Numerical calculations are carried out with use of the R-matrix propagation method to evaluate the scattering phases δ_g and δ_u, from which the reaction probability below the threshold of vibrational excitation is given by

$$P(E) = \sin^2(\delta_g - \delta_u) \,. \tag{10.35}$$

The numerical results of $P(E)$ are plotted in Fig. 10.7 as a function of energy [147]. There are nine Feshbach type resonances with the widths $10^{-7} \sim 10^{-8}$ a.u. due to the excited $E_+^g(\rho)$ state and eight more much sharper resonances due to the excited $E_+^u(\rho)$ state. The resonance energies and widths ($E_{\rm res}$ and Γ) are evaluated with use of the Wigner formula for an isolated resonance [148], which in terms of the phase shift reads

$$\delta_\alpha(E) = \delta_{\rm nonres} + \arctan\left[\frac{2(E - E_{\rm res})}{\Gamma}\right] \tag{10.36}$$

Table 10.3. The resonance energy and width (both in hartree) for the gerade symmetry (see Fig. 10.6).

n	Exact	Semiclassical	P_{L}
0	$[-0.1496736, 0.64 \times 10^{-8}]$	$[-0.1497188, 0.52 \times 10^{-8}]$	0.52×10^{-4}
1	$[-0.1490782, 0.33 \times 10^{-7}]$	$[-0.1491480, 0.31 \times 10^{-7}]$	0.36×10^{-3}
2	$[-0.1485321, 0.76 \times 10^{-7}]$	$[-0.1485997, 0.40 \times 10^{-7}]$	0.48×10^{-3}
3	$[-0.1480178, 0.11 \times 10^{-6}]$	$[-0.1480788, 0.56 \times 10^{-7}]$	0.69×10^{-3}
4	$[-0.1475367, 0.12 \times 10^{-6}]$	$[-0.1475935, 0.66 \times 10^{-7}]$	0.89×10^{-3}
5	$[-0.1470988, 0.11 \times 10^{-6}]$	$[-0.1471504, 0.71 \times 10^{-7}]$	0.11×10^{-2}
6	$[-0.1467117, 0.89 \times 10^{-7}]$	$[-0.1467616, 0.70 \times 10^{-7}]$	0.12×10^{-2}
7	$[-0.1463796, 0.76 \times 10^{-7}]$	$[-0.1464258, 0.66 \times 10^{-7}]$	0.13×10^{-2}
8	$[-0.1461056, 0.54 \times 10^{-7}]$	$[-0.1461463, 0.58 \times 10^{-7}]$	0.14×10^{-2}

"Exact" means the results from Eq. (10.36). "Semiclassical" represents the results from Eqs. (10.10) and (10.11). P_{L} is the overall nonadiabatic transition probability (Eq. (10.37)). Note that $E_X^g = -0.1486376$ [au] and the effective coupling constant $a^2 = 0.0357$ from Eq. (5.89).

Table 10.4.   The same as Table 10.3 but for the ungerade symmetry (see Fig. 10.6).

n	Exact	Semiclassical	P_{L}
0	$[-0.1496364, 0.30 \times 10^{-10}]$	$[-0.1496874, 0.15 \times 10^{-10}]$	0.14×10^{-6}
1	$[-0.1489781, 0.84 \times 10^{-9}]$	$[-0.1490305, 0.71 \times 10^{-9}]$	0.71×10^{-6}
2	$[-0.1483712, 0.18 \times 10^{-9}]$	$[-0.1484242, 0.18 \times 10^{-9}]$	0.20×10^{-5}
3	$[-0.1478200, 0.30 \times 10^{-9}]$	$[-0.1478727, 0.34 \times 10^{-9}]$	0.40×10^{-5}
4	$[-0.1473276, 0.36 \times 10^{-9}]$	$[-0.1473792, 0.51 \times 10^{-9}]$	0.69×10^{-5}
5	$[-0.1468961, 0.66 \times 10^{-9}]$	$[-0.1469456, 0.65 \times 10^{-9}]$	0.10×10^{-4}
6	$[-0.1465266, 0.76 \times 10^{-9}]$	$[-0.1465733, 0.74 \times 10^{-9}]$	0.14×10^{-4}
7	$[-0.1462189, 0.60 \times 10^{-9}]$	$[-0.1462618, 0.76 \times 10^{-9}]$	0.17×10^{-4}

Note that $E_x^u = -0.1524674$ (au) and the effective couling constant $a^2 = 0.0068$ from Eq. (10.42) with $\alpha = 0.55$.

with $\alpha = g$ or u. The width Γ can be detemined with $\sim 1\%$ accuracy. The results are shown in Tables 10.3 and 10.4.

The resonance states supported by the excited potential curves $E_+^{g,u}(\rho)$ decay to the ground states by the nonadiabatic coupling between the two curves. Although it is not evident from a glance at Fig. 10.6, both gerade and ungerade potential curves have minima in the energy difference and depict the Landau–Zener type of avoided crossing. Thus the Zhu–Nakamura theory of nonadiabatic transition should be applicable. The resonance energy and width, $E_{\text{res}}^{(n)}$ and Γ_n, are given by Eqs. (10.10) and (10.11), where $P_{\text{L}}, \phi_{\text{L}}$, and σ_{L} are explicitly expressed as (see Sec. 5.2.1)

$$P_{\text{L}} = 4p_{\text{ZN}}(1 - p_{\text{ZN}}) \sin^2(\psi_{\text{ZN}}), \tag{10.37}$$

$$\phi_{\text{L}} = \frac{1}{2} \arctan\left(\frac{(1 - p_{\text{ZN}}) \sin(2\psi_{\text{ZN}})}{p_{\text{ZN}} + (1 - p_{\text{ZN}}) \cos(2\psi_{\text{ZN}})}\right) - \sigma_{\text{ZN}} \tag{10.38}$$

and

$$\sigma_{\text{L}} = -\frac{\pi}{2}. \tag{10.39}$$

The basic parameters a^2 and b^2 are evaluated from Eqs. (5.89)–(5.91). In the ungerade case, this cannot be done straightforwardly, unfortunately, because the crossing energy $E_X^u = (E_+^u(\rho_0) + E_-^u(\rho_0))/2 = -0.1524674$ a.u. is lower than the bottom of $E_-^u = -0.150031$. Here we used the following modification: First, the reference points x_+^0 and x_-^0 are defined by

$$E_+(x_+^0) = E_-(x_-^0) = E_0 \Rightarrow x_+^0, x_-^0 \tag{10.40}$$

with

$$E_0 = \frac{E_+(\rho_0) + E_-(\rho_0)}{2} + \alpha \frac{E_+(\rho_0) - E_-(\rho_0)}{2}, \tag{10.41}$$

where α is a certain constant in the range $(-1, 1)$. If we use the same technique as before, we obatin

$$a^2 = \sqrt{\frac{d_c^2 - 1}{1 + \alpha^2}} \frac{1 + d_c^2 \alpha^2}{1 + \alpha^2} \frac{\hbar^2}{\mu(x_+^0 - x_-^0)^2(E_+(\rho_0) - E_-(\rho_0))} \tag{10.42}$$

and

$$b^2 = \sqrt{\frac{d_c^2 - 1}{1 + \alpha^2}} \frac{E - (E_+(\rho_0) + E_-(\rho_0))/2}{(E_+(\rho_0) - E_-(\rho_0))/2} \tag{10.43}$$

with

$$d_c^2 = \frac{[E_+(x_-^0(-E_-(x_-^0)][E_+(x_+^0) - E_-(x_+^0)]}{[E_+(\rho_0) - E_-(\rho_0)]^2}. \tag{10.44}$$

It should be noted that in the case of two-state linear model the above quantities do not depend on α. In the general case, the parameter α should be chosen to be as close to zero as possible; and in the present ungerade case it should be in the range $1 > \alpha > 0.4$, where 0.4 corresponds to $E_0 = E^u_{\text{bottom}}$. Here we have used $\alpha = 0.55$ which corresponds to $E^u_0 = -0.148655$ which is just above the second resonance position as in the gerade case. Fortunately, the resonance positions are almost unchanged and the widths change only within the factor two or so in this range. The results are shown in Tables 10.3 for the gerade and 10.4 for the ungerade symmetry.

As can be seen from the tables, overall agreement between the semiclassical results and the quantum mechanical numerical results is very good, indicating the accuracy of the present semiclassical theory. Especially, it is remarkable that such small widths as $10^{-10} \sim 10^{-11}$ a.u. are well reproduced. As can be understood from the analysis here, ordinary chemical reaction on a single potential energy surface can be considered as vibrationally nonadaiabatic processes and can be analyzed by the semiclassical theory. In the next subsection, we will analyze realistic three-dimensional electronically adiabatic chemical reactions.

10.2.3. *Three-dimensional chemical reaction*

In view of the reduction to one-dimensional multi-channel problem and the uniform treatment of various arrangement channels, the hyperspherical coordinate system presents the most convenient method, as was demonstrated already in the collinear reaction in the previous subsection. The hyperspherical coordinate approach implemented for tri- and tetra-atomic reaction systems has shown remarkable progress recently [149–153]. Especially, in the three-dimensional (3D) heavy-light-heavy (HLH) reactions on a single potential energy surface, a newly introduced coordinate system called "hyperspherical elliptic coordinates" has made it possible to define the vibrationally adiabatic ridge lines and clarified the reaction mechanisms nicely in terms of the concept of vibrationally nonadiabatic transitions [152, 153]. Zhu, Nakamura and Nobusada [154] have treated, for the first time, such 3D HLH reactions analytically with use of the Zhu–Nakamura semiclassical theory. Needless to say, the conventional understanding of nonadiabatic transition is a transition between two electronically adiabatic states. This traditional adiabatic separation is, of course, guaranteed by the big mass disparity between electron and nucleus.

This idea may be extended to reactive scattering on a single electronically adiabatic potential energy surface. Especially in the case of 3D HLH systems, the light atom has some analogy to an electron, and the adiabatic separation becomes the separation between hyperradius and hyperangles.

The Schrödinger equation for a triatomic system in hyperspherical coordinates can be written as (in atomic units)

$$\left[\frac{1}{2\mu}\frac{1}{\rho^5}\frac{\partial}{\partial\rho}\rho^5\frac{\partial}{\partial\rho}+\frac{\Lambda^2(\omega)}{2\mu\rho^2}+V(\rho,\omega_{\mathrm{H}})-E\right]\Psi(\rho)=0\,,\qquad(10.45)$$

where $\Lambda^2(\omega)$ is the so called grand angular momentum operator, ρ is the hyperradius and ω represents the five hyperangles which are divided into three Euler angles ω_{E} (not appearing in the interaction potential $V(\rho,\omega_{\mathrm{H}})$) and two geometric angles ω_{H} of a triatomic system. The reduced mass μ is defined by

$$\mu=\sqrt{\frac{m_{\mathrm{A}}m_{\mathrm{B}}m_{\mathrm{C}}}{m_{\mathrm{A}}+m_{\mathrm{B}}+m_{\mathrm{C}}}}\,,\qquad(10.46)$$

where m_{A}, m_{B}, and m_{C} are the mass of atoms A, B and C, respectively. The explicit expression of the grand angular momentum operator, $\Lambda^2(\omega)$, is not given here, since that is not necessary in the present discussion. Furthermore, hereafter we consider only the case of $J=0$ for simplicity, where J is the total angular momentum quantum number. Thus the Euler angles ω_{E} do not show up.

The following expansion is used for solving Eq. (10.45):

$$\Psi(\rho)=\rho^{-5/2}\sum_{\nu}F_{\nu}(\rho)\Phi_{\nu}(\omega_{\mathrm{H}}\colon\rho)\,,\qquad(10.47)$$

where the adiabatic channel functions $\Phi_{\nu}(\omega\colon\rho)$ satisfy the hyperspherical adiabatic eigenvalue problem,

$$[H_{\mathrm{ad}}(\omega_{\mathrm{H}},\rho)-\mu\rho^2\tilde{U}_{\nu}(\rho)]\Phi_{\nu}(\omega_{\mathrm{H}}\colon\rho)=0\qquad(10.48)$$

with

$$H_{\mathrm{ad}}(\omega_{\mathrm{H}},\rho)=\frac{1}{2}\Lambda^2(\omega_{\mathrm{H}})+\mu\rho^2V(\rho,\omega_{\mathrm{H}})\,,\qquad(10.49)$$

in which $\tilde{U}_{\nu}(\rho)$ is the eigenvalue to be determined at each fixed ρ, since $H_{\mathrm{ad}}(\omega,\rho)$ depends on ρ parametrically. The scattering wave function $F_{\nu}(\rho)$

in Eq. (10.47) turns out to satisfy

$$\left\{ \frac{d}{d\rho^2} + 2\mu[E - U_\nu(\rho)] \right\} F_\nu(\rho) + \sum_\mu W_{\nu\mu}(\rho) F_\mu(\rho) = 0 \qquad (10.50)$$

with

$$U_\nu(\rho) = \tilde{U}_\nu(\rho) + \frac{15}{8\mu\rho^2}, \qquad (10.51)$$

where $W_{\nu\mu}(\rho)$ is the nonadiabatic coupling (not given here explicitly (see Ref. [152]). Now, the problem becomes an ordinary multi-channel scattering problem in the same way as in Sec. 9.2. We need only $U_\nu(\rho)$ in Eq. (10.51) to formulate the analytical solution of the reduced scattering matrix within the framework of the present semiclasiccal theory. Namely, we do not need any information about the nonadiabatic coupling $W_{\nu\mu}(\rho)$.

Figure 10.8 shows an example of adiabatic potential curve diagram of the $O(^3P)HCl$ system [153] based on the model LEPS potential energy surface (PES) [155]. Each curve corresponds to a certain ro-vibrational state of a reactant or product molecule at $\rho = \infty$. The dash line is the lowest ($v = 0$) vibrationally adiabatic ridge line, which effectively represents the reaction zone. The concept of ridge line is clear in the case of collinear reaction [156]. Namely, that is nothing but the mountain ridge deviding reactant and product valleys, and generalizes the concept of transition state or saddle point as its limiting point (see Fig. 10.9). Reactions can be regarded as vibrationally adiabatic nonadiabatic transitions in the vicinity of the ridge line.

In the three-dimensional case the ridge becomes a surface and cannot be projected out as a line onto the potential curve diagram, unless we take the highest point, for instance. In the treatment based on the hyperspherical elliptic coordinates, however, we can nicely extract the vibrationally adiabatic ridge lines, assuming the vibrational adiabaticity which holds well in HLH reactions. This is because the two angle variables χ and η in the hyperspherical elliptic coordinate system well represent vibrational and rotational motions, respectively [152]. By solving the vibrational motion at fixed hyperradius ρ first, we can easily find potential barriers in the η-space, which defines the ridge line. The significance of the ridge line can be demonstrated by the following examples [153]. Figure 10.10 shows the reaction probabilities for $O+HCl(v_i = 0, j_i) \rightarrow OH(v_f = 0, \sum_{j_f}) + Cl$. These are the results obtained by solving coupled Schrödinger equations numerically. The reaction probabilities

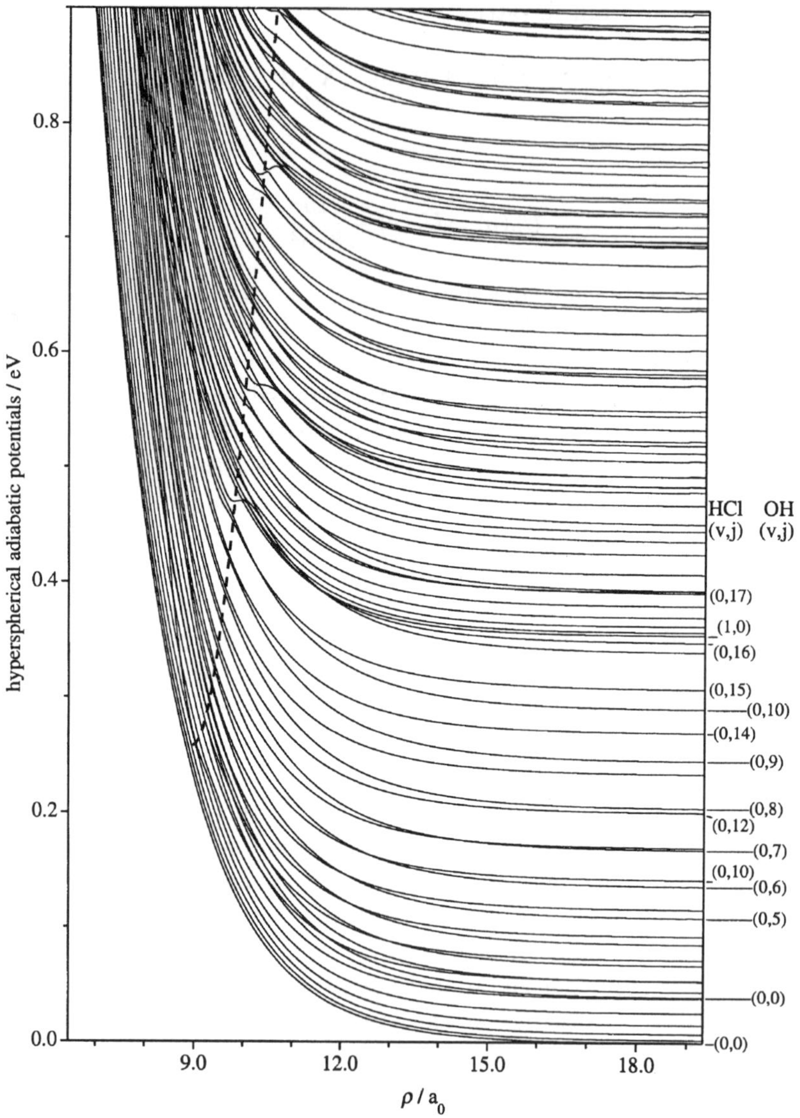

Fig. 10.8.   Adiabatic potential curves as a function of hyperradius ρ in the case of LEPS of OHCl. The energy is measured from the ground state ($v_i = j_i = 0$) of the reactant HCl. The dash line represents the $v = 0$ ridge line. The numbers in brackets at the right edge indicate the vibrational and rotational quantum numbers of HCl and OH. (Taken from Ref. [157] with permission.)

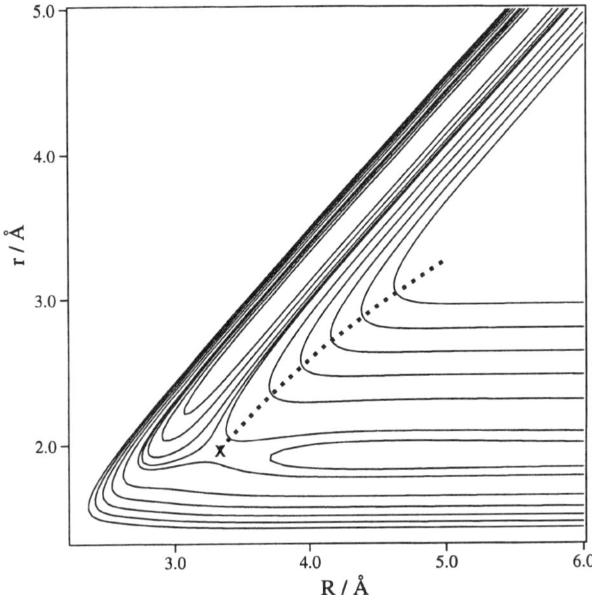

Fig. 10.9. Contour map of a model collinear potential energy surface. The cross and the dotted line represent the saddle point (or transition state) and the ridge line. (Taken from Ref. [153] with permission.)

for small $j_i \leq 6$ rapidly increase at $E \sim 0.22$ eV almost irrespective of j_i. On the other hand, onsets of the reaction probabilities for higher j_i shift toward higher energies with increasing j_i. Although the height of the saddle point of this surface is 0.168 eV, these reactions seldom occur effectively at $E \lesssim 0.2$ eV. These features can be explained by Fig. 10.8. The ridge line ends at $E \sim 0.25$ eV. Since the reactions occur effectively only when the potential ridge becomes energetically accessible, the onset is almost the same for small j_i. The shift of the onset energies for high j_i can be explained in the following way. For instance, in the case of $j_i = 10$ the reaction probability rapidly increases at $E \sim 0.3$ eV. If we follow the adiabatic potential curve of $j_i = 10$ in Fig. 10.8, a sharp avoided crossing appears at $\rho \sim 12.5a_0$ between this level and the level asymptotically corresponding to $j_f = 6$. This avoided crossing is located far right away from the ridge line and do not contribute to any transition. These states are localized in physically distant regions, i.e. in product and reactant regions, and do not interact strongly in spite of the energetical degeneracy.

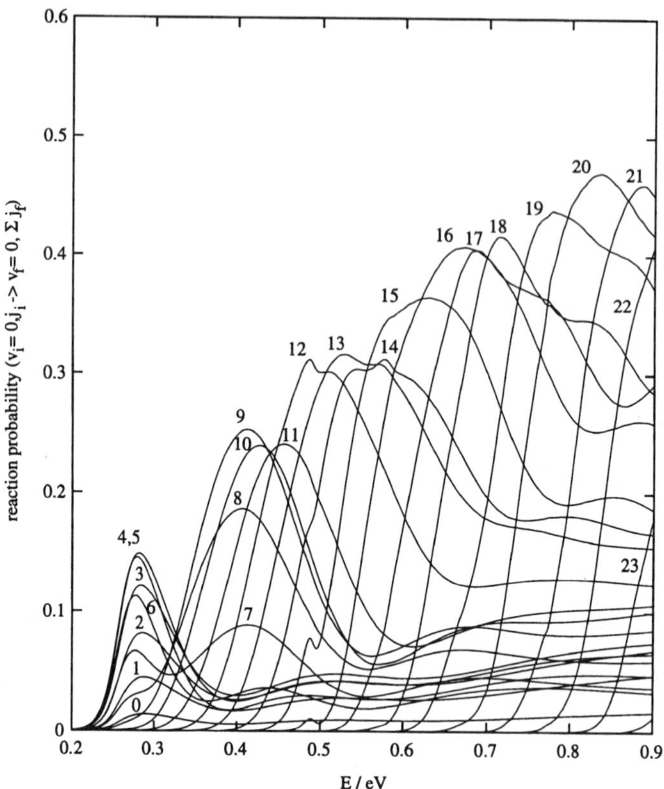

Fig. 10.10.   Reaction probabilities summed over the final rotational states for LEPS PES as a function of energy. (Taken from Ref. [153] with permission.)

This is actually the reason why the states have a very sharp avoided crossing. The system develops diabatically without any transition there. Thus potential curve which asymptotically correlates to $j_i = 10$ reaches the potential ridge at $\rho \sim 9.5a_0$ at $E \sim 0.3$ eV. This energy corresponds to the onset of the reaction probability. Another example is shown in Fig. 10.11. This shows the final rotational state distribution in the reaction $O+HCl(v_i = 0, j_i = 12) \rightarrow OH(v_f = 0, j_f) + Cl$. The final state $j_f = 7$ gives the highest peak. This can be again understood in terms of the ridge line and vibrational nonadiabatic transition. The potential curve corresponding to $j_i = 12$ encounters the ridge line at $\rho \sim 9.7a_0$, where it has an avoided crossing with the lower state which diabatically correlates to the asymptotic state $j_f = 7$.

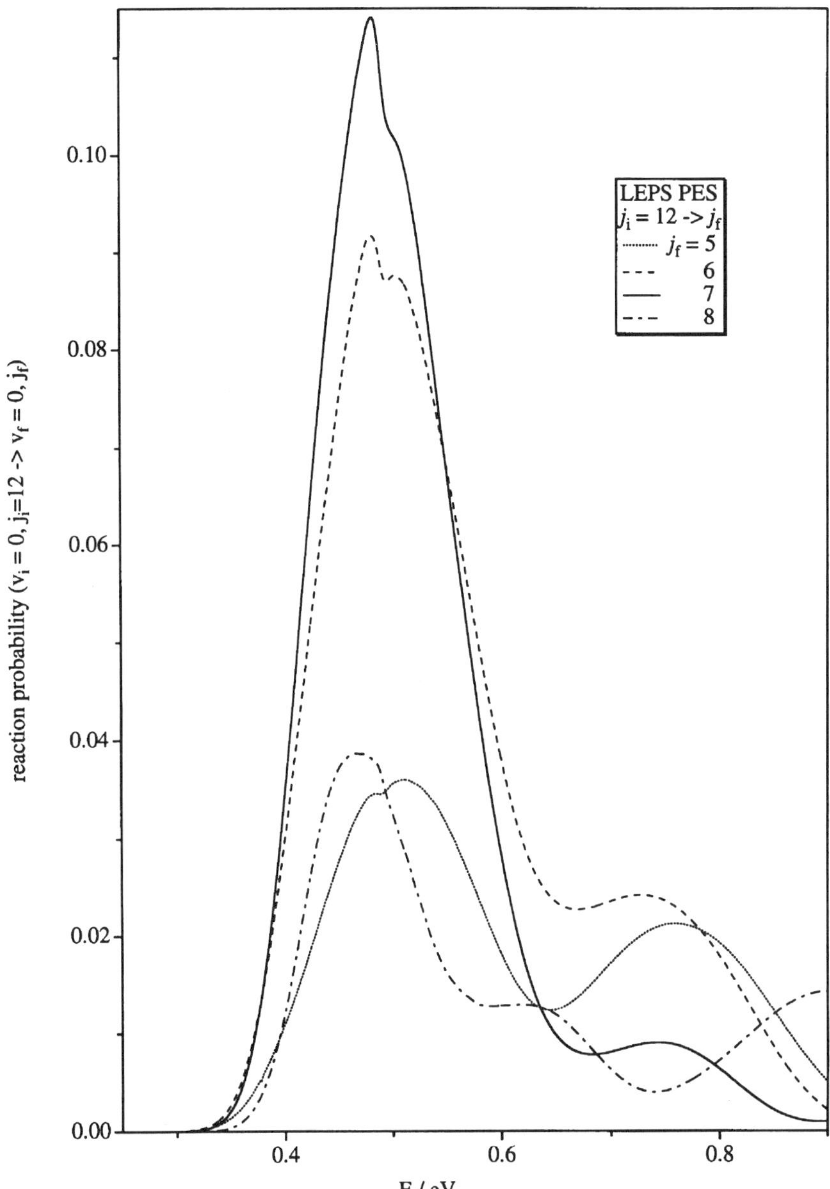

Fig. 10.11.   State-to-State reaction probabilities $(v_i = 0, j_i = 12 \rightarrow v_f = 0, j_f)$ for the OHCl LEPS PES as a function of energy. (Taken from Ref. [153] with permission.)

The above examples have qualitatively demonstrated the usefulness of the concepts of ridge line and of the vibrationally nonadiabatic transitions in the vicinity of the ridge line. In order to confirm this further, Zhu *et al.* analyzed the HLH reactions analytically by using the semiclassical Zhu–Nakamura theory [154]. The scattering wave function $F_\nu(\rho)$ in Eq. (10.50) can be written in the WKB form as

$$F_\nu(\rho) = \frac{A_\nu}{\sqrt{K_\nu(\rho)}} e^{i \int_{T_\nu}^{\rho} K_\nu(\rho)d\rho - i\frac{\pi}{4}} + \frac{B_\nu}{\sqrt{K_\nu(\rho)}} e^{-i \int_{T_\nu}^{\rho} K_\nu(\rho)d\rho + i\frac{\pi}{4}} \qquad (10.52)$$

for $\rho \longrightarrow \infty$, where T_ν is the rightmost turning point on the adiabatic potential $U_\nu(\rho)$ and

$$K_\nu(\rho) = \sqrt{2\mu(E - U_\nu(\rho))}. \qquad (10.53)$$

The reduced scattering matrix S^{R} is defined as

$$\begin{pmatrix} A_1 \\ A_2 \\ \vdots \\ A_n \end{pmatrix} = S^{\mathrm{R}} \begin{pmatrix} B_1 \\ B_2 \\ \vdots \\ B_n \end{pmatrix}, \qquad (10.54)$$

where n represents the number of open channels at a given total collision energy E.

The I-matrix propagation method is directly implemented to obtain the reduced scattering matrix,

$$S^{\mathrm{R}} = (I_1 I_2 \cdots I_N)^t (I_1 I_2 \cdots I_N), \qquad (10.55)$$

where N is the number of avoided crossings that can be as many as thousand among the sea of massive number of adiabatic potential curves $U_\nu(\rho)$. Those avoided crossings represent rovibrationally nonadiabatic transitions which represent reactive as well as non-reactive transitions. For 3D HLH systems vibrationally adiabatic ridge lines can be extracted, and the most important avoided crossings which represent reactive transitions are found to be located along or near these ridge lines. The lowest (ground vibrational, $\nu = 0$) ridge line defines the boundary of reaction zone. The avoided crossings outside this ridge line represent only nonreactive inelastic transitions. Those avoided crossings which are distributed far inside the ridge line represent a mixture of reactive and nonreactive transitions.

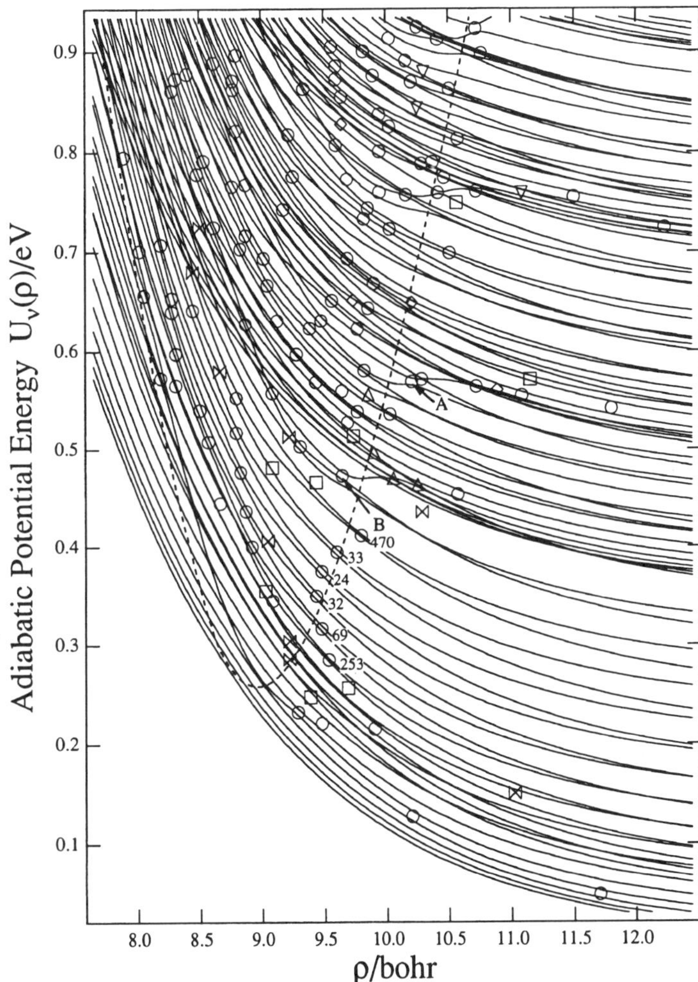

Fig. 10.12. Magnification of the reaction zone of the adiabatic potential curves of OHCl. The dashed line represents the $v = 0$ ridge line. Circles represent important avoided crossings among the adjacent adiabatic potentials and some of the circles assigned with the values of a^2 are the most important ones. The circle designated by letter A (B) indicates the avoided crossing responsible for the peak of certain vibrationally specified cumulative reaction probabilities (see Ref. [154]). The symbols $\square, \triangle, \triangledown, \bowtie$, and $\diamond$ represent the avoided crossings among the nonadjacent adiabatic potentials. The symbols $\square, \triangle$, and $\triangledown$ are from the diabatic potential manifolds with $v = 0, 1, 2$, respectively, for HCl+O arrangement (see Ref. [154]). The symbols $\bowtie$ and $\diamond$ are from the diabatic potential manifolds, respectively, for OH+Cl arrangement (see Ref. [154]). (Taken from Ref. [181] with permission.)

The effective coupling parameter a^2 defined in Eq. (5.89) for LZ-type and Eq. (5.123) for NT-type provides a very nice quantitative index of nonadiabatic coupling strength at each avoided crossing. Most of hundreds of avoided crossings correspond to $a^2 > 1000$ which represent very sharp avoided crossings, and do not play meaningful roles in dynamics. Only about one hundred

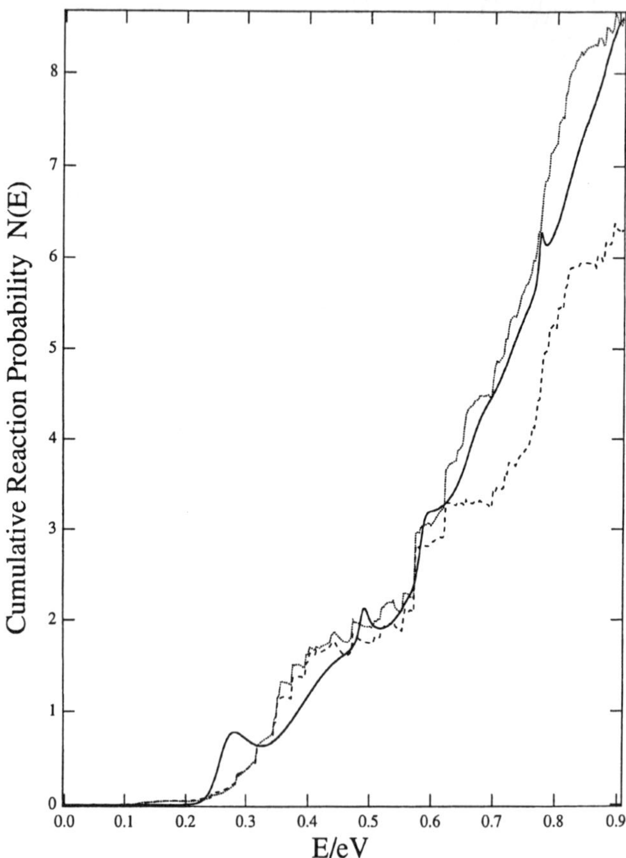

Fig. 10.13. Total cumulative reaction probability $N(E)$ as a function of the total energy E measured from the ground state of the reactant in the case of OHCl system: solid line (exact numerical result); dash line (semiclassical calculation with the avoided crossings among adjacent adiabatic potentials); dotted line (semiclassical calculation with the avoided crossings among both adjacent and nonadjacent adiabatic potentials included). Taken from Ref. [181] with permission.)

avoided crossings with $0.001 \lesssim a^2 \lesssim 1000$ among one $\sim$ two hundreds adiabatic potential curves contribute significantly to the reaction.

In the hyperspherical coordinate approach, all arrangement channels are treated equally and represented as adiabatic potential curves as a function

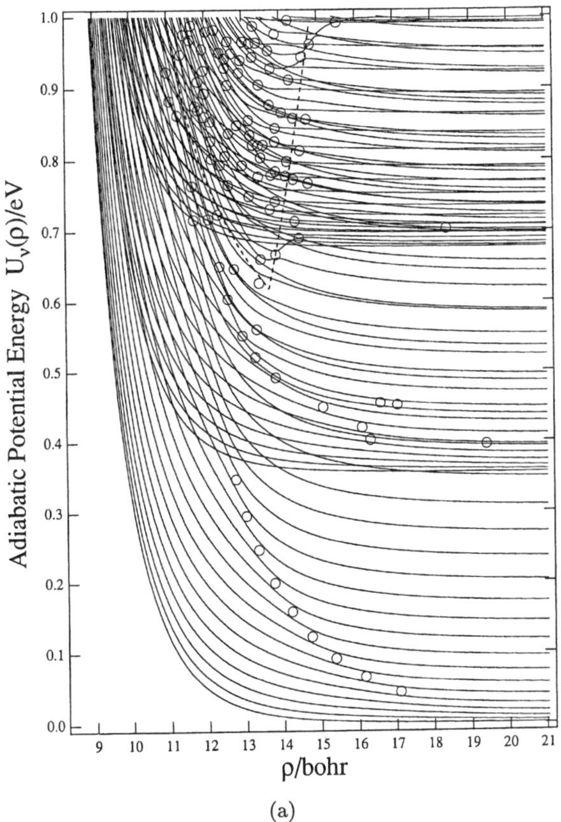

(a)

Fig. 10.14. Adiabatic potential energy curves as a function of hyperradius for BrHCl system, where the dashed line is the ridge line of vibrational quantum number $v = 0$ and the circles represent important avoided crossings with $0.01 < a^2 < 1000$ among the adjacent adiabatic potentials, (a) for the whole energy range considered here, (b) magnification of the reaction zone. Three avoided crossings marked with $\square$ indicate the accidental avoided crossing without any coupling. Two avoided crossings marked with $\triangle$ are those among nonadjacent adiabatic potentials (one is nonreactive between $v = 0$ and $v = 1$ of HCl arrangement and the other designated by A is reactive between $v = 1$ of HCl and $v = 0$ of HBr). For details see Ref. [154]. (Taken from Ref. [154] with permission.)

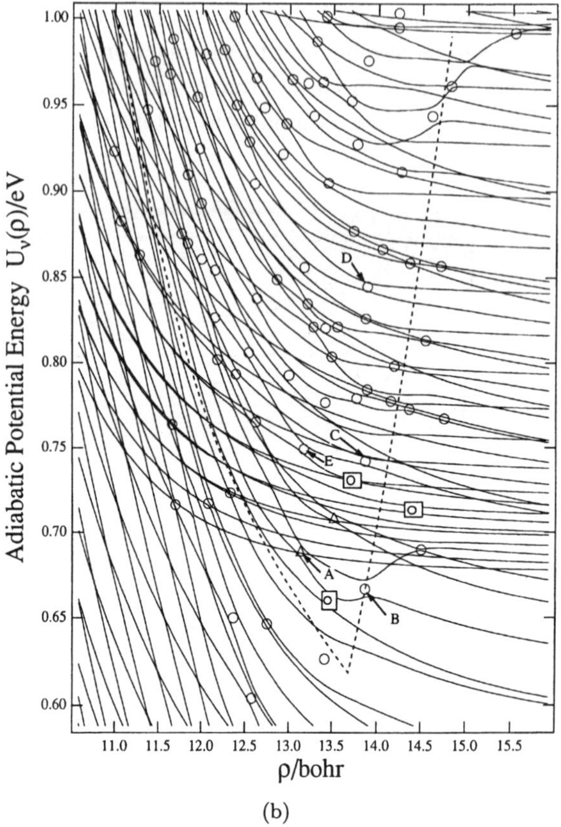

(b)

Fig. 10.14. (*Continued*)

of the hyperradius ρ. Therefore, important avoided crossings exist not necessarily only between adjacent adiabatic potentials, but can appear among non-adjacent adiabatic potentials. Besides, adiabatic potential curves belonging to physically separated arrangement channels avoid crossings very sharply outside the reaction zone and are better connected diabatically without any transitions. In order to extract these avoided crossings among non-adjacent adiabatic potentials, we have developed a certain diabatic decoupling method [154, 157]. We follow $U_\nu(\rho)$ inward from the asymptotic ρ, where each channel can be well assigned. If avoided crossings between adjacent adiabatic potentials on the way in have $a^2 > a_0^2$, then we switch $U_\nu(\rho)$ to $U_{\nu-1}(\rho)$, or $U_{\nu+1}(\rho)$. By repeating this diabatic switching procedure even inside the reaction zone, we can finally obtain a diabatic potential manifold and pick up important avoided crossings among originally non-adjacent potential curves. The dependence of this decoupling procedure on the critical value a_0^2 is not so strong and $a_0^2 \cong 100$ was chosen. All the important avoided crossings are treated analytically to evaluate the scattering matrix. In Ref. [154] we have studied the two examples of the 3D HLH reactions: $O(^3P)+HCl \longrightarrow OH+Cl$ and $Br+HCl \longrightarrow HBr+Cl$. Figure 10.12 shows a magnification of the reaction zone of adiabatic potential curve diagram of OHCl. A comparison between the exact quantum

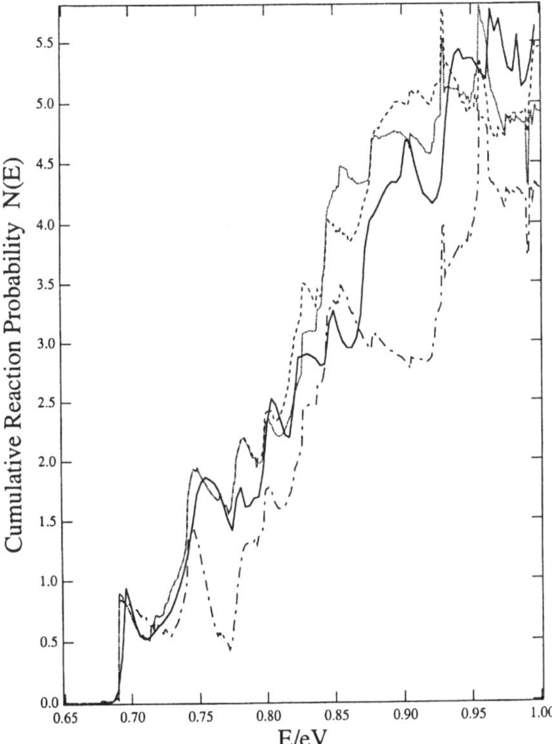

Fig. 10.15. Total cumulative reaction probability $N(E)$ as a function of the total energy E measured from the ground state of the reactant in the case of BrHCl system: solid line (exact numerical result); dash-dot-dash line (semiclassical calculation with the avoided crossings among adjacent adiabatic potentials); dotted (dashed) line (semiclassical calculation with the avoided crossings for the criterion $a_0^2 = 10$ ($a_0^2 = 100$) among both adjacent and non-adjacent adiabatic potentials included). (Taken from Ref. [154] with permission.)

calculation and the present semiclassical result is shown in Fig. 10.13 for cumulative reaction probability. Figures 10.14 and 10.15 show the similar results for BrHCl system [154, 158] based on the LEPS surface [159]. We can see that the agreement is quite good, considering that the semiclassical treatment here is a completely analytical one. The state-to-state reaction probabilities are not quantitatively very well reproduced, however. This is because non-reactive inelastic transitions are not necessarily well represented by avoided crossings.

10.3.   Semiclassical Propagation Method

Although the reduction of a multi-dimensional problem to a one-dimensional multi-channel system is accurate as it is and the latter system can be accurately treated by the one-dimenional semiclassical theories to some extent, as demonstrated in the previous section, this cannot be very practical especially for high dimensional systems. It is more appropriate to directly deal with the multi-dimensional dynamics. Intrinsically multi-dimensional semiclassical mechanics is, however, a very difficult subject and has not yet been established. The theory developed by Maslov would be the best one [118, 160], but it is still in a very primitive stage. However, there have been developed, not completed yet though, some practically usable methodologies which utilize classical trajectories and enable us to incorporate the semiclassical theories of nonadiabatic transitions into the formalism. In view of the importance of electronically nonadiabatic transitions in chemical physics, let us consider here molecular processes accompanying electronic transitions. Electronically nonadiabatic chemical reaction is one of the best examples. Since classical trajectory defines a curvi-linear one-dimensional space, we can utilize the one-dimensional theories of nonadiabatic transitions on the trajectory to deal with the electronically nonadiabatic transitions. Important issues are how to use the classical trajectories to describe the multi-dimensional dynamics and how much of quantum mechanical effects can be incorporated. Examples of such semiclassical propagation schemes are TSH (trajectory surface hopping) method [15, 161, 162], the semiclassical IVR (initial value representation) method [163, 164], and CFGA (cellular frozen Gaussian approximation) method [165–167].

The TSH is actually nothing but an ordinary QCT (quasi-classical trajectory) method without any incorporation of effects of phases, and the reaction probabilities are simply evaluated from the relative number of reactive trajectories with respect to the total. The electronically nonadiabatic transition probabilities, i.e. surface hopping probabilities, are furnished by the semiclassical theory. In the past, the simple Landau–Zener formula at the surface crossing has been used [15, 161, 162]. Nowadays, the time-dependent coupled Schrödinger equations are numerically solved [119, 169, 168]. This portion can be very much improved by using the new theory developed in Chap. 5. The Zhu–Nakamura theory can provide accurate yet simple expressions for the probability. The theory can accurately deal with even the cases that the energy is lower than the crossing point and the nonadiabatic tunneling type of transitions are involved, which cannot be treated by the simle Landau–Zener formula. On the other hand, numerical solutions of time-dependent Schrödinger equations which are required in the case of TFSH (fewest switches trajectory surface hopping) method [119, 168] are not necessary at all. Furthermore, the common trajectory on a certain averaged potential determined from the two electronic potentials, which is inevitable in the case of time-dependent treatment of electronically nonadiabatic transitions, is not necessary to be invoked. It should also be noted that knowledge of the nonadiabatic coupling is not at all required in the Zhu–Nakamura theory, and everything can be evaluated by the simple analytical formulas based on the adiabatic potentials on the real axis. One problem in the trajectory surface hopping type of treatment is how to cut the potential energy surfaces, in other words how to start the classical trajectory after hopping. The conventional choice is considered to be reasonable and acceptable [162]. That is, only the momentum component perpendicular to the crossing seam is changed by the amount corresponding to the adiabatic potential energy difference. It should also be noted that as is clear from the discussions in this book, the adiabatic representation is much better than the diabatic representation. Recently, the new implementation of the TSH method with use of the Zhu–Nakamura theory has been made with applications to the collinear H_3^+ and 3D DH_2^+ reaction systems and demonstrated good accuracy [170]. Even the threshold region can be nicely reproduced.

A more sophisticated method is the semiclassical surface hopping theory, which is an extension of the classical S-matrix theory for single surface dynamics developed by Miller and George [115, 116, 171]. The basic idea of the classical S-matrix theory is that those classical trajectories are searched that satisfy not only initial but also final quantization conditions of the internal states (double-ended trajectories) and that the phases along these trajectories are evaluated and incorporated into the transition amplitude by summing up the amplitudes along all these trajectories. In the case of electronic transition, the double-ended trajectories are analytically continued into the complex coordinate space in the region of surface crossing and go around the branch cut between the two adiabatic electronic potential energy surfaces. This detour into the complex plane gives a complex action to the S-matrix. The imaginary part of this action provides the corresponding nonadiabatic transition probability. In spite of the conceptual beauty, however, this theory turned out to be very cumbersome and time-consuming because of the necessity of handling the full complex trajectories, their complicated branching patterns, and searching for the double-ended trajectories. In order to remedy a part of the drawbacks, Kormonicki *et al.* proposed a decoupling scheme in which the evaluation of electronic transition is decoupled from the trajectory calculations [172]. The electronic transition is evaluated by the local complex integral, Eq. (3.22), with the nonadiabatic transition probability given by Eq. (3.21). This method saves a lot of troublesome matters, but still suffers from some problems. As is well recognized now, Eq. (3.21) cannot be used at energies lower than the crossing point. They have also neglected the dynamical phases, σ_0 and $\phi_{\rm S}$. These defects can be easily remedied by using the Zhu–Nakamura theory. Furthermore, in order to avoid the cumbersome double-ended trajectories and the divergence of the Van Vleck determinant at caustics, the initial value representation (IVR) has been figured out [163, 164, 166, 173]. Instead of specifying the initial and final quantized momenta, only the initial conditions are provided for trajectories. In this formalism the overall transition amplitude can be expressed as follows:

$$\langle \psi_f | e^{-iHt/\hbar} | \psi_i \rangle = \int d\mathbf{p_0} \int d\mathbf{q_0} \left(\left| \frac{\partial \mathbf{q_t}}{\partial \mathbf{p_0}} \right| \Big/ (2\pi i\hbar)^F \right)^{1/2}$$

$$\times \, e^{iS_t(\mathbf{p_0},\mathbf{q_0})/\hbar - i\nu\pi/2} \psi_f^*(\mathbf{q_t}) T_{21} \psi_i(\mathbf{q_0}) , \qquad (10.56)$$

where $\mathbf{q_t} \equiv \mathbf{q_t}(\mathbf{p_0}, \mathbf{q_0})$ is the coordinate at time t for the trajectory with the initial condition $(\mathbf{p_0}, \mathbf{q_0})$, $S_t = S(\mathbf{q_t}(\mathbf{p_0}, \mathbf{q_0}), \mathbf{q_0})$ is the action integral along that trajectory, F is the number of nuclear degrees of freedom, and ν is the number of zeros (Maslov index) experienced by the Jacobian determinant (Van Vleck determinant) in the interval $(0, t)$. The amplitude T_{ji} represents the electronic transition from the electronic state i to j. In the case of electronically adiabatic processes on a single potential surface, this T-matrix is simply replaced by a unit matrix. As is clear now, the theory developed in Chap. 5 can be usefully utilized to deal with the electronically nonadiabatic transition. Especially, the matrix I_X can be directly incorporated into the above formalism. When the energy is lower than the crossing region in the LZ-type of crossing, the reduced scattering matrix S^R can be directly utilized. When the nonadiabatic tunneling (NT) type of crossing is encountered, the transfer matrix N (see Eq. (5.128)) can be used. By using these formulas, not only the transition probabilities can be easily evaluated accurately, but also all the necessary phases including the dynamical phases can be correctly incorporated. Sun and Miller [164] says that the semiclassical analytical theory such as those developed in Chap. 5 is not necessary as an actual computational tool. This is not correct at all. In addition to the obvious strong advantages of the analytical theories to provide physical insights about the mechanisms, they can furnish even numerically effective tools, since everything, probabilties and phases, can be evaluated from simple analytical expressions.

The idea of the initial value representation has also been combined with the idea of frozen Gaussian approximation [174–176] and the cellularized frozen Gaussian approximation (CFGA) has been figured out [165, 166]. In this method, the initial wave function is represented as a superposition of large number of small frozen Gaussian wave packets and all of them are propagated along the classical trajectories of the center of each cell with the Gaussian shape frozen. This method was succcessfully applied to fifteen-dimensional eigenvalue problem on a single potential energy surface [167] and proven to be useful for bigger systems. The method seems immune even to chaotic trajectories. Again, the semiclassical theory developed in Chap. 5 can be incorporated into this formalism to deal with electronically nonadiabatic processes. With use of the idea of the dynamical state representation described in Sec. 3.4, dynamics induced by the Coriolis coupling can be treated in the same way as in the ordinary case induced by radial nonadiabatic couplings.

As discussed above, the combination of the theory developed in Chap. 5 and the various semiclassical propagation methods are expected to be very useful to deal with realistic multi-dimensional dynamics involving electronically nonadiabatic transitions and their actual applications to various practical systems are looked for.

Chapter 11

Complete Reflection and Bound States in the Continuum

11.1. One NT-Type Crossing Case

The nonadiabatic tunneling (NT) type transition presents quite a unique and intriguing mechanism, since this creates a potential barrier which cannot be treated in the same way as the ordinary potential barrier even at energies lower than the barrier top and the phenomenon of complete reflection occurs at energies higher than the bottom of the upper adiabatic potential (cf. Sec. 5.2.2.4). The nonadiabatic tunneling probability is always smaller than the transmission probability of the corresponding single potential barrier with the upper adiabatic potential neglected. This means that the upper adiabatic potential cannot be neglected even at energies lower than the top of the lower adiabatic potential. Furthermore, as is seen in Fig. 5.10, the nonadiabatic tunneling probability oscillates as a function of energy and at certain discrete energies the transmission probability becomes exactly zero. This is called complete reflection. As is given in Eq. (5.156), the transmission probability $|T|^2$ at energies higher than the bottom of the upper adiabatic potential, i.e. $E \geq E_b$, is given by

$$|T|^2 = \frac{4\cos^2\psi_{\rm ZN}}{4\cos^2\psi_{\rm ZN} + p_{\rm ZN}^2/(1 - p_{\rm ZN})}, \tag{11.1}$$

where $p_{\rm ZN}$ is the nonadiabatic transition probability given by Eq. (5.147) for one passage of the crossing point and $\psi_{\rm ZN}$ is the phase given by Eq. (5.148). This equation clearly tells that the complete reflection ($|T| = 0$) occurs when the following condition is satisfied (see also Eq. (5.159)):

$$\psi_{ZN} = \left(l + \frac{1}{2}\right)\pi \quad \text{for } l = 0, 1, 2, \ldots . \tag{11.2}$$

The transmission amplitude T is nothing but the reduced scattering matrix $(S_{NT}^{R})_{12}$ (see Sec. 5.2.2), which is explicitly expressed as

$$T = (S_{NT}^{R})_{12} = \frac{2\sqrt{1 - p_{ZN}}\,e^{i\sigma_{ZN}}}{1 + (1 - p_{ZN})e^{2i\psi_{ZN}}}\cos\psi_{ZN}, \tag{11.3}$$

which can be easily reduced to Eq. (11.1). The above equation (11.3) can be easily rewritten as

$$T = \sqrt{1 - p_{ZN}}\,e^{i\phi_s}\left(1 - \frac{p_{ZN}}{p_{ZN} - 1 - e^{-2i\phi_{ZN}}}\right)$$

$$= (I_X)_{11} + \frac{(I_X)_{12}L_r^2(O_X)_{22}L_l^2(I_X)_{21}}{1 - (O_X)_{22}L_l^2(I_X)_{22}L_r^2}, \tag{11.4}$$

where $(I_X)_{nm}$ is the $n \times m$-element of the matrix I_X given in Eq. (5.158) and represents the nonadiabatic transition amplitude from the adiabatic state m to the state n at the avoided crossing point when the wave propagation proceeds inward, namely from the right to the left. The matrix O is the transpose of I, and represents the transition in the outgoing (left to right) segment. L_l and L_r are defined by (see Fig. 11.1)

$$L_l = \exp\left[i\int_{T_1}^{x_b} k_2(x)dx + i\frac{\pi}{4}\right] \tag{11.5}$$

and

$$L_r = \exp\left[i\int_{x_b}^{T_2} k_2(x)dx + i\frac{\pi}{4}\right]. \tag{11.6}$$

The second equation of Eq. (11.4) can be obtained with the help of diagrammatic technique described in Sec. 7.1. This equation and the first equation in Eq. (11.4) enable us to clarify the physical meaning of this complete reflection phenomenon. The first term I_{11} represents the wave which simply crosses the barrier without any transition to the upper adiabatic state $E_2(x)$, and the second term corresponds to the transmitting wave which is trapped by the upper adiabatic potential. At the energies of complete reflection, these waves

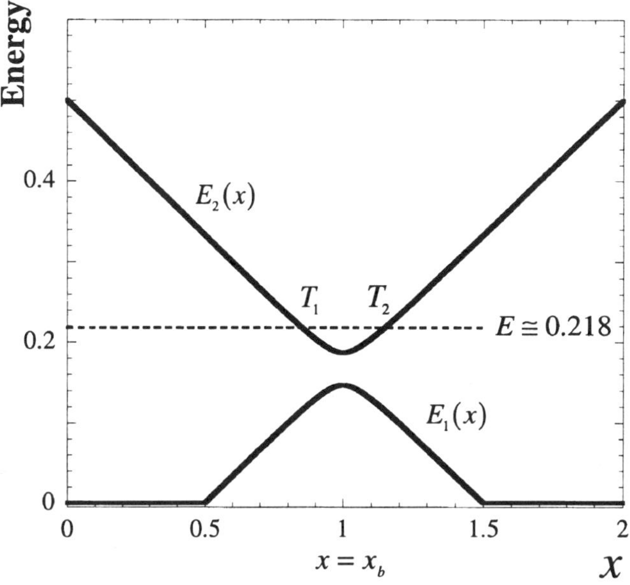

Fig. 11.1. One unit of NT-type potentials defined by Eqs. (11.17)–(11.19).

interfere destructively at the exit whatever the nonadiabatic transition probability p_{ZN} is and the incident wave is completely reflected back. When the condition in Eq. (11.2) is satisfied, the second term in the bracket of the first equation of Eq. (11.4) becomes exactly unity and cancels the first term whatever the probability p_{ZN} is. This condition is similar to the Bohr–Sommerfeld quantization condition; in the present case, however, the effect of nonadiabatic coupling naturally appears as the additional phase ϕ_S, which tends to $\pi/4$ (0) in the weak diabatic coupling limit $\delta_{ZN} \to 0$ (the strong diabatic coupling limit $\delta_{ZN} \to \infty$).

This unique phenomenon of complete reflection suggests some intriguing possibilities such as bound states in the continuum, molecular switching in a periodic potential system and control of molecular dynamic processes by laser. The latter two will be discussed in the subsequent Chaps. 12 and 13. Here we discuss the possibility of creating bound states in the continuum in between two units of NT-type potential curve crossings (see Fig. 11.2). If the two NT-type potential units are the same, a wave can be trapped between them forever; that is to say, a bound state in the continuum is created. This is realized when

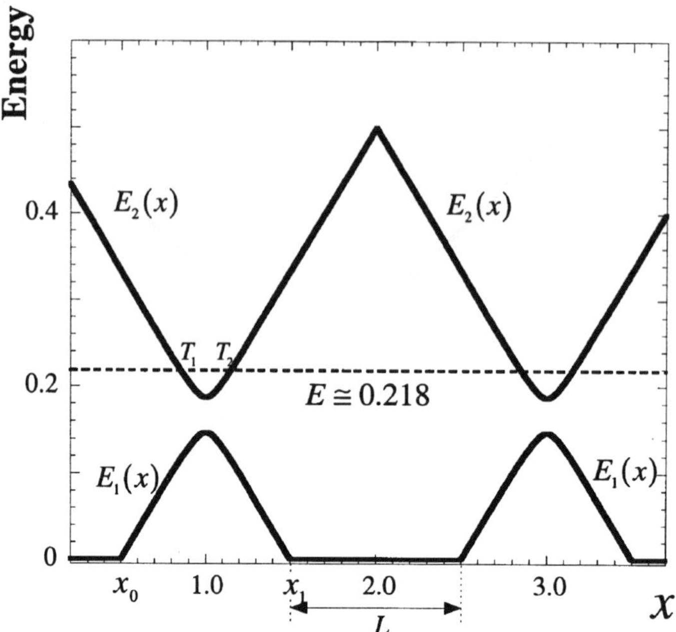

Fig. 11.2.   Two units of the same NT-type potentials.   One unit is the same as that in Fig. 11.1.

the following two conditions are fulfilled:

$$\psi_{ZN} = \left(s + \frac{1}{2}\right)\pi \quad \text{for } (s - 0, 1, 2, \ldots) \tag{11.7}$$

and

$$\delta_{T} + kL = \left(s' + \frac{1}{2}\right)\pi \quad \text{for } (s' = 0, 1, 2, \ldots), \tag{11.8}$$

where δ_{T} is the phase of the transmission amplitude $(T = |T|e^{i\delta_{T}})$ and is explicitly given by

$$\delta_{T} = \int_{x_0}^{x_1} k_1(x)dx + \sigma_{ZN} - \arctan\left[\frac{(1 - p_{ZN})\sin(2\psi_{ZN})}{1 + (1 - p_{ZN})\cos(2\psi_{ZN})}\right]$$
$$+ \Theta[-\text{sgn}(\cos\psi_{ZN})], \tag{11.9}$$

k and L are the wave number at zero potential and the distance between the two units (see Fig. 11.2). The condition, Eq. (11.7), is the condition of complete reflection, exactly the same as Eq. (11.2), and the second condition, Eq. (11.8), represents the condition of quantization. Interestingly, this condition coincides with the condition of complete transmission (see Eq. (12.25) in the next chapter). Thus, the bound states are created when the conditions of complete reflection ($E = E_{\mathrm{r}}$) and complete transmission ($E = E_{\mathrm{t}}$) coincide. The transmission probability $|T^{(2)}|^2$ for the two-unit system is expressed as

$$|T^{(2)}|^2 = \frac{|T|^4}{|T|^4 + 4|R|^2 \cos^2(\delta_{\mathrm{T}} + kL)} \, , \tag{11.10}$$

where R is the reflection amplitude for the one-unit system. Expanding ψ_{ZN} and $\delta_{\mathrm{T}} + kL$ at $E \simeq E_{\mathrm{r}}$ and $E \simeq E_{\mathrm{t}}$, respectively, as

$$\psi_{\mathrm{ZN}} \simeq \left(s + \frac{1}{2}\right)\pi + \left[\frac{\partial \psi_{\mathrm{ZN}}}{\partial E}\right]_{E_{\mathrm{r}}} (E - E_{\mathrm{r}}) \quad \text{at } E \simeq E_{\mathrm{r}} \tag{11.11}$$

$$\delta_{\mathrm{T}} + kL \simeq \left(s' + \frac{1}{2}\right)\pi + \left[\frac{\partial \delta_{\mathrm{T}}}{\partial E}\right]_{E_{\mathrm{t}}} (E - E_{\mathrm{t}}) \quad \text{at } E \simeq E_{\mathrm{t}} \, , \tag{11.12}$$

then the overall transmission probability $|T^{(2)}|^2$ behaves as

$$|T^{(2)}|^2 \simeq \frac{\tau^4}{4} \left[\frac{\partial \psi_{\mathrm{ZN}}}{\partial E}\right]_{E_{\mathrm{r}}}^{-2} (E - E_{\mathrm{r}})^2 \quad \text{at } E \simeq E_{\mathrm{r}} = E_{\mathrm{t}} \tag{11.13}$$

with

$$\tau = 2\left[\frac{\sqrt{1 - p_{\mathrm{ZN}}}}{p_{\mathrm{ZN}}}\right]_{E_{\mathrm{r}}} \left[\frac{\partial \psi_{\mathrm{ZN}}}{\partial E}\right]_{E_{\mathrm{r}}} . \tag{11.14}$$

It should be noted that if $E_{\mathrm{r}} \neq E_{\mathrm{t}}$, then

$$|T^{(2)}|^2 \simeq o(|E - E_{\mathrm{r}}|^4) \quad \text{at } E \simeq E_{\mathrm{r}} \, . \tag{11.15}$$

From a pole of $T^{(2)}$, we can obtain the decay width Γ of the resonance state created in between the two units as

$$\Gamma \simeq \frac{\tau^2}{2} \left[\frac{\partial \psi_{\mathrm{ZN}}}{\partial E}\right]_{E_{\mathrm{r}}}^{-1} (E_{\mathrm{r}} - E_{\mathrm{t}})^2 \, , \tag{11.16}$$

which goes to zero when $E_{\mathrm{r}} \to E_{\mathrm{t}}$.

If one of the potential units, say the left one, is replaced by a simple repulsive wall, the system becomes the same as that in a dissociating diatomic molecule. Bound states in the continuum can also be created in this system. In the case of IBr, the above conditions seem to be satisfied and the corresponding predissociation dip was analyzed [177].

Another example is resonances in the transition $^2\Sigma_g^+ \to {}^2\Sigma_u^+$ of H_2^+ due to the laser-induced avoided crossing [178]. The condition stated in these analyses is the coincidence of vibrational energy levels supported by the attractive diabatic potential and by the upper adiabatic potential with the correction ϕ_S due to nonadiabatic coupling. But the complete reflection in the one-unit system, as in Fig. 11.1, was not explicitly recognized. The latter example mentioned above is an intriguing one, because we can manipulate molecules by lasers so as to satisfy the necessary conditions.

The existence of complete reflection and the bound states in the continuum are numerically demonstrated below. The following one-unit diabatic potentials are employed [58]:

$$V_1(x) = \begin{cases} \dfrac{2v_b}{(a+b)}\left(x - \dfrac{b-a}{2}\right) & \text{for } (b-a)/2 \le x \le b+l \\ 0 & \text{for } -l \le x \le (b-a)/2\,, \end{cases} \tag{11.17}$$

$$V_2(x) = \begin{cases} -\dfrac{2v_b}{(a+b)}\left(x - \dfrac{b+a}{2}\right) & \text{for } -l \le x \le (a+b)/2 \\ 0 & \text{for } (a+b)/2 \le x \le b+l\,, \end{cases} \tag{11.18}$$

$$V_{12}(x) = \begin{cases} [1 - (2x-b)^2/c^2]v_p & \text{for } (b-c)/2 \le x \le (b+c)/2 \\ 0 & \text{otherwise}\,, \end{cases} \tag{11.19}$$

where the parameters a, b, c, l, v_b, and v_p are 1.0, 2.0, 0.5, 0.0, 0.5, and 0.02 atomic units. The adiabatic potentials are shown in Fig. 11.1. The particle mass is taken to be 1800 atomic units. There are six complete reflection dips in the energy region $0.10 < E < 0.35$ in atomic units. To illustrate a complete reflection wave function, we have chosen the first zero of $|T|^2$, that is $E \sim 0.21809$. The corresponding reflection coefficient, $R \equiv |R|\exp(i\delta_R)$ is equal to $R \sim (1 - 2 \times 10^{-13})$ and $\delta_R \sim 0.42223$. The calculated wave functions $\psi_1(x)$ and $\psi_2(x)$ corresponding to the diabatic states, chosen real, are shown

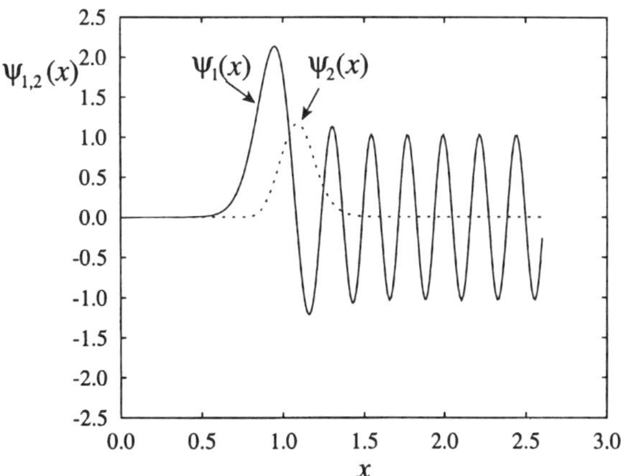

Fig. 11.3. Two component wave functions at the energy of complete reflection at $E \simeq$ 0.21809. The potential unit used is the one given in Fig. 11.1. The function $\psi_j(x)$ $(j = 1, 2)$ corresponds to the diabatic state $V_j(x) \cdot \psi_2(x)$ decays at $x \lesssim 0.5$ despite the fact that channel 2 is open there. (Taken from Ref. [58] with permission.)

in Fig. 11.3, the normalization being such that

$$\psi_1(x) = \cos(kx + \theta) \quad \text{for } x \geq \frac{b+a}{2}, \tag{11.20}$$

where $\theta = 18\pi - kb + \delta_{\rm R}/2 \sim 0.71980$ in this particular case. The wave function $\psi_2(x)$ dies out at $x \lesssim 0.5$ despite the fact that this channel is energetically open in this region. For the demonstration of a bound state in the continuum, we take the two potential units symmetrical with respect to the crossing point, as is shown in Fig. 11.2. Because of the symmetry of the system, $R = \pm 1$ at the center of the whole array, and accordingly either an even $[R = 1]$ or an odd $[R = -1]$ function of $(x - b - l)$ is obtained. In this model $L = 2l + b - a$ and we can adjust l to satisfy the conditions, Eqs. (11.7) and (11.8). One bound state corresponding to $s' = 0$ of Eq. (11.8) is shown in Fig. 11.4. The adiabatic potentials are shown in Fig. 11.2.

The wave function $\psi_2(x)$ dies out in the regions $x \lesssim 0.5$ and $x \gtrsim 3.5$ despite the fact that this channel is open in these regions. The wave function $\psi_1(x)$, on the other hand, decays naturally there, and has several nodes in the central region. In spite of these nodes, this bound state in the continuum is energetically the lowest in the corresponding system. The number of nodes is determined essentially by l and the energy E.

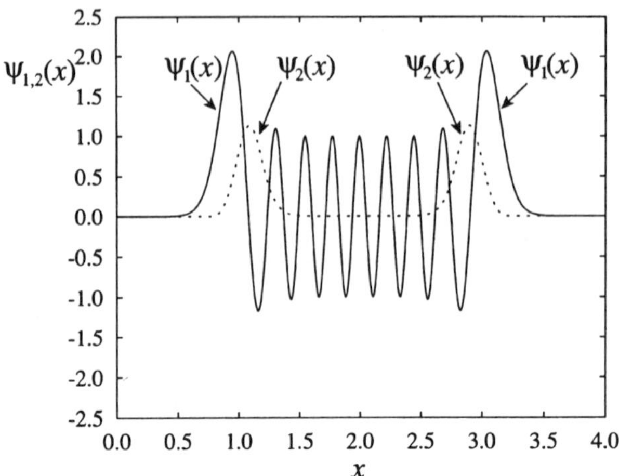

Fig. 11.4. Two component wave functions when the two conditions of complete reflection and quantization are satisfied at the same time for the potential system shown in Fig. 11.2. $\psi_2(x)$ is trapped inside despite the fact that the diabatic channel 2 is open outside. (Taken from Ref. [58] with permission.)

11.2. Diabatically Avoided Crossing (DAC) Case

In Sec. 5.3.2 analytical solutions of the diabatically avoided crossing model depicted in Fig. 5.14 are briefly discussed, the potentials of which are defined by Eq. (5.164). It is interesting to note that the complete reflection and complete transmission phenomena can occur in this potential system [66]. First, let us try to analyze the behavior of the transmission coefficient $|\alpha|^2$ given by Eq. (5.176) at high energies,

$$q_1 \gg 1, \quad q_2 \gg 1 \quad \text{and} \quad |q_1 - q_2| \gg 1, \tag{11.21}$$

where q_j are defined by Eq. (5.167). We keep only the leading order term with respect to energy E, use the expansion formula of the G-function given below in Eqs. (11.26)–(11.28), and define the new phase,

$$\psi = \frac{q_1}{2}\ln|z| - \arg\Gamma(iq_1) + \arg\Gamma\left(\frac{1}{2} + i\frac{q_2 - q_1}{2}\right)$$

$$+ \arg\Gamma\left(\frac{1}{2} - i\frac{q_1 + q_2}{2}\right) \tag{11.22}$$

together with $\psi' = \psi(q_1 \leftrightarrow q_2)$. Then we have

$$\phi_1 \simeq \arctan \left[\frac{\sin(\psi + \psi') + \eta \sin(\psi - \psi')}{2 \cos \psi \cos \psi'} \right] \tag{11.23}$$

and

$$\phi_2 \simeq \arctan \left[\frac{\sin(\psi + \psi') - \eta \sin(\psi - \psi')}{2 \sin \psi \sin \psi'} \right], \tag{11.24}$$

where

$$\eta = \tanh \left(\pi \frac{q_1 - q_2}{4} \right). \tag{11.25}$$

The expansion of the G-function used [65] is

$$G_{04}^{40}(z|\{b\}) = \sum_{h=1}^{4} \left(\prod_{\substack{j=1..4 \\ j \neq h}} \Gamma(b_j - b_h) z^{b_h} \right)$$
$$\times {}_0F_3[1 + b_h - b_1, \cdots * \cdots, 1 + b_h - b_4, z] \tag{11.26}$$

with

$${}_0F_3[d_1, d_2, d_3, z] = \sum_{n=0}^{\infty} \frac{z^n}{n!(d_1)_n(d_2)_n(d_3)_n} \tag{11.27}$$

and

$$(d)_n = d(d+1) \cdots (d + n - 1). \tag{11.28}$$

Setting $x_i = \tan \phi_i$, the result for the transmission and reflection coefficients reads

$$|\alpha|^2 \simeq \frac{(x_1 + x_2)^2}{(1 + x_1^2)(1 + x_2^2)} \quad \text{and} \quad |\gamma|^2 = 1 - |\alpha|^2. \tag{11.29}$$

Hence the complete reflection, $\alpha = 0$, is given by the condition

$$\tan(\psi + \psi') = \eta \tan(\psi - \psi'). \tag{11.30}$$

If we use the simple relations in the high energy approximation,

$$|q_1 - q_2| \sim E^{-1/2} \sim \psi - \psi' \sim \eta, \tag{11.31}$$

in the zeroth order we have

$$\psi = \psi', \qquad \phi_1 = \psi, \qquad \phi_2 = \frac{\pi}{2} - \psi, \qquad |\alpha|^2 = 1. \tag{11.32}$$

Namely, the transmission coefficient tends to unity at energies asymptotically high. The necessary correction to this oversimplified estimate is to re-express the transmission coefficient near the complete reflection point. Expanding

$$\psi(E) = \frac{\pi}{2} + \omega + \xi(E - E_0), \tag{11.33}$$

$$\psi'(E) = \frac{\pi}{2} - \omega + \xi(E - E_0), \tag{11.34}$$

and substituting these expressions into Eq. (11.29), we obtain

$$|\alpha|^2 = \frac{\epsilon^2}{\epsilon^2 + [\xi(\epsilon - \epsilon_+)(\epsilon - \epsilon_-)]^2}, \qquad \epsilon \equiv E - E_0, \qquad \epsilon_\pm = \frac{\omega}{\xi}(\eta \pm 1). \tag{11.35}$$

Here

$$\xi = \frac{1}{2}\frac{d}{dE}(\psi + \psi')\bigg|_{E=E_0} \quad \text{and} \quad \omega = \frac{1}{2}(\psi - \psi')\bigg|_{E=E_0}. \tag{11.36}$$

The resonance profile in Eq. (11.35) smoothly connects the zero dip in $|\alpha|^2$ to the unity background within the accuracy of the high energy expansion. The resonance position, E_0, follows from Eqs. (11.30) and (11.31), and is given by

$$\psi(E_0) \sim \sqrt{E_0}\left[\frac{1}{2}\ln|z_0| + 2 - \ln(E)\right] = (2n - 1)\frac{\pi}{2}. \tag{11.37}$$

This condition coincides with the Bohr–Sommerfeld quantization condition in the upper well,

$$\int_{x_L^t}^{x_R^t} \sqrt{E - V_{11}(x)}\, dx = \left(n - \frac{1}{2}\right)\pi, \qquad n = 1,\ldots \tag{11.38}$$

In order to make the physical interpretation of the above equations clear we carry out the semiclassical analysis of the diabatically avoided crossing model. Suppose that the diabatic and adiabatic potentials at $x \to 0$ almost coincide, i.e. $V \ll |U_1 - U_2|$, or in other words the region $x \sim 0$ is far enough from the nonadiabatic transition regions which can be represented by the real parts of the complex crossing points. Then the scattering can be decomposed into adiabatic wave propagation and two nonadiabatic transitions (left and right). Each of them is described by a nonadiabatic transition matrix

$$
I(x_s) = \begin{pmatrix} \sqrt{1-p_s}\exp(-i\phi_s) & \sqrt{p_s}\exp(i\psi_s) \\ -\sqrt{p_s}\exp(-i\psi_s) & \sqrt{1-p_s}\exp(i\phi_s) \end{pmatrix}, \quad s = L \text{ or } R. \quad (11.39)
$$

These matrices connect from left to right the WKB wave functions centered at the transition points, p_s is the nonadiabatic transition probability and ϕ_s and ψ_s are the dynamical phases. It should be noted that the upper (lower) adiabatic channel is indexed by 1 (2) (see Fig. 5.14(a)). There are three co-ordinate regions, (x_L^t, x_L), (x_L, x_R), and (x_R, x_R^t), where $x_s^t(x_s)$ represents the turning point (transition point). The transition points x_L and x_R are the real parts of the corresponding complex crossing points. We denote these regions as i ($i = \text{I}, \text{II}, \text{III}$, $x_i < x < x_{i+1}$ with $x_1 = x_L^t, \ldots, x_4 = x_R^t$) and introduce the following phase factors,

$$
c_i \equiv e^{id_i} = \exp\left(i\Re \int_{x_i}^{x_{i+1}} \sqrt{E - u_1^{(a)}(x)}\, dx\right), \qquad c \equiv c_1 c_2 c_3, \quad (11.40)
$$

and

$$
d \equiv e^{i\delta} = \frac{1}{c_2}\exp\left(i\Re \int_{x_2}^{x_3}\sqrt{E - u_2^{(a)}(x)}\, dx\right) \equiv \frac{1}{c_2}e^{id_2^2}. \quad (11.41)
$$

The semiclassical diagram corresponding to this problem is shown in Fig. 5.14(b). The circle, rectangular and arrow indicate turning point, nonadiabatic transition and adiabatic wave propagation, respectively. In order to simplify the expressions derived below, we introduce a term which represents the interference of a wave propagating from the adiabatic state j (left) to the adiabatic state i (right),

$$
z_{ij} \equiv R_{i1}L_{1j} + R_{i2}L_{2j}d, \quad (11.42)
$$

where L (R) stands for the left (right) nonadiabatic transition matrix. The matching of corresponding wave functions between the three regions follows. According to Fig. 5.14(b), the waves in region I are determined by the boundary conditions,

$$
Ac_1\overrightarrow{1} - iAc_1^*\overleftarrow{1} + B\overleftarrow{2}. \quad (11.43)
$$

Then we propagate the total wave function to region II,

$$L_{11}Ac_1c_2\overrightarrow{1} + c_2^*(L_{11}^*c_1^*(-iA) + L_{12}^*B)\overleftarrow{1} + L_{21}c_1c_2dA\overrightarrow{2}$$

$$+ c_2^*d^*(L_{21}^*c_1^*(-iA) + L_{22}^*B)\overleftarrow{2}\,, \tag{11.44}$$

and region III,

$$X\overrightarrow{1} + Y\overleftarrow{1} + z_{21}c_1c_2A\overrightarrow{2} + [z_{21}^*c_1^*c_2^*(-iA) + z_{22}^*c_2^*B]\overleftarrow{2}\,. \tag{11.45}$$

The notations c_i, d and z_{ij} are introduced in Eqs. (11.40)–(11.42); arrows denote the direction of adiabatic wave,

$$X \equiv z_{11}cA\,, \tag{11.46}$$

$$Y \equiv z_{11}^*c^*(-iA) + c_2^*c_3^*z_{12}^*B\,. \tag{11.47}$$

In Eq. (11.43), the so far arbitrary constant A (B) denotes the amplitude of a wave in the upper (lower) channel which satisfies the boundary conditions for scattering from right to left. The right turning point imposes a condition on these constants (see Eq. (11.43)),

$$X(A) = -iY(A, B)\,. \tag{11.48}$$

The reflection coefficient, $|R|^2$, is easily obtained as

$$|R|^2 = \frac{|z_{12}z_{21}|^2}{|z_{12}z_{21}c - (z_{11}c + z_{11}^*c^*)z_{22}|^2}\,. \tag{11.49}$$

There are two special cases that solve Eq. (11.48). Taking $B = 0$, we obtain the complete reflection condition,

$$\int_{x^{(a)\,t}_{\mathrm{L}}}^{x^{(a)\,t}_{\mathrm{R}}} \sqrt{E - u_1^{(a)}(x)}\,dx = \left(n - \frac{1}{2}\right)\pi + \Delta(E) \tag{11.50}$$

with an additional shift $\Delta(E)$ to the Born–Sommerfeld quantization condition in the upper well,

$$\Delta(E) = \arctan\left[\frac{\zeta\sin(\phi_{\mathrm{L}} + \phi_{\mathrm{R}}) + \sin(\psi_{\mathrm{L}} - \psi_{\mathrm{R}} + \delta)}{\zeta\cos(\phi_{\mathrm{L}} + \phi_{\mathrm{R}}) - \cos(\psi_{\mathrm{L}} - \psi_{\mathrm{R}} + \delta)}\right)\,, \tag{11.51}$$

where

$$\zeta \equiv \sqrt{\frac{(1 - p_{\rm L})(1 - p_{\rm R})}{p_{\rm L} p_{\rm R}}} \tag{11.52}$$

and δ is given in Eq. (11.41). Now, it is clear that Eqs. (11.38) is just the high energy limit of Eq. (11.50). If both $p_{\rm L}$ and $p_{\rm R}$ are small, the quantization in Eq. (11.50) occurs naturally in the adiabatic well with the imposed shift from diagonal dynamical phases, $\phi_{\rm L}$ and $\phi_{\rm R}$ (see the above equation). If both $p_{\rm L}$ and $p_{\rm R}$ tend to unity, on the other hand, the quantization occurs in the diabatic well and the imposed phase shift is due to the off-diagonal dynamical phases, $\psi_{\rm L}$ and $\psi_{\rm R}$. Applying the complete transmission condition to Eq. (11.48), which turns out to be equivalent to $A = 0$, we obtain

$$p_{\rm L} = p_{\rm R} \equiv p \quad \text{and} \quad \delta + \phi_{\rm L} + \phi_{\rm R} + \psi_{\rm L} - \psi_{\rm R} = \pi. \tag{11.53}$$

If p tends to zero or unity, the second condition in Eq. (11.53) can be omitted. The above equation has a very simple and interesting physical meaning: the incident wave interferes between the two transition points in such a way that no part of it reaches the turning points in the upper adiabatic channel. This is because in region I (left turning point), $A = 0$, and consequently in region III (right turning point), $X(A) = 0$ and $Y(A, B) = 0$ [see Eqs. (11.46) and (11.48)]. The parameters of nonadiabatic transition matrices in Eq. (11.39) in the case of above model are given in Eqs. (3.37)–(3.41). Equation (11.49) with Eqs. (3.37)–(3.41) give the semiclassical transmission and reflection coefficients. Let us finally note that Eq. (11.51) is invariant with respect to the following transformation

$$\text{L} \leftrightarrow \text{R} \quad \text{and} \quad \psi \leftrightarrow \pi - \psi, \tag{11.54}$$

as it should be due to the symmetry.

Finally, we give some numerical examples [66]. In order to avoid unnecessary parameters let us use the units

$$[E] = (\hbar\alpha)^2 (2M)^{-1}, \qquad [x] = \alpha^{-1}. \tag{11.55}$$

Then, as it follows from Eq. (5.164), there are only two substantial parameters, V and C. These parameters affect the general behavior of the transmission coefficient (see Fig. 11.5), which is: (1) the exponential decrease at energies far below the top of the lower potential, then (2) overall monotonous increase

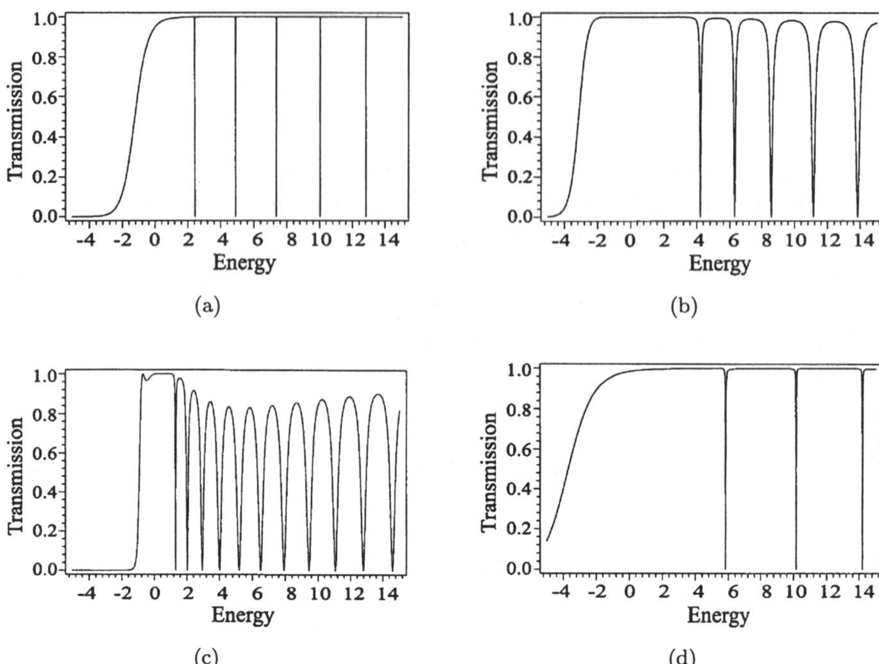

Fig. 11.5. Transmission probability as a function of energy in the DAC model of Eq. (5.164) with $(U = 0)$. The bottom E_b (top — E_b) of the upper (lower) adiabatic potential is given by $\sqrt{V^2 + C^2}$. The dimensionless parameters are as follows: (a) $C = 0.2$ and $V = 1$, (b) $C = 6$ and $V = 1$, (c) $C = 2$ and $V = 0.1$, (d) $C = 2$ and $V = 3$. (Taken from Ref. [66] with permission.)

up to the first complete reflection point, and (3) complete reflection dips with an envelope that converges to unity.

In the limit of small coupling, C (see Fig. 11.5(a)), the transmission coefficient corresponds to that of a single barrier penetration, except that the complete reflection dips survive with very narrow widths. For large values of C (see Fig. 11.5(b) in comparison with Fig. 11.5(a)) the first step broadens and the first resonance moves to higher energies because the bottom of the upper adiabatic potential shifts up with growing C. Also the complete reflection dips become wider.

In the limit of small pre-exponential constant, V (see Fig. 11.5(c) in comparison with Fig. 11.5(d)), the potential curves become flat, the step is sharper and the resonances become more dense, because the semiclassical phase in the

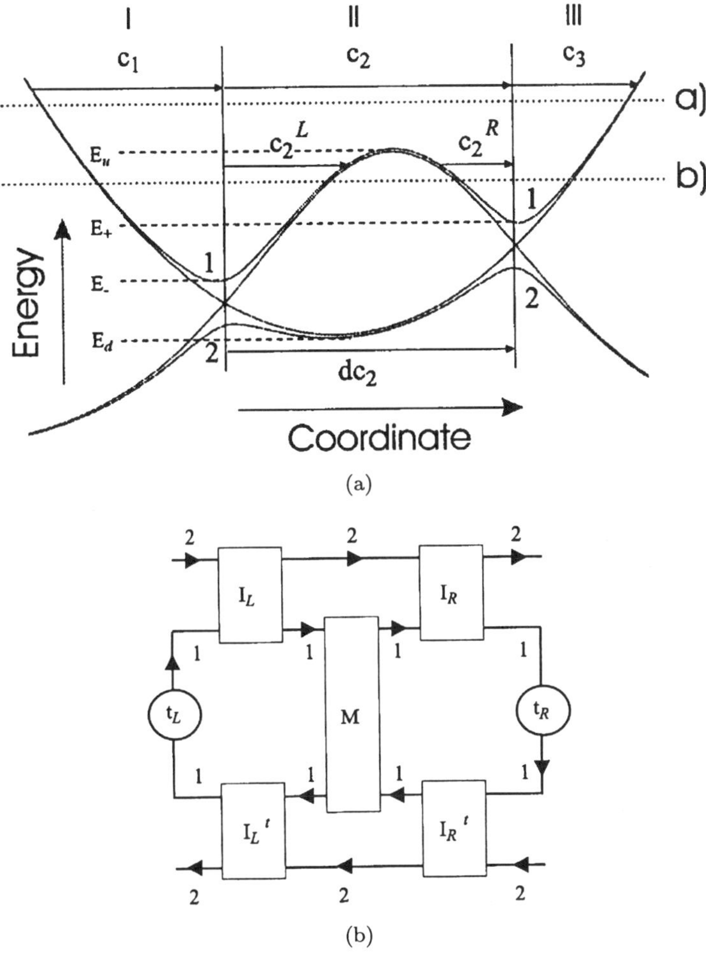

Fig. 11.6. Double-crossing NT-type model. (a) The potential curves. The roman numbers denote the three important coordinate regions; arabic numbers index the adiabatic potentials. (b) The semiclassical diagram (energy below the top of the barrier). The matrix M stands for the tunneling through the central potential barrier. (Taken from Ref. [66] with permission.)

upper well accumulates easily. The semiclassical theory based on a sequence of two RZ-type transitions works well (see Fig. 11.6). For large values of V (see Fig. 11.5(d)), on the other hand, the dips get narrower again and more separated, and the transmission coefficient decreases slowly at low energies.

In addition to this, there is a nontrivial envelope of the transmission coefficient as clearly seen in Fig. 11.5(c). Such envelope may even have a deep dip before getting converged to unity at large energies.

In the ordinary single unit of NT-type of transition, the qualitative behavior of the transmission coefficient is quite different. Since the nonadiabatic transition probability tends to unity as energy increases, the incoming wave is effectively switched to the upper well at the crossing point, then reflected back at the turning point and returns to the initial channel. Thus the envelope of the transmission coefficient monotonically decreases to zero with increasing energy. In the above DAC model, on the other hand, the envelope behaves in the non-monotonous way before getting converged to unity. This is basically due to a sequence of two symmetric RZ-type nonadiabatic transitions, and is in accordance with the semiclassical analysis made above. Another difference between DAC and NT-cases is the coupling strength dependence of the transmission coefficient. While in the DAC model the zero coupling limit corresponds to a single barrier penetration problem with nonzero transmission, in a single NT-case the transmission is not possible, since each diabatic potential diverges at one side.

11.3.　Two NT-Type Crossings Case

Here we discuss the conditions for the complete reflection and transmission in potential systems shown in Fig. 11.6. We derive the semiclassical conditions for the complete transmission and complete reflection to occur and also compare these NT-models with the DAC model.

11.3.1.　*At energies above the top of the barrier: (E_u, ∞)*

Since the sequence of turning and transition points is the same in both Fig. 5.14(a) and 11.6(a), Eq. (11.50) with the phase shift given in Eq. (11.51) holds. The same applies to Eq. (11.53). However, the nonadiabatic transition matrices are, of course, different. The expressions for the nonadiabatic transition probability and the dynamical phases are given by the Zhu–Nakamura theory given in Sec. 5.2. Referring to the results of the previous section we conclude that the complete reflection and complete transmission are possible, when the following conditions are satisfied:

$$\int_{x^{(a)t}_{\text{L}}}^{x^{(a)t}_{\text{R}}} \sqrt{E - u_1^{(a)}(x)}\, dx = \left(n - \frac{1}{2}\right)\pi + \Delta(E)\,, \tag{11.56}$$

(Δ given in Eq. (11.51)) for the complete reflection, and

$$p_{\text{L}} = p_{\text{R}} \equiv p \quad \text{and} \quad \delta + \phi_{\text{L}} + \phi_{\text{R}} + \psi_{\text{L}} - \psi_{\text{R}} = \pi\,, \tag{11.57}$$

for the complete transmission.

11.3.2. *At energies between the barrier top and the higher crossing:* (E_+, E_u)

Here, we consider energy above the bottom of upper adiabatic potential (which is denoted here as E_+, see Fig. 11.6(a)), because the Zhu–Nakamura theory presents different formulas for $E \geq E_+$ and $E \leq E_+$. Since the energy is below the barrier top of the higher adiabatic potential (case (b) in Fig. 11.6(a)), we use the tunneling matrix M which connects the in/out-going WKB waves from left to right,

$$\begin{pmatrix} \rightarrow \\ \leftarrow \end{pmatrix} = \begin{pmatrix} \sqrt{1 + \kappa^2}\, e^{-i\Phi} & i\kappa e^{i\Theta} \\ -i\kappa e^{-i\Theta} & \sqrt{1 + \kappa^2}\, e^{i\Phi} \end{pmatrix} \begin{pmatrix} \rightarrow \\ \leftarrow \end{pmatrix}, \tag{11.58}$$

where $1/\kappa = e^{-\pi\epsilon}$ represents the Gamov factor with ϵ equal to the tunneling action integral when the energy is lower than the barrier top (for further details see e.g. Ref. [5]). This situation is schematically shown in the semiclassical diagram, Fig. 11.6(b). Using similar arguments as before, the complete reflection is possible and the condition in Eq. (11.50) still holds with some modifications. In particular, we have

$$I(E) \equiv \text{Re}\left\{\int_{x^{(a)t}_{\text{L}}}^{x^{(a)t}_{\text{R}}} \sqrt{E - u_1^{(a)}(x)}\, dx\right\} = \left(n - \frac{1}{2}\right)\pi + \Delta(E)\,, \tag{11.59}$$

where $\Delta(E)$ reads one of the following forms:

$$\Delta_{\text{L}}(E) = -\arg\{de^{i(\psi_{\text{L}} - \psi_{\text{R}})} - \zeta\sqrt{1 + \kappa^2}\, e^{-i(\phi_{\text{R}} + \phi_{\text{L}} + \Phi)}$$

$$- \zeta\kappa e^{-i(\Theta + \phi_{\text{L}} - \phi_{\text{R}})}(c_3^* c_2^{\text{R}*})^2\}\,, \tag{11.60}$$

or

$$\Delta_{\mathrm{R}}(E) = -\arg\{de^{i(\psi_{\mathrm{L}}-\psi_{\mathrm{R}})} - \zeta\sqrt{1+\kappa^2}e^{-i(\phi_{\mathrm{R}}+\phi_{\mathrm{L}}+\Phi)}$$

$$- \zeta\kappa e^{i(\Theta+\phi_{\mathrm{L}}-\phi_{\mathrm{R}})}(c_1^* c_2^{\mathrm{L}*})^2\}. \tag{11.61}$$

That is to say, the complete reflection occurs when $I(E) - \Delta_{\mathrm{L}}(E) = (n - 1/2)\pi$ or $I(E) - \Delta_{\mathrm{R}}(E) = (n - 1/2)\pi$ is satisfied. The factor $\Delta_{\mathrm{L}}(E)(\Delta_{\mathrm{R}}(E))$ represents the effects of both nonadiabatic coupling and tunneling [and obeys the symmetry relation in Eq. (11.54)]. If the tunneling probability is small, $\Delta_{\mathrm{L}}(E)(\Delta_{\mathrm{R}}(E))$ corresponds to the Bohr–Sommerfeld quantization condition in the left (right) upper adiabatic potential well. If $\kappa = 0 = \Phi$, namely the matrix in Eq. (11.58) turns to be a unit matrix, then the parameters $\Delta(E)$ in Eqs. (11.60), (11.61), and (11.51) are the same (the complete reflection condition naturally agrees with that in the case 1 above). Taking the diabatic limit in Eqs. (11.60) and (11.61), i.e. $p_{\mathrm{L}}, p_{\mathrm{R}} \to 1$ and thus $\zeta \to 0$, yields the Bohr–Sommerfeld quantization condition in the diabatic potential well.

The complete transmission is also possible but its mechanism is quite different from the above-the-barrier case. It occurs when

$$R_{21}L_{12}^* M_{12}(u+u^*) + i[ic_2^{\mathrm{R}*}c_3^*(R_{11}^* L_{12}^* M_{22} + R_{12}^* L_{22}^* d^*) + R_{11}L_{12}^* M_{12}c_2^{\mathrm{R}}c_3]$$

$$\times [ic_1 c_2^{\mathrm{L}}(R_{21}L_{11}M_{11} + R_{22}L_{21}d) + R_{21}L_{11}^* M_{21}^* c_1^* c_2^{\mathrm{L}*}] = 0 \tag{11.62}$$

with

$$u \equiv R_{11}(L_{11}M_{11}c_1 c_2 - iL_{11}^* M_{12}c_1^* c_2^{\mathrm{L}*}c_2^{\mathrm{R}})c_3 + R_{12}L_{21}cd \tag{11.63}$$

is satisfied. If the tunneling matrix M is replaced by a unity matrix, we obtain Eq. (11.57). The constants from Eq. (11.43), A (amplitude of the wave reflected from the left turning point) and B (amplitude of the transmitted wave), must be both nonzero. Physically it means that the complete transmission occurs only when the half-standing waves both left and right from the central barrier interfere destructively with the reflected part of the incident wave. Because of this, Eq. (11.62) is, unfortunately, quite complicated and difficult to solve analytically. Let us also note that depending on the potential parameters, the complete reflection and transmission can occur at energies close to each other, in which case we have the Fano type resonance [179].

When the maximum of the upper adiabatic potential is so high that the tunneling can be neglected ($\kappa \to \infty$), the formalism simplifies considerably. As also follows from Eqs. (11.60) and (11.61) in this limit, the complete reflection occurs, if

$$\cos(d_1 + d_2^{\mathrm{L}} - \phi_{\mathrm{L}}) = 0 \quad \text{or} \quad \cos(d_3 + d_2^{\mathrm{R}} - \phi_{\mathrm{R}}) = 0 \qquad (11.64)$$

is satisfied, i.e. its energy is given by the Bohr–Sommerfeld quantization rule with the additional phase correction, ϕ_{L} or ϕ_{R}. This is simply equivalent to the case discussed before [9, 57–59]. The complete transmission condition follows

$$p_{\mathrm{R}}(1 - p_{\mathrm{L}})\cos(d_1 + d_2^{\mathrm{L}} - \phi_{\mathrm{L}})\exp(-i(\phi_{\mathrm{L}} + \psi_{\mathrm{L}} - \psi_{\mathrm{R}} + d_3 + d_2^2 - d_2^{\mathrm{L}}))$$

$$+ p_{\mathrm{L}}(1 - p_{\mathrm{R}})\cos(d_3 + d_2^{\mathrm{R}} - \phi_{\mathrm{R}})\exp(i(\phi_{\mathrm{R}} + \psi_{\mathrm{L}} - \psi_{\mathrm{R}} + d_1 + d_2^2 - d_2^{\mathrm{R}}))$$

$$+ p_{\mathrm{L}}p_{\mathrm{R}}\cos(d_1 + d_2^{\mathrm{L}} + d_2^{\mathrm{R}} + d_3 + \delta + \psi_{\mathrm{L}} - \psi_{\mathrm{R}}) = 0 . \qquad (11.65)$$

(confer with Eq. (11.54)). The three cosine functions in Eq. (11.65) turn to zero when the scattering energy coincides with a shifted bound state in the left, right or the global diabatic well (see Fig. 11.6). Let us also note that Eq. (11.65) is a limiting case of Eq. (11.62). When we set $\Theta = 0 = \Phi$ and expand Eq. (11.62) in powers of κ, the leading term κ^2 vanishes identically, while the term proportional to κ yields Eq. (11.65).

11.3.3. *At energies in between the two crossing regions:* (E_-, E_+)

We can use the above formalism also when the energy is in between E_+ and E_-, where E_- is the bottom of the upper adiabatic potential in the lower crossing region (see Fig. 11.6(a))). In this case, the lower crossing can still be treated by the I-matrix (see Eq. (11.39)) as previously, but the upper crossing should be described by the nonadiabatic tunneling matrix (transfer matrix) N [9, 180, 181, 183] which connects the waves on both sides of the crossing in the lower adiabatic channel (see also Eq. (11.58)). Furthermore, we have to assume that the barrier on the upper adiabatic channel (see Fig. 11.6(a)) is high enough so that the tunneling through it can be neglected. Then, the above formalism can still be used with the replacements of the tunneling matrix M and the I-matrix for the higher crossing by the N-matrix and a unit matrix, respectively. Finally we derive the following results:

The complete reflection occurs, if the Bohr–Sommerfeld condition in the open adiabatic well is satisfied, i.e. it is also given by Eq. (11.64). The condition for the complete transmission reads

$$2R_{22}N_{12}dc_2\cos(d_2^{\mathrm{R}} + d_3) = R_{21}c_2^{\mathrm{R}}(R_{12}N_{12}dc_2c_3 + iR_{12}^*N_{22}^*d^*c_2^*c_3^*) \quad (11.66)$$

(here we suppose for simplicity that the crossing point energy is higher on the left than on the right).

11.3.4.   *At energies below the crossing points:* $(-\infty, E_-)$

Using similar assumptions as in the case 3 above, the semiclassical wave propagation for $E < E_-$ should be described by the two nonadiabatic tunneling matrices (transfer matrices), N (left crossing) and N' (right crossing). These are parametrized with κ, Φ and Θ, similarily as in Eq. (11.58). Then we find that the semiclassical complete reflection is not possible. Yet the complete transmission can still occur, provided (1) that energy is above the bottom of the lower adiabatic potential, $E_{\mathrm{d}} < E < E_-$, and (2) the following equation is satisfied,

$$\kappa = \kappa' \qquad (dc_2)^2 = -e^{\Phi+\Phi'+\Theta-\Theta'}. \qquad (11.67)$$

Since it is possible to control κ and κ', for instance by changing the intensity of the laser field, the complete transmission condition above could be useful for enhancing chemical reactions, especially those which are otherwise unlikely due to the tunneling [182]. At energies below the bottom of the lower adiabatic potential, E_{d}, the whole scattering can be described by semiclassical analysis only as a single barrier penetration. The complete reflection and the complete transmission are not possible.

11.3.5.   *Numerical examples*

Atomic units are used throughout this section. The reduced mass is chosen to be $M = 1000$ a.u., if not stated otherwise. First, we illustrate the theoretical results for energies below the top of the barrier. The accuracy of the semiclassical theory (cf. Eq. (11.49)) and the Zhu–Nakamura theory for energies above the barrier top is also demonstrated. Finally, we briefly discuss the Fano type of resonance [179, 182] using our semiclassical analysis.

In Fig. 11.7, numerical examples of the transmission probability are depicted for an asymmetric double NT-type crossing model. These figures

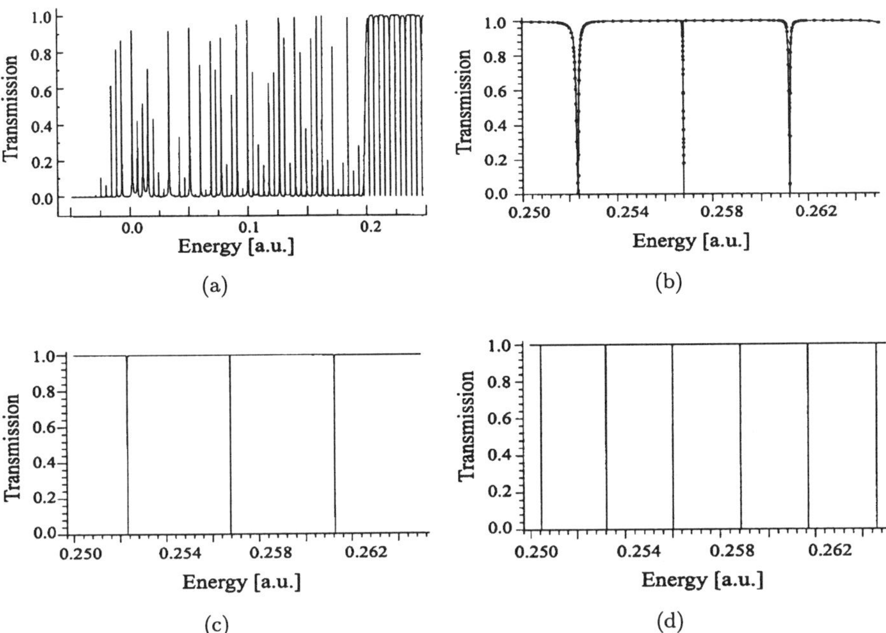

Fig. 11.7. Transmission in double-crossing NT-case. The asymmetric model potential is defined by Eq. (11.68). (a) Transmission coefficient as a function of energy (overall feature). (b) Magnification of (a) ($V_{12} = 5 \times 10^{-3}$). Solid circles represent the semiclassical results from Eq. (11.49) and the Zhu–Nakamura theory presented in Chap. 5. (c) The same as (b) except for $V_{12} = 1 \times 10^{-3}$. (d) The same as (b) except for $V_{12} = 8 \times 10^{-2}$. (Taken from Ref. [66] with permission.)

demonstrate the occurence of complete reflection and transmission in the energy regions discussed in the previous section. Figure 11.7(a) shows the overall behavior of the transmission coefficient in the case of asymmetric potential model defined as

$$V_{11} = 0.2 - 0.01x^2, \qquad V_{22} = 0.01(x - 1)^2 - 0.3, \qquad V_{12} = 5.10^{-3}. \quad (11.68)$$

There appear as expected complete reflection dips at $E > E_-$ and complete transmission peaks at $E > E_d$. The five energy intervals are divided by $E_u = 0.2$, $E_+ = 5 \times 10^{-3}$, $E_- = -0.095$, and $E_d = -0.3$. In the energy region below the barrier top, both the complete reflection and transmission occur close to each other (see also Fig. 11.8). At energies above the barrier top, the very flat transmission peaks are separated by the complete reflection

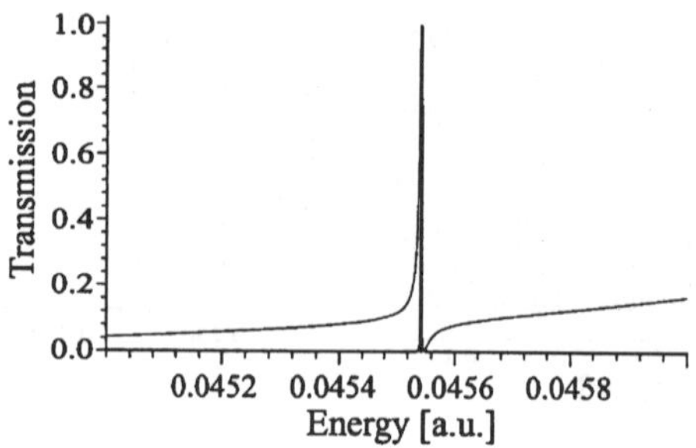

Fig. 11.8.   Fano type resonance in the Vardi–Shapiro case Ref. [182]. The model potentials are given by Eqs. (11.69)–(11.72). Complete reflection and transmission appear below the top of the central barrier (maximum of $V_{11} \simeq 4.71 \times 10^{-2}$). The thin line corresponds to the potential given in the text, while the bold line is the transmission coefficient for an artificially magnified central barrier to such an extent that the tunneling can be neglected. (Taken from Ref. [66] with permission.)

dips. This follows from Eq. (11.53), since $p_{\mathrm{L}} \simeq p_{\mathrm{R}} \to 1$ and the phase condition therein is also satisfied at large energies. We note that the dependance of the resonance width on the diabatic coupling strength is not monotonous. This is because both for the weak coupling, $p \to 1$, and for the strong coupling, $p \to 0$, the two subsequent transitions do not allow the wave to reach the turning points and get reflected back. Figure 11.7(b) shows a magnification of some portion of Fig. 11.7(a), demonstrating the accuracy of the Zhu–Nakamura semiclassical theory (solid circles). Figures 11.7(c) and 11.7(d) show the transmission coefficient for different coupling strengths, i.e. $V_{12} = 1 \times 10^{-3}$ (c) and $V_{12} = 8 \times 10^{-2}$ (d). Since also here the semiclassical and exact results practically coincide, only one curve is plotted. The Zhu–Nakamura semiclassical theory works very well (also in the tunneling range which is not shown here). Figures 11.7(b)–11.7(d) demonstrate the non-monotonous character of the dip widths on the coupling strength, too.

Finally, it is interesting to note that the Fano type resonance [179] discussed by Vardi and Shapiro under the name of laser catalysis [182] can be reproduced by the present analysis. We take the following two diabatic potentials [182]:

$$V_{11} = U_1 - \frac{B\xi}{(1-\xi)^2}, \qquad V_{22} = U_2 + \frac{B\xi}{(1-\xi)^2} \qquad (11.69)$$

with

$$\xi = -e^{\pi x/2}, \qquad (11.70)$$

where

$$B = 6.247 \times 10^{-2}, \qquad U_1 = 3.15 \times 10^{-2}, \qquad U_2 = 5.917 \times 10^{-2}. \qquad (11.71)$$

The coupling is given by the dipole moment and the laser field intensity,

$$V_{12} = 6.8 \times 10^{-5}, \qquad M = 1060.83. \qquad (11.72)$$

Although in the above mentioned reference the authors used a laser pulse and propagate wave packets, the constant coupling in Eq. (11.72) is adequate, since the field changes slowly. Figure 11.8 shows the transmission coefficient as a function of energy. This figure is very similar to Fig. 8 in the reference [182]. This is because an infinitesimal shift of a Floquet state by $d\omega$ is roughly equivalent to a shift in scattering energy, $dE = -d\omega$. As a result of a very small coupling, the wave passes through the crossing points almost diabatically (p_{L}, $p_{\mathrm{R}} \sim 1$). Thus the background in Fig. 11.8 is just a single barrier penetration. Since the phases d_1 and d_3 are very small, Eq. (11.64) can never be satisfied in this particular energy range. Consequently, the complete reflection is not the same as that in the single NT-case. It can also be seen from Eqs. (11.60) and (11.61) that the limit of infintely high barrier ($\zeta\kappa \gg 1$) is not justified. Hence, the tunneling through the central potential barrier is responsible for the complete reflection. Taking $\zeta \sim 0$, $\Delta \sim \delta + \psi_{\mathrm{L}} - \psi_{\mathrm{R}}$ (see Eqs. (11.61) and (11.60)), the complete reflection condition (11.59) reduces to a quantization in the global diabatic well, $d_1 + d_2^{\mathrm{L}} + d_2^{\mathrm{R}} + d_3 + \delta + \psi_{\mathrm{L}} - \psi_{\mathrm{R}} = \pi/2$. This explains the complete reflection in Fig. 11.8, since the energy is close to the first bound state supported by the diabatic potential V_{22}. For the complete transmission, on the other hand, the tunneling does not play so important role. This can be checked both analytically and numerically. When we increase artificially the height of the barrier, the complete reflection disappears while the complete transmission remains stable (bold line in Fig. 11.8). Thus we can use Eq. (11.65) instead of Eq. (11.62). The first two terms in this equation, proportional to $1 - p$, are very small and the last term turns to zero when $d_1 + d_2^{\mathrm{L}} + d_2^{\mathrm{R}} + d_3 + \delta + \psi_{\mathrm{L}} - \psi_{\mathrm{R}} = \pi/2$. Thus just a small change in energy suffices to switch between the complete reflection and complete transmission and the Fano type of resonance can be nicely explained by the semiclassical picture and theory.

Chapter 12

New Mechanism of Molecular Switching

12.1. Basic Idea

Transmission properties and switching in one-dimensional periodic potential systems have been continuously attracting much attention since the original works done by Kronig and Penney [184], and by others [185–187]. Not only the transmission problems in periodic delta-function potentials and square well potentials have been analyzed [188–191], but also molecular switches have been discussed on the basis of a tight-binding model from the viewpoint of molecular electronics [192]. With use of an array of periodic potential barriers, Carter proposed also an idea of molecular switching [193]. In the array of potential barriers, by somehow modifying some of the potential barriers one may switch off the complete transmission which exists always in the periodic system. He proposed to use, for instance, a certain kind of transition metal intercalated coordination compound as a control unit to modify the shape of a potential barrier in the array, since the compound induces intramolecular charge transfer by absorbing light and thus creates an electric field. Here we discuss the possibility of new molecular switching with use of the phenomenon of complete reflection described in detail in the previous chapter. Since potential barriers in molecular systems are highly probable to be the NT-type of curve crossing and thus the latter is expected to appear commonly in the nature, the peculiar phenomenon of complete reflection must be taking a certain crucial role in various physical, chemical and biological phenomena. On the other hand, it should be possible to utilize this phenomenon effectively as a sort of molecular device in future high technology. As is clear now, the NT-type transition is far more versatile than the ordinary barrier transmission phenomenon, and the variety of molecules could provide us with a much broader possibility. Namely, we could effectively use the properties of a single molecule. In this chapter such a new speculative possibility of molecular switching is discussed and analyzed theoretically for not only a one-dimensional model [57, 58] but also for a two-dimensional constriction system [59].

12.2.   One-Dimensional Model

12.2.1.   *Transmission in a pure system*

First, we consider a one-dimensional array of the same M units as that of Fig. 12.1.

The total transfer matrix $N^{(M)}$ is simply given by

$$N^{(M)} = N^M ,\qquad(12.1)$$

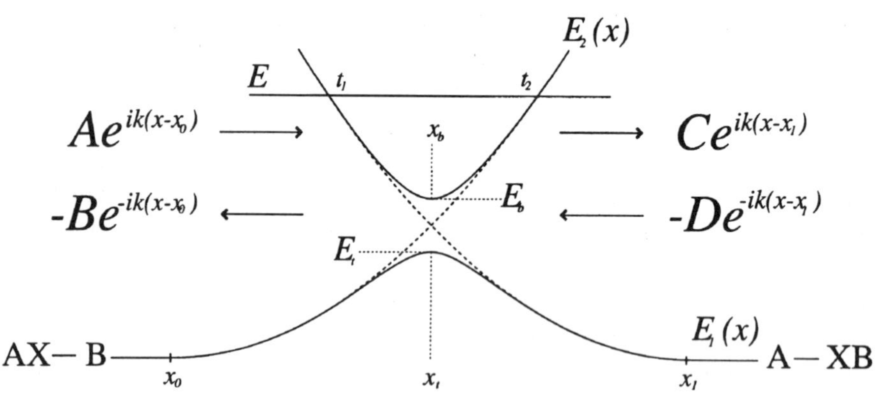

Fig. 12.1.   One unit of the nonadiabatic tunneling type potential. Potentials for $x \leq x_0$ and $x \geq x_1$ are assumed to be flat. The top (bottom) of the lower (upper) adiabatic potential $E_1(x)(E_2(x))$ is denoted by x_t and E_t (x_b and E_b). The transmission from the left $(AX-B)$ to the right $(A-XB)$ corresponds to an effective transfer of X. $X \equiv H$ is a hydrogen atom transfer and $X \equiv e$ is an effective electron transfer $(A^- \to B^-)$. (Taken from Ref. [58] with permission.)

where N is the transfer matrix for one-unit and is given explicitly by (see Eq. (5.128))

$$N_{11} = N_{22}^* = \frac{1}{T^*} \equiv \frac{1}{|T|} e^{i\delta_{\rm T}} \tag{12.2}$$

$$N_{21} = N_{12}^* = \frac{R}{T} \equiv \frac{|R|}{|T|} e^{i(\delta_{\rm R} - \delta_{\rm T})} \,. \tag{12.3}$$

If necessary, we can add an interval L between the units. In this case, the transfer matrix N for one-unit is replaced by

$$N \to \begin{pmatrix} e^{ikL} & 0 \\ 0 & e^{-ikL} \end{pmatrix} N \,. \tag{12.4}$$

As in the Kronig–Penney model in solid state pysics [184–187], the following parameter μ plays an important role:

$$\mu \equiv {\rm Re}\left(\frac{1}{T}\right) = \cos\frac{\delta_{\rm T}}{T} \,. \tag{12.5}$$

The eigenvalues of the N-matrix are given by

$$\lambda_{1,2} = \mu \pm \sqrt{\mu^2 - 1} = \begin{cases} e^{\pm i\phi}(\phi = \cos^{-1}\mu) & |\mu| \le 1, \\ \pm e^{\pm\phi}(\phi = \cosh^{-1}\mu) & |\mu| > 1. \end{cases} \tag{12.6}$$

The region of $|\mu| \le 1(|\mu| > 1)$ corresponds to a conduction (forbidden band) band in the $M \to \infty$ limit. The way of forming these bands shows an interesting chaotic behavior described by the so called logistic equation [57, 194].

For an M unit periodic array, we employ the Lagrange–Sylvester formula for a matrix function,

$$f(A) = \sum_j f(\lambda_j) \prod_{i \ne j} \frac{(A - \lambda_i I)}{(\lambda_j - \lambda_i)} \,, \tag{12.7}$$

where λ_j's are the eigenvalues of a matrix A, and I is a unit matrix. In the case of a 2×2 matrix, we have

$$f(A) = \frac{\lambda_1 f(\lambda_2) - \lambda_2 f(\lambda_1)}{\lambda_1 - \lambda_2} I + \frac{f(\lambda_1) - f(\lambda_2)}{\lambda_1 - \lambda_2} A \,. \tag{12.8}$$

In our present problem, $A = N$ and

$$f(N) = N^M \equiv \begin{pmatrix} \dfrac{1}{T^{(M)*}} & \dfrac{R^{(M)*}}{T^{(M)*}} \\[2ex] \dfrac{R^{(M)}}{T^{(M)}} & \dfrac{1}{T^{(M)}} \end{pmatrix}. \tag{12.9}$$

Then, from Eq. (12.7) we can obtain the following expressions:

$$T^{(M)} = \begin{cases} \dfrac{\sin(\phi)}{1/T\sin(M\phi) - \sin[(M-1)\phi]} & \text{for } |\mu| \leq 1 \\[3ex] \dfrac{\sinh(\phi)}{(\pm 1)^{M-1}/T\sinh(M\phi) - (\pm 1)^M \sinh[(M-1)\phi]} & \text{for } |\mu| > 1, \end{cases} \tag{12.10}$$

and

$$R^{(M)} = \begin{cases} \dfrac{R}{T}\dfrac{\sin(M\phi)}{\sin(\phi)} & \text{for } |\mu| \leq 1 \\[3ex] (\pm 1)^{(M-1)}\dfrac{R}{T}T^{(M)}\dfrac{\sinh(M\phi)}{\sinh(\phi)} & \text{for } |\mu| > 1. \end{cases} \tag{12.11}$$

The corresponding probabilities and phases for $|\mu| \leq 1$ are given by

$$|T^{(M)}|^2 = 1 - |R^{(M)}|^2 = \frac{\sin^2(\phi)}{\sin^2(\phi) + \sin^2(M\phi)(1/|T|^2 - 1)}$$

$$= \frac{1}{\cos^2(M\phi) + (\tan^2(\delta_T)/\tan^2(\phi))\sin^2(M\phi)}, \tag{12.12}$$

$$\delta_{T^{(M)}} = \tan^{-1}\left[\frac{\tan(M\phi)}{\tan(\phi)}\tan(\delta_T)\right], \tag{12.13}$$

and

$$\delta_{R^{(M)}} = \delta_{T^{(M)}} + \delta_R + \delta_T - \pi\Theta\left[-\frac{\sin(M\phi)}{\sin(\phi)}\right]. \tag{12.14}$$

The parameter $\mu^{(M)}$ corresponding to Eq. (12.5) is given by

$$\mu^{(M)} \equiv \text{Re}(N_{11}^{(M)}) = \frac{\cos(\delta_{T^{(M)}})}{T^{(M)}} = \begin{cases} \cos(M\phi) & |\mu^{(M)}| \leq 1 \\[2ex] \pm\cosh(M\phi) & \text{for } |\mu^{(M)}| > 1. \end{cases} \tag{12.15}$$

The semiclassical wave function at $x \sim x_j$ is expressed as

$$\psi_j(x) \simeq A_j e^{ik(x-x_j)} - B_j e^{-ik(x-x_j)}, \qquad (12.16)$$

where the jth unit $(j = 1 \sim M)$ is located at $x_j \leq x \leq x_j$. The waves at the exit (A_M, B_M) and those at the entrance (A_0, B_0) are connected by

$$\begin{pmatrix} A_M \\ B_M \end{pmatrix} = N_T^{(M)} \begin{pmatrix} A_0 \\ B_0 \end{pmatrix} \qquad (12.17)$$

and

$$\begin{pmatrix} B_0 \\ A_M \end{pmatrix} = S^{(M)} \begin{pmatrix} A_0 \\ B_M \end{pmatrix}, \qquad (12.18)$$

with

$$N_T^{(M)} = \prod_{j=1}^{M} N_j = N_M \cdot N_{M-1} \cdots N_1, \qquad (12.19)$$

where $S^{(M)}$ is the scattering matrix and N_j is the transfer matrix of the jth unit. If all the units are the same, we naturally have $N_T^{(M)} = N^{(M)} = N^M$. The physical boundary condition leads to

$$A_0 = C(\text{constant}), \qquad B_M = 0, \qquad (12.20)$$

and thus

$$B_0 = S_{11}^{(M)} C, \qquad A_M = S_{21}^{(M)} C. \qquad (12.21)$$

The wave function in Eq. (12.16) is thus given by

$$\begin{pmatrix} A_j \\ B_j \end{pmatrix} = N_{j,1} \begin{pmatrix} A_0 \\ B_0 \end{pmatrix} = N_{j,1} \begin{pmatrix} 1 \\ S_{11}^{(M)} \end{pmatrix} C, \qquad (12.22)$$

with

$$N_{p,q} = \prod_{j=q}^{p} N_j. \qquad (12.23)$$

Equation (12.12) shows that complete transmission $|T^{(M)}| = 1$ occurs when $|T| = 1$ or $\sin(M\phi) = 0$ is satisfied. If one unit potential is like the one shown in Fig. 12.1, then $|T| = 1$ is never satisfied. The second condition leads to

$$\mu = \frac{\cos \delta_{\mathrm{T}}}{|T|} = \cos\left(\frac{m\pi}{M}\right) \qquad (m = 1, 2, \ldots, M - 1). \tag{12.24}$$

When $M = 2$, this condition reads

$$\delta_{\mathrm{T}} = \left(s' + \frac{1}{2}\right)\pi \qquad (s' = 0, 1, 2, \ldots). \tag{12.25}$$

This coincides with the quantization condition, since here we are considering the case of $L = 0$. Each level of Eq. (12.25) splits into $(M - 1)$ levels, and eventually the regions of $|\mu| \leq 1$ become conduction bands in the limit $M \to \infty$ (see Figs. 12.2 and 12.3).

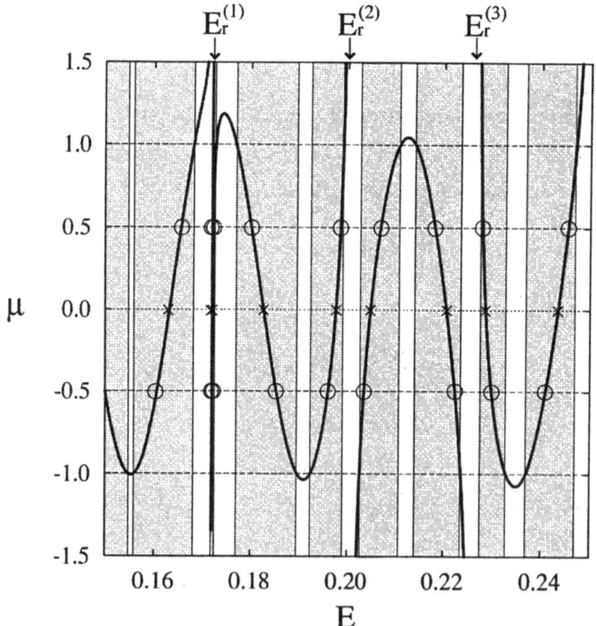

Fig. 12.2.   The quantity μ of Eq. (12.5) as a function of energy E. The crosses (circles) correspond to the complete transmission peaks for $M = 2$ ($M = 3$). At $E = E_r^{(1)}$, $E_r^{(2)}$, and $E_r^{(3)}$ complete refraction occurs. The shaded regions correspond to conduction bands in the limit $M \to \infty$. (Taken form Ref. [58] with permission.)

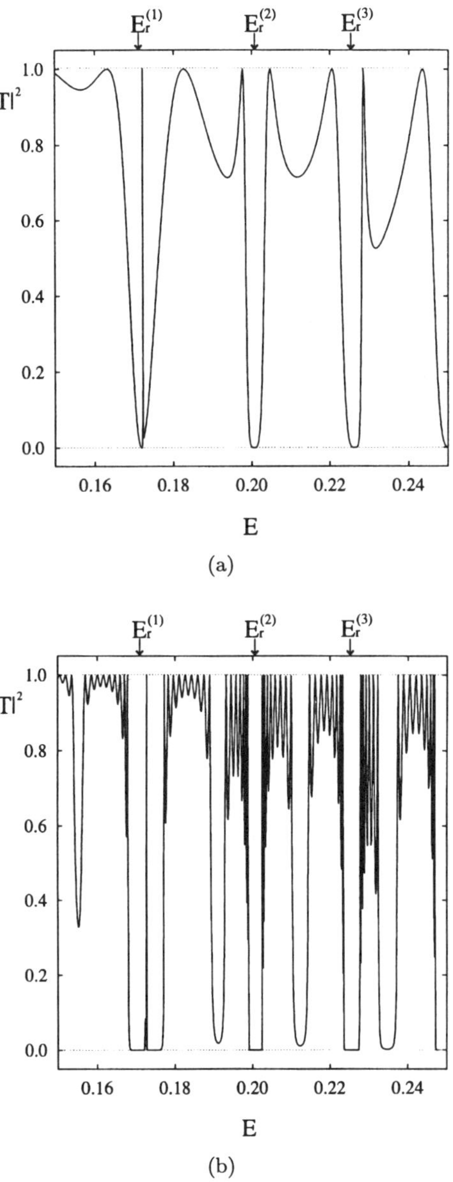

Fig. 12.3. Transmission probabilities as a function of energy E for $M = 2(a)$ and $10(b)$. The model potentials used are the same as in Eqs. (12.26) and (12.27). Creation of conduction bands and complete reflection dips are seen clearly. (Taken form Ref. [58] with permission.)

Here we have used the following diabatic potentials:

$$V_1(x) = a\left(x - \frac{b}{2}\right)^2 \qquad \text{for } -1.0 \le x \le 1.0,$$

$$V_2(x) = a\left(x + \frac{b}{2}\right)^2 \qquad \text{for } -1.0 \le x \le 1.0 \tag{12.26}$$

and

$$V_{12}(x) = V_0 e^{-cx^2} \tag{12.27}$$

with $a = 0.125, b = 2.0, c = 10.0$, and $V_0 = 0.025$ in atomic units. At $E \sim E_t$, where the complete transmission occurs, the transmission peak is roughly expressed as

$$|T^{(M)}|^2 \simeq \frac{[\Gamma_t^{(M)}]^2}{[\Gamma_t^{(M)}]^2 + (E - E_t)^2}, \tag{12.28}$$

with

$$\Gamma_t^{(M)} = \frac{|\sin(m\pi/M)|T_t}{M|\phi_t'|R_t} \qquad \text{for } m \sim \frac{M}{2}, \tag{12.29}$$

where

$$\phi_t' \equiv \left[\frac{\partial \phi}{\partial E}\right]_{E_t} = \left[\frac{\partial \mu}{\partial E}\right]_{E_t} \bigg/ \sin\left(\frac{m\pi}{M}\right), \tag{12.30}$$

$$|T| = T_t + \left[\frac{\partial |T|}{\partial E}\right]_{E_t} (E - E_t) + \cdots, \tag{12.31}$$

and

$$|R| = R_t + \left[\frac{\partial |R|}{\partial E}\right]_{E_t} (E - E_t) + \cdots. \tag{12.32}$$

The complete reflection is a property of one potential unit of the NT-type and its position is determined by Eq. (11.2). This dip width, however, depends on the number, M, of units. Since the complete reflection occurs at $|\mu| \to \infty$, we must use the expression for $|\mu| > 1$. From Eqs. (12.10) and (12.11), we have

$$|R^{(M)}|^2 = \left[1 + \frac{\sinh(|\phi|)}{\sinh^2(M|\phi|)} \left|\frac{T}{R}\right|^2\right]^{-1},$$ (12.33)

with

$$|\mu| = \cosh|\phi|.$$ (12.34)

At $E \sim E_{\mathrm{t}}$, where the complete reflection occurs, we can obtain

$$|R^{(M)}|^2 \simeq \frac{[\Gamma_{\mathrm{r}}^{(M)}]^{2M}}{[\Gamma_{\mathrm{r}}^{(M)}]^{2M} + (E - E_{\mathrm{r}})^{2M}}$$ (12.35)

with

$$\Gamma_{\mathrm{r}}^{(M)} = \frac{1}{T_{\mathrm{r}}}[2|\cos\delta_T^{\mathrm{r}}|]^{1-1/M},$$ (12.36)

where

$$|T|^2 \simeq 4\left[\frac{1-p}{p^2}\right]_{E_{\mathrm{r}}}\left[\frac{\partial\psi}{\partial E}\right]_{E_{\mathrm{r}}}(E - E_{\mathrm{r}})^2 \equiv T_{\mathrm{r}}^2(E - E_{\mathrm{r}})^2 \qquad \text{at } E \sim E_{\mathrm{r}}$$ (12.37)

and

$$\delta_{\mathrm{T}}^{\mathrm{r}} = \delta_{\mathrm{T}}(E_{\mathrm{r}}).$$ (12.38)

The smaller T_{r} is, the larger $\Gamma_{\mathrm{r}}^{(M)}$ is. With increasing M, the shape of the complete reflection dip becomes closer to a square, the size of which is well represented by $\Gamma_{\mathrm{r}}^{(M)}$. This M-dependence of the shape comes from the fact that the reflection probability at off-resonance ($E \neq E_{\mathrm{r}}$) depends on M.

Finally, let us look into the time delay or gain in the propagation through a pure potential system. It is an interesting matter whether the propagation can be accelerated or not when the complete transmission occurs. There are a lot of discussions about the tunneling time, i.e., the time necessary to tunnel through a potential barrier [196]. Unfortunately, or rather naturally, there is no definite answer to this. One of the reasonable measures for that is, however, given by the phase time delay, which is defined by the derivative of the total phase $\delta_{T(M)}$ with respect to k (wave number). This is calculated directly from

Eq. (12.13) as [57]

$$\frac{d\delta_{T(M)}}{dk} = \frac{1}{1 + \{[\tan(M\phi)/\tan(\phi)]\tan\delta_T\}^2} \left(\frac{d\delta_T}{dk} \left\{ \frac{1}{\cos^2(\delta_T)} \frac{\tan(M\phi)}{\tan(\phi)} \right. \right.$$

$$\left. + \frac{\tan(\delta_T)\sin(\delta_T)}{|T|\cos^2(M\phi)\sin^3(\phi)} [M\cos(\phi)\sin(\phi) - \cos(M\phi)\sin(M\phi)] \right\}$$

$$\left. + \frac{d|T|}{dk} \frac{\sin(\delta_T)}{|T|^2} \frac{M\cos(\phi)\sin(\phi) - \cos(M\phi)\sin(M\phi)}{\cos^2(M\phi)\sin^3(\phi)} \right). \tag{12.39}$$

If we are interested in the time delay at the complete transmission, then the above equation can be simplified as follows: In the case of $|T| = 1$, we easily obtain

$$\left(\frac{d\delta_{T(M)}}{dk} \right)_{E_t} = M \left(\frac{d\delta_T}{dk} \right)_{E_t}, \tag{12.40}$$

since in this case $(d|T|/dk)_{E_t} = 0$. In the case of Eq. (12.24), we have

$$\left(\frac{d\delta_{T(M)}}{dk} \right)_{E_t} = \frac{M\sin(\delta_T)}{|T|^2 - \cos^2(\delta_T)} \left[\sin(\delta_T)\frac{d\delta_T}{dk} + \frac{1}{|T|}\cos(\delta_T)\frac{d|T|}{dk} \right]_{E_t}. \tag{12.41}$$

Then the relative dimensionless measure of deceleration/acceleration at the resonant transmission is given by

$$Q \equiv \left(\frac{1}{v}\frac{d\delta_{T(M)}}{dk} - \frac{1}{v}Ml \right)_{E_t} \bigg/ \left(\frac{Ml}{v} \right)$$

$$= \begin{cases} \dfrac{1}{l}\left(\dfrac{d\delta_T}{dk} \right)_{E_t} - 1 \quad \text{(in the case of Eq. (12.39))} \\[2ex] \dfrac{1}{l}\left\{ \dfrac{\sin(\delta_T)}{|T|^2 - \cos^2(\delta_T)}\left[\sin(\delta_T)\dfrac{d\delta_T}{dk} + \dfrac{1}{|T|}\cos(\delta_T)\dfrac{d|T|}{dk} \right] \right\}_{E_t} - 1 \\[2ex] \quad \text{(with } |T| > |\cos\delta_T| \text{ in the case of Eq. (12.40))}, \end{cases} \tag{12.42}$$

where l and v represent the length of one unit of potential and the velocity corresponding to k. Interestingly, this relative measure Q does not depend on M. If we could design a potential in such a way that Q becomes negaive, then we could have an accelerated complete resonant transmission. This means that the resonant transmission could proceed faster than the translation of a free particle at the same asymptotic velocity v. In general, Q is expected to be positive for the system composed only of potential barriers, although it is difficult to prove this analytically for a general case. Deceleration at resonant transmission was shown explicitly in the case of a rectangular potential barrier system [195]. It is, however, interesting to note that Q defined above can be negative, if a potential system involves a strong enough attractive potential. This is simply because the attractive potential accelerates the particle and can compensate the delay due to barrier penetration. Here let us try to estimate the time delay directly by propagating a wave packet. This is done by evaluating the probability flux as a function of time. The actual potential system used is a 15-unit pure array, the potential parameters of which are the same as those given by Eqs. (12.26) and (12.27). The initial wave packet is given by

$$\psi(x; t = 0) = \pi^{-1/4} \omega^{-1/2} \exp\left[-ik(x - x_0) - \frac{(x - x_0)^2}{2\omega^2}\right], \qquad (12.43)$$

where ω measures the packet width,

$$\omega = 2\langle(\Delta x)^2\rangle = \frac{1}{2\langle(\Delta p)^2\rangle}. \qquad (12.44)$$

The computation procedure of wave packet propagation will be explained later. The packet with the parameters $\omega = 7.1$ and $E_0 = 0.19546$ is sent in from $x = 50$, and the completed transmitted flux is evaluated at $x = -41$ as a function of time. The result is shown in Fig. 12.4 by the dashed line. The solid line is the result of free propagation of the same wave packet without any potentials. The time delay, i.e., the difference between the two peaks, is equal to $\Delta t \simeq 264$. The phase time delay $\Delta\tau_{\mathrm{ph}}$, on the other hand, is defined by

$$\Delta\tau_{\mathrm{ph}} = \frac{1}{v}\left(\frac{d\delta_{T(M)}}{dk} - l_{\mathrm{T}}\right), \qquad (12.45)$$

where v is the group velocity, l_{T} is the total length of the potential array. Under the same conditions used in the wave packet propagation, we obtain $\Delta\tau_{\mathrm{ph}} \simeq 251$, which is in good agreement with the above estimate. This coincidence does not hold always, but the above result indicates that the phase time delay may provide a rough reasonable estimate.

12.2.2. *Transmission in a system with impurities*

In the previuos subsection, we have shown how the complete transmission peaks and complete reflection dips are formed when the number M of potential units

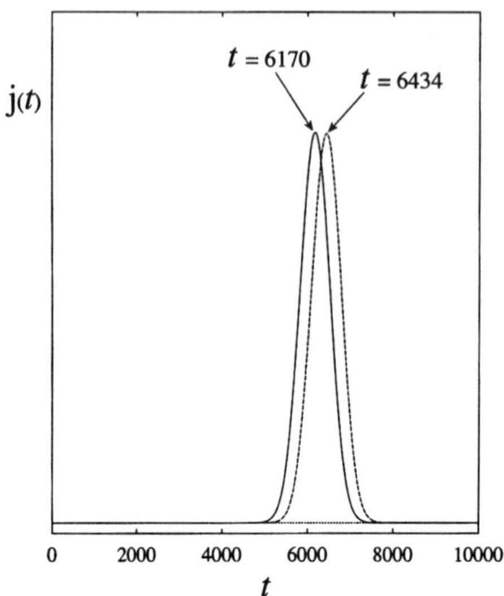

Fig. 12.4. Probability flux as a function of time to demonstrate the transmission time delay due to potentials. The potential system used is the 15-unit pure array, the parameters of which are the same as those of Eqs. (12.26) and (12.27). The solid line represents the free propagation without potentials. The dashed line corresponds to the complete transmission through the potential system, and the difference between the two peaks indicates the delay. (Taken from Ref. [58] with permission.)

is increased in the pure system. Here we demonstrate how this feature changes with an introduction of impurities into the system. Impurities, here, mean such potential units that have diabatic potentials and/or diabatic couplings different from those of other units.

First of all, it can be shown easily that the complete transmission is destroyed by an impurity. Suppose we have an impurity at the right end of a periodic system. Denoting the transfer matrix of the periodic system as N_A and that of the impurity as N_I, we obtain the $(1,1)$ element of the transfer matrix of the whole $(A + I)$ system as

$$N_{11}^{AI} = \frac{1}{|T_A T_I|} \exp[i(\delta_{T_A} + \delta_{T_I})][1 + |R_A R_I|e^{-i(2\delta_{T_A} + \delta_{R_I} - \delta_{R_A})}]. \quad (12.46)$$

By a simple manipulation, we obtain

$$N_{11}^{AI} = \frac{|R_A R_I|}{|T_A T_I|} \left[\frac{|R_A|}{|R_I|} + \frac{|R_I|}{|R_A|} + 2\cos(2\delta_{T_A} + \delta_{R_I} - \delta_{R_A}) \right] + 1. \quad (12.47)$$

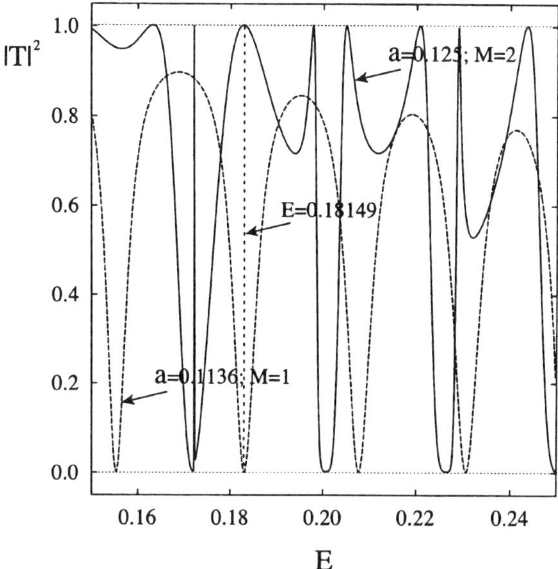

Fig. 12.5. An example showing the coincidence of a complete transmission peak in a pure system and a complete reflection dip in a system with an impurity. The model potentials used are as in Eqs. (12.26) and (12.27). (Taken form Ref. [58] with permission.)

This equation tells that unless $|R_A| = |R_I|$ and $2\delta_{T_A} + \delta_{R_I} - \delta_{R_A} = (2s+1)\pi$ ($s = 0, \pm 1, \pm 2, \cdot$) are satisfied simultaniously, the overall tunneling probability cannot be unity (always smaller than unity); i.e., no complete transmission is possible.

It is interesting to note that if we can somehow change the shape and/or coupling of one potential unit in the array so that a new complete reflection position coincides with one of the complete transmission peaks in the original pure system, then we can switch off the complete transmission completely. Figure 12.5 shows an example of the coincidence of a complete transmission peak in the pure system and a complete reflection in the system with an impurity. The model ptentials used are the same as Eqs. (12.26) and (12.27), except that the parameters a and V_0 are changed to 0.1136 and 0.0175, respectively, for the impurity. If the creation of such an impurity and the restoration to the original pure system could be made reversible, we could switch complete transmission off and on completely. This is the fundamental idea of the molecular switching. The positions of complete reflections are more sensitive to the shape of

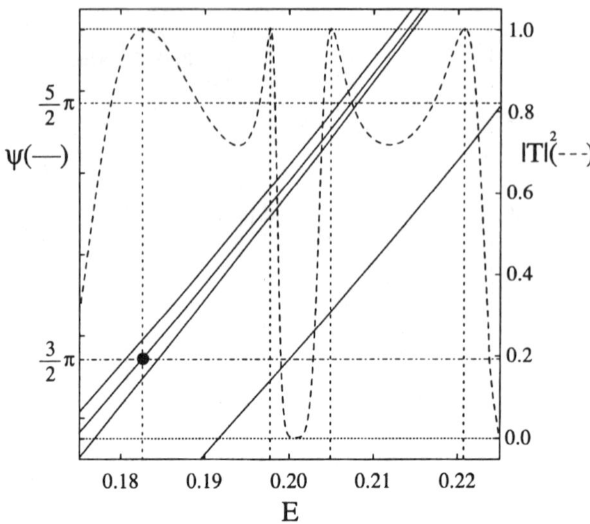

Fig. 12.6.   The phase ψ against energy E for various potential shapes and coupling strengths. A curve of $|T|^2$ versus E for the pure system with $M = 2$ is superimposed. The model potentials used are as in Eqs. (12.26) and (12.27). (Taken form Ref. [58] with permission.)

the potential than to the diabatic coupling. The widths, on the other hand, are more sensitive to the coupling. In order to find a system which satisfies the position matching condition, it is convenient to draw a figure of ψ_{ZN} [of Eq. (5.148)] versus E for various potential shapes and coupling strengths. If the curve of $|T|^2$ versus E for the pure system is superimposed, as is shown in Fig. 12.6, we can find proper conditions for the impurity which meet the complete switching condition.

So far we have considered only one impurity in the system. We can actually think of a group of impurities, the potential parameters of which are slightly different from each other and distributed in a certain range. In that case a complete reflection dip band with a desired width Γ_{dip} may be created. Since the dip width of one unit is given by $\Gamma_{\mathrm{r}}^{(M=1)}$ of Eq. (12.36), the number of impurity units is roughly equal to $\Gamma_{\mathrm{dip}}/\Delta E_{\mathrm{r}} - 1$, where ΔE_{r} is an interval between the complete reflection dips of two adjacent impurities and is adjusted to be equal to $\sim \Gamma_{\mathrm{r}}^{(1)}$, so that the two adjacent dips sufficiently overlap. When $M \gtrsim 10 \sim 20$, the complete transmission peaks are closer together in each region of $|\mu| \leq 1$, and almost constitute conduction bands. In such a case, switching may be made more effectively, if the above mentioned complete reflection band is matched to the transmission band. Actually, such a complete reflection band may be used as an energy filter. We may make the system completely translucent in a certain energy range by using the energy filter, which may be designed as desired with respect to the energy position and width. First, we investigate effects of distributions of impurities in the array. And then, after that we will discuss the energy filter.

Figure 12.7 shows the transmission probability as a function of energy for ten-unit arrays which contain one [Fig. 12.7(a)] and five [Fig. 12.7(b)] impurities. The model potentials used are the same as Eqs. (12.26) and (12.27) with $a = 0.125$, $b = 2.0$, $c = 10.0$, $V_0 = 0.025$. The parameters a and V_0 are changed to $a = 0.11627$ and $V_0 = 0.01875$ in the case of impurity. This impurity is designed so that a new complete reflection dip appears inside the region of the complete transmission band of the original pure sustem [dashed line in Fig. 12.7(a)]. In the case of Fig. 12.7(a) [12.7(b)], the last one unit (five units) in the ten-unit array is (are) replaced by the one impurity (five impurities). As is seen from these figures, complete transmission is not possible anymore, if the system contains an impurity. (There are some peaks reaching close to 1.0, but they are actually lower than 1.0). There appear new complete reflection dips due to the impurity, two in the present energy range at $E \simeq 0.15897$ and 0.18736. Original complete reflection dips of the pure system, only one denoted as P in the present energy range, survive and stay at the same positions irrespective of the number of impurities M_{imp}, as is expected. On the other hand, the peak positions and their shapes are slightly changed from the original ones depending on M_{imp}. The effect of M_{imp} on the complete reflection dips due to impurities can be seen from the two figures. The dips become wider and square with increasing M_{imp}. This is in accordance with the general discussion given before.

Let us next look into the effects of distribution of impurities by arranging the same five impurities in the 15-unit array (see Fig. 12.8). Figure 12.8(a) is a pure system and depicts 14 complete transmission peaks in each region of $|\mu| \leq 1$. Figures 12.8(b)–12.8(d) correspond to the cases in which the five impurities are put in the front (b), in the middle (c), and randomly (d), respectively. The complete reflection dip at $E \sim 0.1625$ is created even in the case (d). Since case (c) is more regular than the others, higher transmission peaks appear more than in cases (b) and (d). In the random case (d), the transmissivity is the worst. This is in accordance with the Anderson's localization [197].

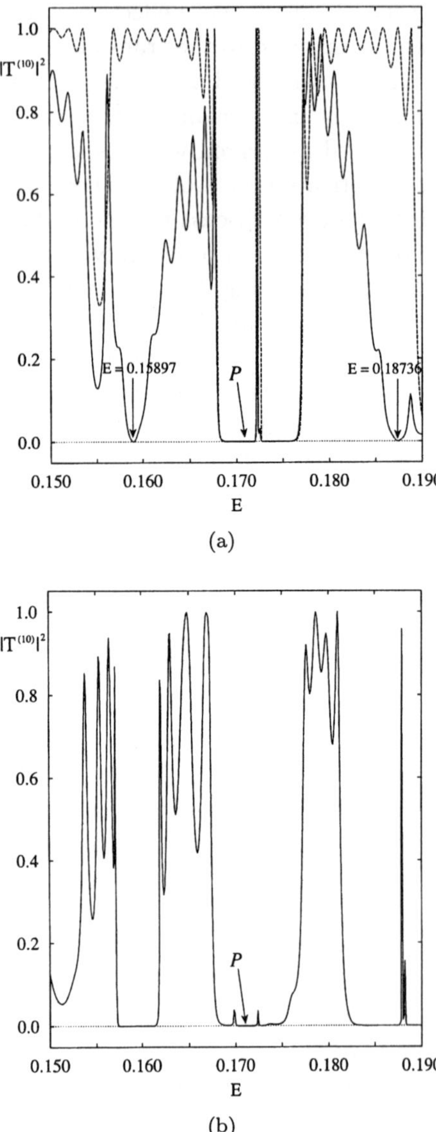

(a)

(b)

Fig. 12.7. Transmission probability as a function of energy in the case of ten-unit arrays which contain impurities. The model potentials used are given by Eqs. (12.26) and (12.27). The dashed line corresponds to the pure system. (a) One impurity case. Two complete reflection dips at $E \cong 0.15897$ and 0.18736 are due to this impurity. (b) Five same impurities case. (Taken from Ref. [58] with permission.)

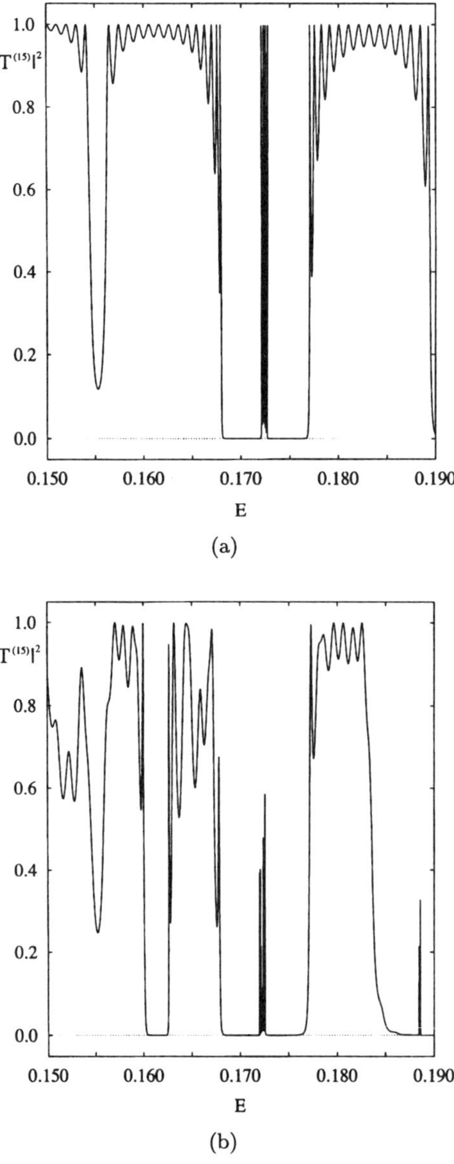

Fig. 12.8. Effects of the distribution of five same impurities in the 15-unit array. The potentials used are the same as those given by Eqs. (12.26) and (12.27). (a) Pure system, (b) Five impurities in the front, (c) Five impurities in the middle, (d) Five impurities are distributed randomly. (Taken from Ref. [58] with permission.)

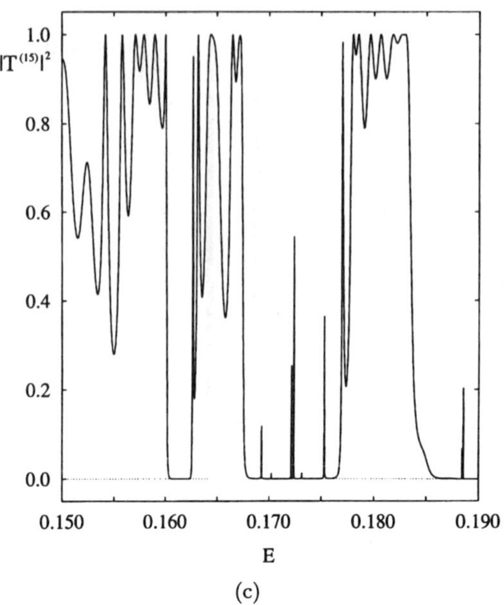

(c)

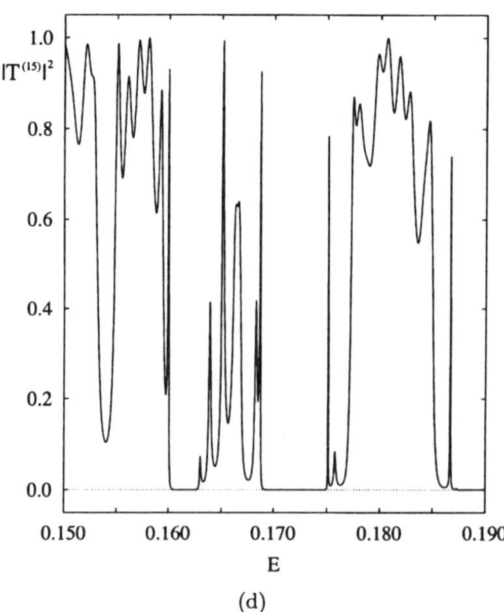

(d)

Fig. 12.8.   (*Continued*)

Complete reflection bands of a desirable width at any desirable energy can be created, in principle, by introducing impurities in a multi-unit system. This is quite important in order to switch the transmission efficiently. Here we discuss two methods to realize this, and prepare an appropriate system for the wave packet propagation discussed in the next subsection. First, we create a complete reflection dip at a desired energy E_{dip} by changing potential parameters of an impurity. This can be done effectively by plotting the phase ψ_{ZN} of Eq. (5.148) as a function of energy for various potential parameters. Next, we want to make the complete band has a desired width Γ_{dip}, so that the portion of the incoming wave packet corresponding to $E_{\mathrm{dip}} - \Gamma_{\mathrm{dip}}/2 \leq E \leq E_{\mathrm{dip}} + \Gamma_{\mathrm{dip}}/2$ will be reflected. The band width can be controlled by either one of the following two methods: (i) by introducing a group of the same impurities, or (ii) by introducing a group of impurities which have potential parameters slightly different from each other. In the first case, the maximum possible width Γ_{dip} is determined by the range of $|\mu_{\mathrm{imp}}| > 1$, where μ_{imp} is the μ parameter of the impurity defined by Eq. (12.5). Thus it is necessary to find a proper impurity which can give rise to a dip band of the desired width Γ_{dip}. Once this is found, it is easy to create the band. On the other hand, the second method is quite flexible, because we can introduce various kinds of impurities. If the dip width of one impurity is equal to $\Gamma_{\mathrm{r}}^{(M=1)}$ of Eq. (12.36), the number of impurities necessary to cover the desired width Γ_{dip} is roughly given by $\Gamma_{\mathrm{dip}}/\Delta E_{\mathrm{r}} - 1$, where ΔE_{r} is an interval between the complete reflection dips of two adjacent impurities and should be adjusted to be equal to $\sim \Gamma_{\mathrm{r}}^{(M=1)}$. Two examples corresponding to the above two cases, (i) and (ii), are shown in Figs. 12.9(a) and 12.9(b), where the system is composed of 20-units, including 9 units of impurities at the rear end. The model potentials used are again the same as in Eqs. (12.26) and (12.27), with $a = 0.10989$, $b = 2.0$, $c = 10.0$, and $V_0 = 0.0425$. Figure 12.9(a) corresponds to case (i), for which the impurity parameters are $a = 0.10363$, $b = 2.0$, $c = 10.0$, and $V_0 = 0.03$. The complete reflection band obtained is at $E_{\mathrm{dip}} = 0.1949$, with the width $\Gamma_{\mathrm{dip}}(M_{\mathrm{imp}} = 9) \simeq 0.001714$. There appear a wavy structure just outside the band, as is seen in Fig. 12.9(a). Figure 12.9(b) depicts the result of nine slightly different impurities. The parameters $a \sim c$ are taken to be the same for the nine impurities ($a = 0.11765$, $b = 2.0$, $c = 10.0$), but the coupling strength V_0 is varied as follows: $V_0 = 0.0415$, 0.041075, 0.04065, 0.040225, 0.0398, 0.039375, 0.03895, and 0.0381. The dip width of one impurity $\Gamma_{\mathrm{r}}^{(M_{\mathrm{imp}}=1)}$ is roughly equal to 0.00017, and thus the total band

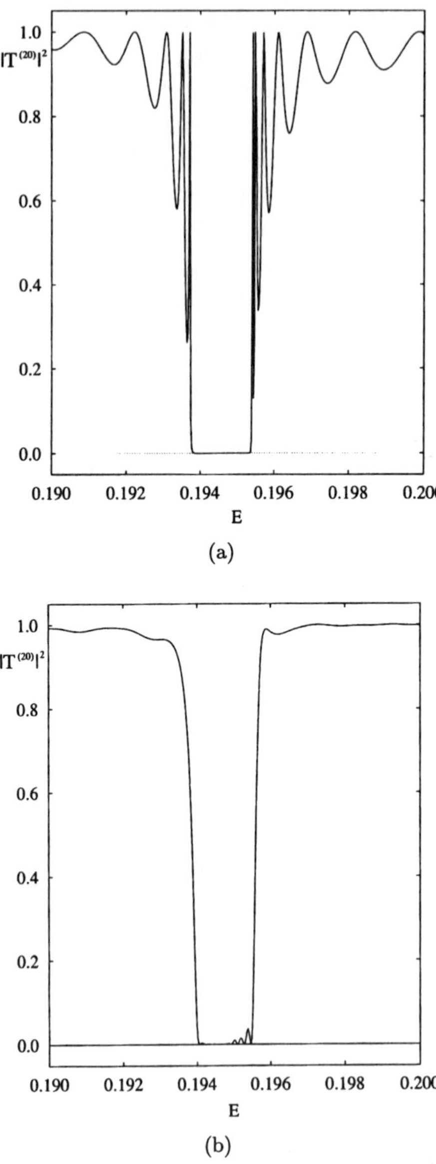

(a)

(b)

Fig. 12.9.   Complete reflection band (energy filter) in the 20-unit array with nine impurities. The potential functions are the same as in Eqs. (12.26) and (12.27). (a) The same impurities are put in the end. (b) Nine impurities at the end with slightly different potential parameters (see the text). (Taken from Ref. [58] with permission.)

width is expected to be ~ 0.0017. Actually, a dip band of the width ~ 0.00167 is obtained, as is seen in Fig. 12.9(b). There is no wavy structure outside the band. Switching in the wave packet propagation will be discussed in the next subsection with use of these reflection bands.

12.2.3. Molecular switching

The switching of transmission is directly demonstrated by utilizing the wave packet propagation in time. In the diabatic representation, our one-dimensional two-state problem is described by the following time-dependent Schrödinger equation:

$$H\psi = i\frac{\partial}{\partial t}\psi \tag{12.48}$$

with

$$
H = \begin{pmatrix} -\dfrac{1}{2\mu}\dfrac{\partial^2}{\partial x^2} + V_1 & V_{12} \\[2mm] V_{12} & -\dfrac{1}{2\mu}\dfrac{\partial^2}{\partial x^2} + V_{12} \end{pmatrix}
$$

$$
= \begin{pmatrix} -\dfrac{1}{2\mu}\dfrac{\partial^2}{\partial x^2} & 0 \\[2mm] 0 & -\dfrac{1}{2\mu}\dfrac{\partial^2}{\partial x^2} \end{pmatrix} + \begin{pmatrix} V_1 & V_{12} \\[1mm] V_{12} & V_2 \end{pmatrix}
$$

$$
= T + V \tag{12.49}
$$

and

$$\psi = \begin{pmatrix} \psi_1 \\ \psi_2 \end{pmatrix}, \tag{12.50}$$

where V_1 and V_2 are the diabatic potentials and V_{12} is the coupling between the diabatic states ψ_1 and ψ_2. The diabatic model potentials defined by Eqs. (12.26) and (12.27) are used here. The diabatic potential matrix V is diagonalized by the unitary matrix,

$$U^\dagger V U = \lambda = \begin{pmatrix} \lambda_+ & 0 \\ 0 & \lambda_- \end{pmatrix}, \tag{12.51}$$

then the wave function in the adiabatic state representation ϕ is given by

$$\phi = U^\dagger \psi. \tag{12.52}$$

The time propagation of the wave function is carried out by repeating the following procedure:

$$\begin{aligned}
\psi(x : t = \delta t) &= e^{-iH\delta t}\psi(x : t = 0) \\
&= e^{-iT\delta t}e^{-iV\delta t}\psi(x : t = 0) \\
&= e^{-iT\delta t}Ue^{-iU^\dagger VU\delta t}U^\dagger\psi(x : t = 0) \\
&= e^{-iT\delta t}Ue^{-i\lambda\delta t}U^\dagger\psi(x : t = 0),
\end{aligned} \tag{12.53}$$

where δt is a time step which should be small enough to keep the unitarity. When the kinetic energy term operates on the wave function, we utilize the fast Fourier transform (FFT) between the spatial and frequency domains. If we restrict ourselves to a finite spatial domain, the wave packet eventually reaches the boundary and appears from the other side of the boundaries in the FFT treament. In order to avoid this, the following imaginary absorbing potential is introduced:

$$V_{\mathrm{damp}} = -iCx^4, \tag{12.54}$$

where the coefficient C is determined according to the procedure of Vibok and Balint-Kurti [198]. The transmission and reflection probabilities are estimated by integrating the flux of the corresponding wave packet with respect to time. The flux is given by

$$j(x; t) = -\frac{i}{2\mu}\left[\phi^*\frac{\partial\phi}{\partial x} - \left(\frac{\partial\phi^*}{\partial x}\right)\phi\right], \tag{12.55}$$

where $\partial\phi/\partial x$ is numerically evaluated by the fifth order finite difference method.

In order to demonstrate the switching explicitly, we employ the 20-unit system introduced in the previous section and propagate a wave packet which has the width $\omega = 26.7$. The initail wave packet is sent in from the right, as is shown in Fig. 12.10(a). The peak energy of the packet corresponds to $E = 0.19494$. Figure 12.10(b) shows the propagation in the case of a pure system. An inerference structure appears when the wave packet is moving

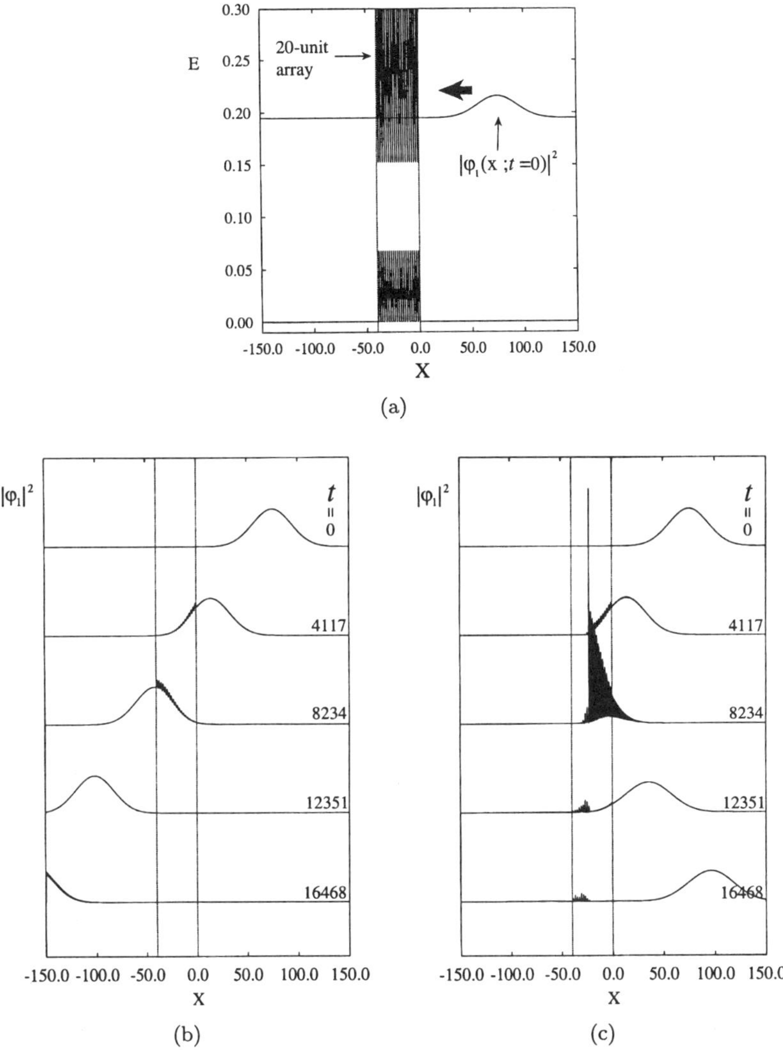

Fig. 12.10.   Wave packet propagation in the case of 20-unit array. Potentials are the same as those given in Eqs. (12.26) and (12.27). The peak energy and the width of the packet are $E \cong 0.19494$ and $\omega = 26.7$. (a) Overall scheme. The initial packet starts from $x = 75$ to the left. The 20-unit potential array is distributed in $-40 \le x \le 0$. (b) Complete transmission through the pure system. (c) Complete reflection by the same system, with impurities as those of Fig. 12.9(a). (d) Transmission probability as a function of switching time t_s, at which the nine impurities are created. (e) Switching of the potential system when the wave packet stays inside the system. (Taken from Ref. [58] with permission.)

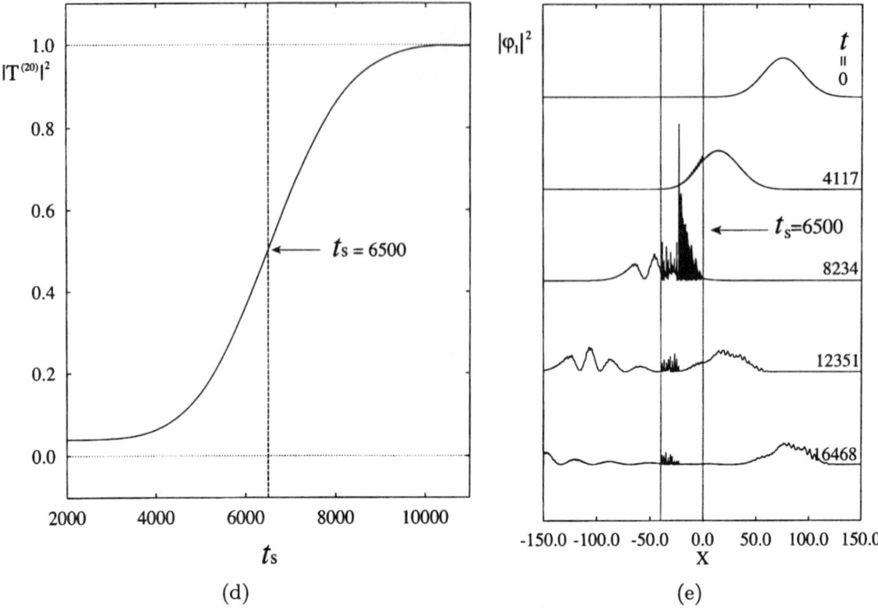

Fig. 12.10.   (*Continued*)

through the potential system, but after that the packet transmits the system completely, since the packet is narrower than the corresponding range of complete transmission band ($0.19375 \leq E \leq 0.19539$). The transmission probability calculated from the flux at $x = -45.0$ is 0.99999. Figure 12.10(c) demonstrates the (almost) complete reflection of the wave packet in the system with impurities corresponding to Fig. 12.9(a). A violent interference structure appears inside the potential system, but the reflected packet retains the original Gaussian shape. The reflection probability evaluated at $x = 149$ at $t \leq 41177$ is 0.96085; the residual portion is trapped by the impurities for quite a long time. Figure 12.10(d) depicts the transmission probability as a function of the switching time t_s, at which the potential system is abruptly switched from the pure system to the impurity system. Figure 12.10(e) shows the feature when the potential switching is made at $t_s = 6500$, at which time the packet just stays inside the potential system. After the violent interference, half of the packet is transmitted and the residual half is reflected. Both waves are quite deformed from the Gaussian within the present scales of space and time.

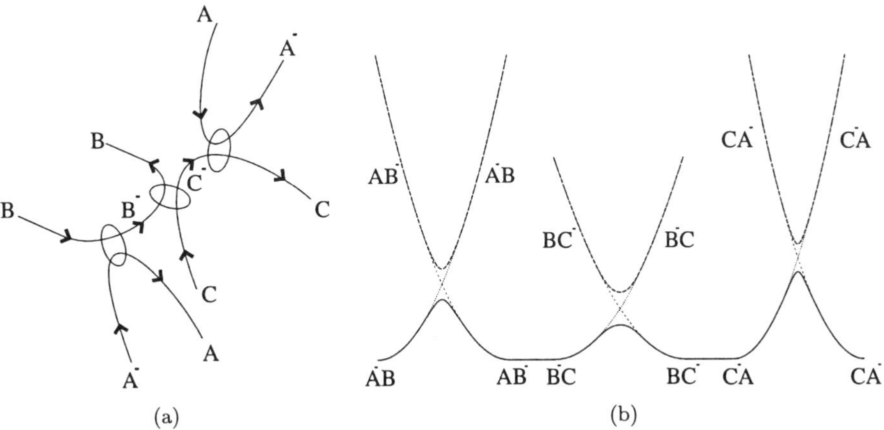

Fig. 12.11.   (a) An effective electron transfer $A^- \to B^- \to C^- \to A^-$. (b) Corresponding schematic potential unit. This type of potential unit may be one unit in the array. (Taken from Ref. [58] with permission.)

The creation of impurities and restoration to the original pure system by changing the shape and coupling of potential units might be realized by using laser or by introducing control units of molecules into the system, as suggested by Carter [193]. One possible candidate for such control units would be metal coordination compounds which induce intramolecular charge transfer reversibly by absorbing light. The phenomenon of complete reflection is due to the quantum mechanical interference effect and is complete only in one-dimensional space. In the present treatment we have considered the single crossing potential system as a unit in the array. One unit may be a more complex one, however. For instance, we can employ the one shown in Fig. 12.11, which describes an effective electron transfer $A^- \to B^- \to C^- \to A^-$. In this case $|T| = 1$ is possible, and many more complete reflections also appear.

12.3. Two-Dimensional Model

It is a big question whether the complete reflection phenomenon can survive in multi-dimensional systems or not, since this is based on a sensitive phenomenon of phase interference. Some traces of this intriguing phenomenon are, however, expected to survive, if it is not complete [199]. In this subsection, a two-dimensional constriction model shown in Fig. 12.12 [59] is considered. This

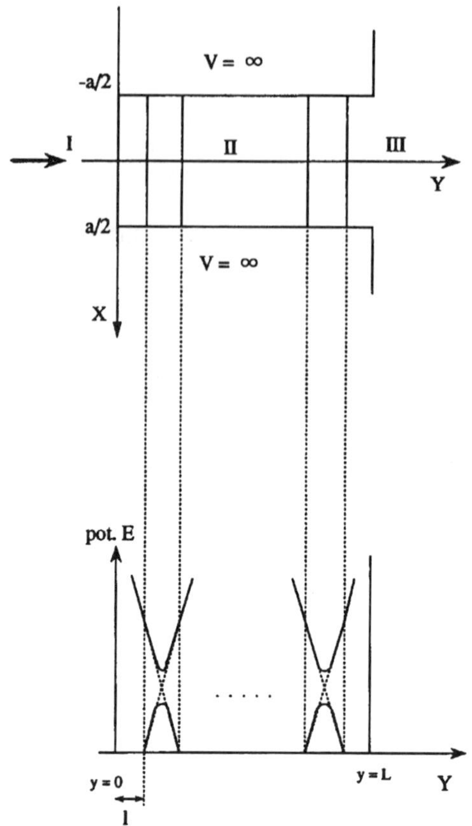

Fig. 12.12.   Two-dimensional constriction model. (Taken from Ref. [58] with permission.)

model is chosen, since it is simple enough and analytical treatment is possible
to a good extent and yet some basic features of a multidimensional case can be
comprehended. A similar type of 2D constriction model has been employed in
the study of electron conduction [200, 201], although there are some incorrect
procedureds in the treatments [59].

12.3.1.   *Two-dimensional constriction model*

In regions I and III in the model shown in Fig. 12.12, potentials are taken to be
zero, and the periodic nonadiabatic tunneling type potential which is the same

as that used in the previous subsection (Eqs. (12.26) and (12.27)) is assumed inside the constriction (region III). In the x-direction in region II the potential is assumed to be constant which is y-dependent. Thus, the one-unit potential in the direction of transmission is given by

$$V_1(y) = d(y - b/2)^2 \qquad \text{for } -b/2 \le y \le b/2,$$

$$v_2(y) = d(y + b/2)^2 \qquad \text{for } -b/2 \le y \le b/2, \tag{12.56}$$

$$V_{12}(y) = V_0 \exp(-cy^2),$$

where the actual values of the parameters used are given later. It should be noted that what is important is to show whether the switching is actually possible or not in a two-dimensional space. Particular forms of potential functions such as those given above are not very important, since the complete reflection phenomenon occurs whenever we have a NT-type potential unit.

Although regions I and III are separable even in the simple Cartesian coordinates (tx, y), we should employ the following elliptic coordinates (ξ, η) in order to correctly satisfy matching conditions at the boundaries between I and II and between II and III:

$$x = \frac{a}{2} \cosh\xi \cos\eta \tag{12.57}$$

and

$$y = \frac{a}{2} \sinh\xi \sin\eta \tag{12.58}$$

with $0 \le \xi < \infty$ and $0 \le \eta \le 2\pi$. This is because the quantum mechanical diffracrion due to the potential edges at $(x = \pm a/2, y = 0, L)$ cannot be treated correctly by the Cartesian coordinates. In this sense the methods used in [200, 201] are not correct and should be modified by following the procedure described here.

The Schrödinger equation for a free particle can be separated into the following two equations:

$$\left[\frac{d^2}{d\eta^2} + (\lambda - h^2 \cos^2 \eta)\right] Z(\eta) = 0, \tag{12.59}$$

$$\left[\frac{d^2}{d\eta^2} + (-\lambda + h^2 \cosh^2 \xi)\right] Y(\xi) = 0, \tag{12.60}$$

where $h \equiv ak/2$ (k is the wave number) and λ is the separation constant. What is needed in physical applications of the angle function $Z(\eta)$ is a periodic function and two such solutions are possible; one is even with respect to $\eta = 0$ and π, and the other is odd. In the notations of Morse–Feshbach [202] these are the Mathieu functions $Se_\mu(h, \cos\eta)$ and $So_\mu(h, \cos\eta)$, respectively, where $\mu = 0, 1, 2, \ldots$. In the limit of $h \to 0$ they tend to $\cos(\mu\eta)$ and $\sin(\mu\eta)$, respectively. The separation constant $\lambda_\mu^{(o,e)}$ go to μ^2 in this limit. Since in our present problem we need only the odd solutions $\{So_\mu(h, \cos\eta)\}_\mu$, hereafter formulas are given only for them. These solutions satisfy the following expansion and normalization conditions:

$$So_s(h, \cos\eta) = \sum_t B_t^o(h, s) \sin(t\eta)\,, \tag{12.61}$$

$$\int_0^{2\pi} So_t(h, \cos\eta) So_{t'}(h, \cos\eta) d\eta = M_t^o \delta_{tt'}\,, \tag{12.62}$$

where

$$t = \begin{cases} 2\nu & \text{(for } s = 2\mu) \\ 2\nu + 1 & \text{(for } s = 2\mu + 1) \end{cases} \tag{12.63}$$

and

$$\sum_t t B_t^o(h, t) = 1\,. \tag{12.64}$$

These make a complete set on $(0, 2\pi)$ separately for $p = $ even and odd. In the present treatment, however, we need only the half region $(0, \pi)$ in which $\{So_p\}_{p=1,2,\ldots}$ constitute a complete set. For later convenience we introduce the following normalized complete set of functions:

$$u_\mu(\eta) = \sqrt{\frac{2}{M_\mu^o}} So_\mu(h, \cos\eta) \tag{12.65}$$

with

$$\int_0^\pi u_\mu(\eta) u_{\mu'}(\eta) d\eta = \delta_{\mu\mu'}\,. \tag{12.66}$$

The above definition of angle functions follows that of Morse–Feshbach [202], and its relation to the definition used by Abramowitz–Stegun [203] may be simply given by (AS is attached to the latter)

$$B_t^{o(\text{AS})}(h,s) = \sqrt{\frac{\pi}{M_s^o}} B_t^o(h,s) \tag{12.67}$$

with

$$\sum_t [B_t^{o(\text{AS})}]^2 = 1 \tag{12.68}$$

and thus

$$So_\mu^{\text{AS}}(a,q) = \sqrt{\frac{\pi}{M_{\mu o}}} So_\mu(h, \cos\eta)\,, \tag{12.69}$$

where $a = \lambda - h^2/2$ and $q = h^2/4$ are the parameters used by Abramowitz and Stegun.

Corresponding to So_μ and Se_μ, the following two sets of radial solutions are defined. They are denoted as $\{Jo_\mu(h, \cosh\xi), No_\mu(h, \cosh\xi)\}_\mu$ and $\{Je_\mu(h, \cosh\xi), Ne_\mu(h, \cosh\xi)\}_\mu$. What we need later is

$$Ho_\mu(\xi) \equiv Ho_\mu(h, \cosh\xi)$$

$$= Jo_\mu(h, \cosh\xi) + iNo_\mu(h, \cosh\xi)$$

$$\xrightarrow[\xi\to\infty]{} \frac{1}{\sqrt{h\cosh\xi}} \exp\left[ih\cosh\xi - i\frac{\pi}{2}\left(\mu + \frac{1}{2}\right)\right]. \tag{12.70}$$

In the limit of $h \to 0$, $(Jo_\mu$ and $Je_\mu)$, $(No_\mu$ and $Ne_\mu)$ and $(Ho_\mu$ and $He_\mu)$ correspond to the ordinary Bessel functions J_μ, N_μ, and H_μ, respectively.

12.3.2. *Wave functions, matching, and transmission coefficient*

We assume that a plane wave comes in from $y = -\infty$ along the y-axis (see Fig. 12.12). Then the 2D wave function in region I is given by

$$\psi_\mathrm{I} = e^{iky} - e^{-iky} + \sum_\mu a_\mu u_\mu(\eta) Ho_\mu(\xi)\,, \tag{12.71}$$

where the first (second) plane wave is the incident wave (wave reflected by the wall at $y = 0$), and the third term represents the wave scattered by the slit at $\xi = 0$. Only the $So_\mu(h, \cos\eta)$ functions survive, because the boundary condition

$$\psi_{\mathrm{I}}(\xi > 0, \eta = \pi, 2\pi) = 0 \tag{12.72}$$

should be satisfied. In $y_{2j+1} \le y \le y_{2j}$ $(j = 0, 1, 2, \ldots, N)$ in region II we put the potential unit defined by Eq. (12.56), namely, $y_{2j}(y_{2j+1})$ corresponds to $-b/2(b/2)$. The total wave function between the units is expressed as

$$\psi_j^{\mathrm{II}} = \sum_{n_o}[-b_{n_oj}^{(-)}e^{-iq_{n_o}(y-y_{2j})} + b_{n_oj}^{(+)}e^{iq_{n_o}(y-y_{2j})}]\phi_{n_o}(x)$$

$$+ \sum_{n_c} b_{n_c}^{(+)}e^{-|q_{n_c}|(y-y_{2j})}\phi_{n_c}(x), \tag{12.73}$$

where $\phi_n(x)$ is the nth normalized eigenfunction in the x-direction given by

$$\phi_n(x) = \frac{2}{a}\sin\left[\frac{n\pi}{a}\left(x + \frac{a}{2}\right)\right] \qquad n = 0, 1, 2, \ldots \tag{12.74}$$

with

$$\int_{-a/2}^{a/2} \phi_n(x)\phi_{n'}(x)dx = \delta_{nn'}. \tag{12.75}$$

The corresponding eigenvalue E_n is equal to

$$E_n = \frac{\hbar^2}{2m}\left(\frac{n\pi}{a}\right)^2, \tag{12.76}$$

and q_n represents the wave number of the motion in the y-direction and is given by

$$q_n = \frac{\sqrt{2m}}{\hbar}\sqrt{E - E_n}, \tag{12.77}$$

where E is the total energy and m is the mass of the transmitting particle. In Eq. (12.73) $n_o(n_c)$ represents an open (closed) channel for which $q_{n_o}(q_{n_c})$ is positive (pure imaginary). The coefficients $b_{n_oj}^{(+)}$, etc. are the quantities to be determined. The connection of wave functions at $y = y_{2j+1}$ and at $y = y_{2j+2}$ is made with use of the transfer matrix which represents nonadiabatic tunneling transition by the $(j + 1)$th unit of potential. As is shown in Fig. 12.12, there are N potential units in region II and the total length of the constriction is

$L = N(b+1)+l$. Repeating the above procedure, we can obtain the coefficients $b_{n_o N}^{(\pm)}$ as follows:

$$\tilde{\mathbf{b}}_N^{(+)} = N_{11}^{(N)}\mathbf{b}_0^{(+)} + N_{12}^{(N)}\mathbf{b}_0^{(-)} , \qquad (12.78)$$

$$\tilde{\mathbf{b}}_N^{(-)} = N_{21}^{(N)}\mathbf{b}_0^{(\pm)} + N_{22}^{(N)}\mathbf{b}_0^{(-)} , \qquad (12.79)$$

where

$$(\tilde{\mathbf{b}}_N^{(\pm)})_{n_o} = b_{n_o N}^{(\pm)}e^{\pm q_{n_o}l} \qquad (12.80)$$

$$(\mathbf{b}_0^{(\pm)})_{n_o} = b_{n_o 0}^{(\pm)} . \qquad (12.81)$$

The overall transfer matrix for the total N-unit system $N_{ij}^{(N)}$ is defined as

$$(N_{ij}^{(N)})_{n_o n_o'} = N_{ij}^{(n_o N)}e^{-iq_{n_o}l}\delta_{n_o n_o'} \qquad \text{for } (ij) = (22) \text{ and } (21), \qquad (12.82)$$

$$(N_{11}^{(N)})_{n_o n_o'} = (N_{22}^{(N)})^*_{n_o n_o'} , \qquad (12.83)$$

$$(N_{12}^{(N)})_{n_o n_o'} = (N_{21}^{(N)})^*_{n_o n_o'} , \qquad (12.84)$$

with

$$N_{22}^{(n_o N)} = \frac{1}{T_{n_o}^{(N)}}$$

$$N_{21}^{(n_o N)} = \frac{R_{n_o}^{(N)}}{T_{n_o}^{(N)}} , \qquad (12.85)$$

where $T_{n_o}^{(N)}$ and $R_{n_o}^{(N)}$ represent the overall transmission and reflection amplitudes corresponding to the n_oth open channel. When all N potential units are the same (pure system), these quantities can be easily obtained from $T_{n_o}^{(1)}$ and $R_{n_o}^{(1)}$ of one-unit. In region III the total wave function can be expanded as

$$\psi_{\text{III}} = \sum_\mu c_\mu u_\mu(\eta)Ho_\mu(\xi) , \qquad (12.86)$$

because of the boudary condition

$$\psi_{\text{III}}(\xi > 0, \eta = 0, \pi) = 0 . \qquad (12.87)$$

In this case the x-axis is shifted to the outlet of the constriction.

The matching of ψ_{I} and $\psi^{\mathrm{II}}_{j=0}$ at $y = 0$ leads to

$$a_\mu H o_\mu(0) = \sum_{n_o}(b^{(+)}_{n_o 0} - b^{(-)}_{n_o 0})C_{n_o\mu} + \sum_{n_c} b^{(+)}_{n_c 0}C_{n_c\mu} \qquad (12.88)$$

and

$$2k - E_\mu - \frac{2i}{a}H o'_\mu(0)$$

$$= \sum_{n_o} q_{n_o}(b^{(+)}_{n_o 0} + b^{(-)}_{n_o 0})D_{n_o\mu} + i\sum_{n_c} |q_{n_c}|b^{(+)}_{n_c 0}D_{n_c\mu}\,, \qquad (12.89)$$

where $C_{n\mu}, D_{n\mu}$, and E_μ are the expansion coefficients of $\phi_n(x), \phi_n(x)\sin\eta$, and $\sin\eta$ in terms of $u_\mu(\eta)$ over the region $(\pi, 2\pi)$, namely,

$$C_{n\mu} = \int_\pi^{2\pi} u_\mu(\eta)\phi_n(x)d\eta\,,$$

$$\qquad (12.90)$$

$$D_{n\mu} = \int_\pi^{2\pi} u_\mu(\eta)\phi_n(x)\sin\eta d\eta\,,$$

and

$$E_\mu = \int_\pi^{2\pi} u_\mu(\eta)\sin\eta d\eta = \begin{cases} 0 & (\text{for } \mu = \text{even}) \\ \dfrac{\pi}{\sqrt{2M^0_\nu}}B_1(h,\mu) & (\text{for } \mu = 2\nu + 1)\,. \end{cases} \qquad (12.91)$$

With use of the orthogonality relation between the coefficients $C_{n\mu}$ and $D_{n\mu}$, we can finally obtain

$$\mathbf{b}^{(+)}_0 = (1 - A_{oo})^{-1}(1 + A_{oo})\mathbf{b}^{(-)}_0 + (1 - A_{oo})^{-1}[A_{oc}\mathbf{B}_c + \mathbf{B}_o]\,, \qquad (12.92)$$

where

$$A_{oo} = (\Gamma_Q)_{oo} + (\Gamma_Q)_{oc}[1 - (\Gamma_Q)_{cc}]^{-1}(\Gamma_Q)_{co}\,, \qquad (12.93)$$

$$A_{oc} = (\Gamma_Q)_{oc}[1 - (\Gamma_Q)_{cc}]^{-1}\,, \qquad (12.94)$$

$$(\mathbf{B}_\alpha)_{n_\alpha} = \frac{i}{2}a^2 k \sum_\mu \frac{H o_\mu(0)}{H O'_\mu}D_{n_\alpha\mu}E_\mu \qquad (\alpha = \mathrm{o}, \mathrm{c})\,, \qquad (12.95)$$

and

$$[(\Gamma_Q)_{\alpha\beta}]_{n_\alpha m_\beta} = -\frac{i}{4}a^2 \sum_\mu \frac{Ho_\mu(0)}{Ho'_\mu(0)} D_{n_\alpha\mu} C_{m_\beta\mu} |q_{m_\beta}| \qquad (\alpha, \beta = \mathrm{o}, \mathrm{c}). \quad (12.96)$$

The matching of $\psi^{\mathrm{II}}_{j=N}$ and ψ_{III} at the right end $y = L$ leads to

$$c_\mu Ho_\mu(0) = -\sum_{n_\mathrm{o}} (\tilde{b}^{(+)}_{n_\mathrm{o}N} - \tilde{b}^{(-)}_{n_\mathrm{o}N}) C_{n_\mathrm{o}\mu} + \sum_{n_\mathrm{c}} \tilde{b}^{(-)}_{n_\mathrm{c}N} C_{n_\mathrm{c}\mu}, \quad (12.97)$$

and

$$c_\mu Ho'_\mu = \frac{i}{2}a \sum_{n_\mathrm{o}} (\tilde{b}^{(+)}_{n_\mathrm{o}N} + \tilde{b}^{(-)}_{n_\mathrm{o}N}) q_{n_\mathrm{o}} D_{n_\mathrm{o}\mu} - \frac{a}{2} \sum_{n_\mathrm{c}} \tilde{b}^{(-)}_{n_\mathrm{c}N} |q_{n_\mathrm{c}}| D_{n_\mathrm{c}\mu}. \quad (12.98)$$

From these expressions we have

$$\mathbf{b}^{(-)}_0 = [(\Omega N^{(N)}_{11} - N^{(N)}_{21})\Omega + \Omega N^{(N)}_{12} - N^{(N)}_{22}]^{-1}(N^{(N)}_{21} - \Omega N^{(N)}_{11}$$
$$\times (1 - A_{\mathrm{oo}})^{-1}(A_{\mathrm{oc}}\mathbf{B}_\mathrm{c} + \mathbf{B}_\mathrm{o}), \quad (12.99)$$

$$\mathbf{b}^{(+)}_0 = \Omega \mathbf{b}^{(-)}_0 + (1 - A_{\mathrm{oo}})^{-1}(A_{\mathrm{oc}}\mathbf{B}_\mathrm{c} + \mathbf{B}_\mathrm{o}), \quad (12.100)$$

$$\tilde{\mathbf{b}}^{(+)}_N = (N^{(N)}_{11}\Omega + N^{(N)}_{12})\mathbf{b}^{(-)}_0 + N^{(N)}_{11}(1 - A_{\mathrm{oo}})^{-1}(A_{\mathrm{oc}}\mathbf{B}_\mathrm{c} + \mathbf{B}_\mathrm{o}), \quad (12.101)$$

$$\tilde{\mathbf{b}}^{(-)}_N = (N^{(N)}_{21}\Omega + N^{(N)}_{22})\mathbf{b}^{(-)}_0 + N^{(N)}_{21}(1 - A_{\mathrm{oo}})^{-1}(A_{\mathrm{oc}}\mathbf{B}_\mathrm{c} + \mathbf{B}_\mathrm{o}), \quad (12.102)$$

where

$$\Omega = (1 - A_{\mathrm{oo}})^{-1}(1 + A_{\mathrm{oo}}). \quad (12.103)$$

The flux at $y = Y$ is defined as usual by

$$j_Y = \frac{\hbar}{m} \int_{-a/2}^{a/2} dx \, \Im\left[\psi^* \frac{\partial}{\partial y}\psi\right]_{y=Y}. \quad (12.104)$$

With use of the wave functions ψ_{I}, ψ^{II}_0, ψ^{II}_N, and ψ_{III} given by Eqs. (12.71), (12.73), and (12.86) we can obtain

$$j_0 = -\frac{\hbar}{m} \sum_\mu \{akE_\mu \Re[a^*_\mu Ho^*_\mu(0)] + |a_\mu|^2 \Im[Ho^*_\mu(0)Ho'_\mu(0)]\}, \quad (12.105)$$

$$j_{0+} = \frac{\hbar}{m} \sum_{n_\mathrm{o}} q_{n_\mathrm{o}}[|b^{(+)}_{n_\mathrm{o}0}|^2 - |b^{(-)}_{n_\mathrm{o}0}|^2], \quad (12.106)$$

$$j_{L^-} = \frac{\hbar}{m} \sum_{n_o} q_{n_o} [|\tilde{b}_{n_o N}^{(+)}|^2 - |\tilde{b}_{n_o N}^{(-)}|^2] \,, \qquad (12.107)$$

$$j_{L^+} = \frac{\hbar}{m} \sum_{\mu} c_\mu \Im[Ho_\mu^*(0)Ho_\mu'(0)] \,. \qquad (12.108)$$

The conservation of flux is quaranteed because of the symmetry of the transfer matrix and the completeness of the functions $\{\phi_n(x)\}$. Since the total initial flux j_{in} coming to the slit is equal to $\hbar ak/m$, the transmission coefficient is defined by

$$t \equiv \frac{j_{0^+}}{J_{in}} = \frac{1}{ak} \sum_{n_o} q_{n_o} [|b_{n_o 0}^{(+)}|^2 - |b_{n_o 0}^{(-)}|^2] \,. \qquad (12.109)$$

In order to comprehend the above formulation qualitatively, the following simplified analysis may be useful. The matrix Γ_Q represents the basis transformations between regions I and II and also between II and III. The above vector $\mathbf{B}$ gives a sort of initial condition. If we assume that there is only one open channel n_o (no closed channel) and $A_{oo} = -1$, then $\Omega = 0$ and from Eqs. (12.99) to (12.102) and (12.81) to (12.85) we have

$$b_0^{(-)} = -\frac{B_0 N_{21}^{(N)}}{2 N_{22}^{(N)}} \propto R_{n_o}^{(N)} \,, \qquad (12.110)$$

$$b_0^{(+)} = \frac{B_o}{2} \,, \qquad (12.111)$$

$$\tilde{b}_N^{(-)} = N_{22}^{(N)} b_0^{(-)} + N_{21}^{(N)} b_0^{(+)} = 0 \,, \qquad (12.112)$$

$$\tilde{b}_N^{(+)} = \frac{B_o}{2}\left(N_{11}^{(N)} - \frac{N_{12}^{(N)}}{N_{22}^{(N)}} N_{21}^{(N)} \right) = \frac{B_o}{2 N_{22}^{(N)}} \propto T_{n_o}^{(N)} \,. \qquad (12.113)$$

Thus, if we assume $B_o = 2a$ and $q_{n_o} = k$, then we can obtain

$$t = |T_{n_o}^{(N)}|^2 \,. \qquad (12.114)$$

12.4.   Numerical Examples

For the periodic potential in the y-direction the same functions given by Eqs. (12.56) as those used before in the case of one-dimensional model are employed. The parameters used here are $d = 0.007$, $b = 2.0$, $c = 10.0$, and $V_0 = 0.0014$ (in atomic units), which give the lowest complete reflection at $E \simeq 0.015$. The transmitting particle mass m and the distance l between the potential units (see Fig. 12.12) are taken to be 1000 and 0 in atomic units, respectively. The slit width a in the x-direction is assumed to be 3.0 a.u. Thus the eigenvalues E_n given by Eq. (12.76) are E_n $(n = 1, 2, \ldots) = 5.483 \times 10^{-4}$, 2.193×10^{-3}, 4.935×10^{-3}, 8.773×10^{-3}, 1.37×10^{-2}, $1.974 \times 10^{-2}, \ldots$, in which because of the symmetry only the odd order states $(n = 1, 3, 5, \ldots)$ are effective. Accordingly, only the odd order Mathieu functions $\{So_{2\mu+1}(h, \cos\eta)\}_\mu$ survive. Although the mass m was taken to be 1000, the transmitting particle is not necessary a proton, but can be an electron. Transfers of these particles take place effectively through the molecular motion, i.e., $AH^+ + B \to A + H^+B$ in the case of proton transfer and $A^- + B \to A + B^-$ in the case of electron transfer, where A and B represent certain atoms or molecules. It should be noted that the spatial coordinates (x, y) correspond to the relative nuclear (not the elctron) coordinates. The potential parameters given above are changed quite a bit from the values in the previous subsection and the potential crossing energy is made much lower. This was made, primarily because the numerical calculations can be simplified without losing the generality. This is not unrealistic at all, however, since excitation energies of big molecules like biological molecules are quite small. The Mathieu functions are calculated with use of the code written by Shirts [204]. Numerical check was made by comparing to references [202, 203] as much as possible. Other checks such as the orthogonalities of the coefficients $C_{n\mu}$ and $D_{n\mu}$ are also made. The maixum μ in the summation was taken to be 10, which was confirmed to be good enough within the range of the total energy considered here, which is $8 - 20 \times 10^{-3}$ a.u. The number of open channels with respect to n is up to 3, i.e., $n = 1, 3, 5$, in this energy range, and two closed channels were confirmed to be good enough for the convergence. The transfer matrix $N^{(N)}$ was evaluated with use of the semiclassical theory of Zhu–Nakamura.

Figures 12.13–12.15 show the results of the transmission coefficient t against the total energy E for one, three, and seven unit-systems. The potential parameters are the same as those given above. The lowest transmission peak, which will be utilized later for switching, appears at $E \sim 1.0 \times 10^{-2}$ a.u. In the three- and seven-unit cases, this peak naturally splits into two and six transmission peaks, which cannot be complete like in the one-dimensional case.

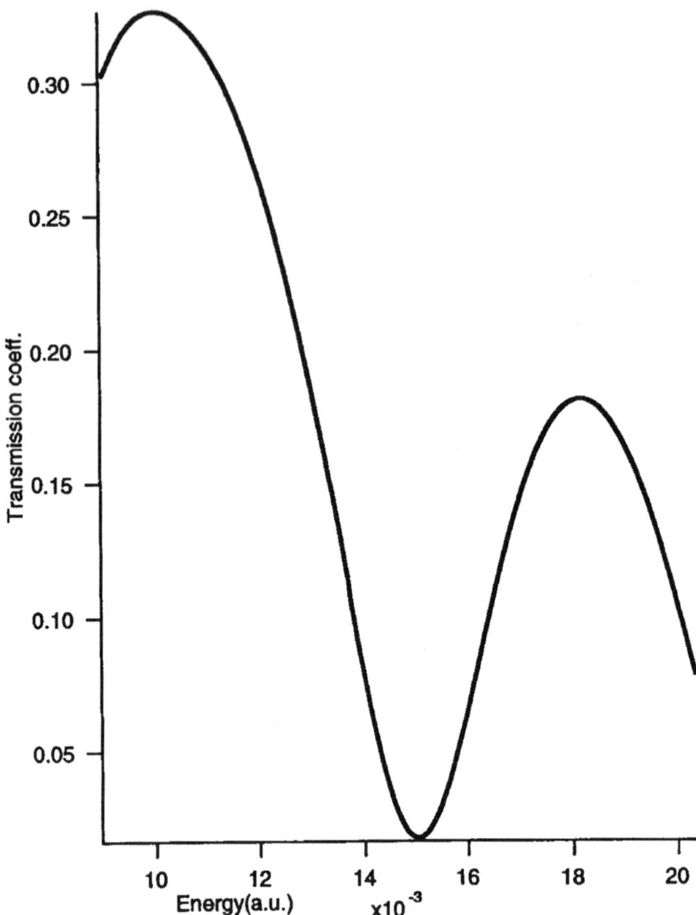

Fig. 12.13.   Transmission coefficient in the case of one-unit as a function of energy (a.u.). The potential parameters are given in the text. (Taken from Ref. [59] with permission.)

Figures 12.16 and 12.17 show the results of one- and two-unit systems composed only of impurity potentials, the parameters of which are chosen to be $d = 0.0044$, $b = 2.0$, $c = 10.0$, and $V_0 = 0.001$. The lowest reflection dip, which again cannot be complete, appears at $E \sim 1.05 \times 10^{-3}$ a.u. the two-unit system has s rather broad reflection band around $E \sim 1.0 \times 10^{-2}$ and may be appropriate to switch off the lowest transmission band in Fig. 12.15. Figure 12.18 shows the results of the seven-unit system out of which the right end two units

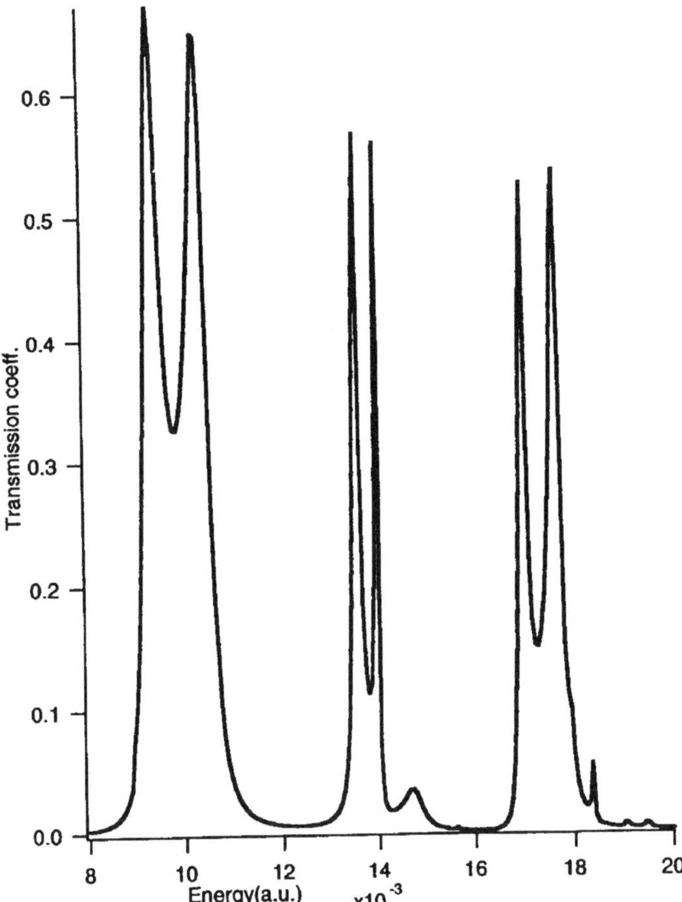

Fig. 12.14. The same as Fig. 12.13 for the seven-unit system. (Taken from Ref. [59] with permission.)

are impurities. The transmission band appearing around $E \sim 1.0 \times 10^{-2}$ in Fig. 12.15 is almost switched off. The impurity parameters were chosen just by a simple trial and error, but can, of course, be found more systematically to achieve much better switching. We can use the same prescription as that given previously, but this kind of fine tuning is not important here at this stage. What is important here is to demonstrate the possibility of molecular switching in a two-dimensional system.

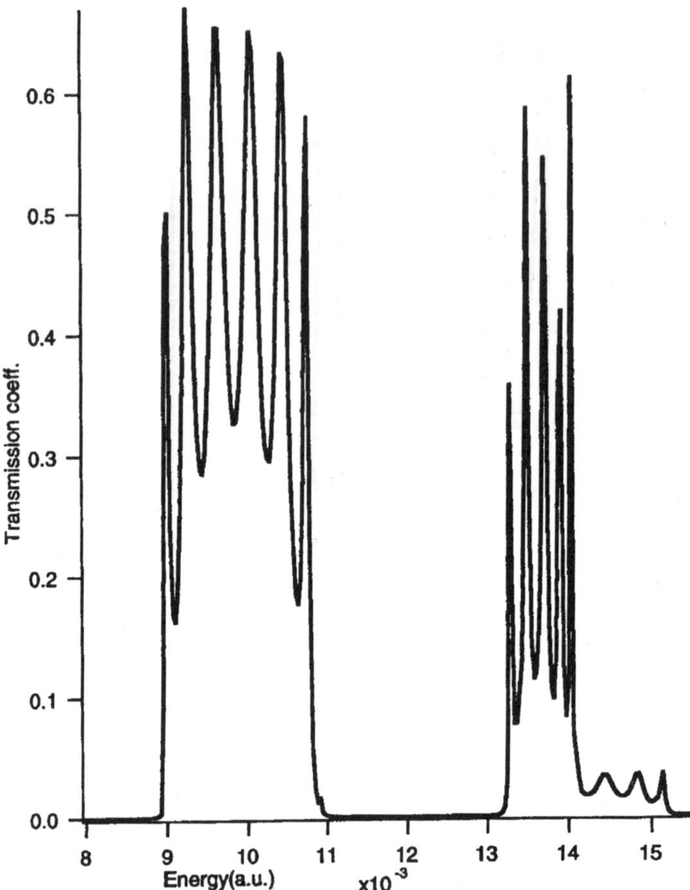

Fig. 12.15. The same as Fig. 12.13 for the seven-unit system. (Taken from Ref. [59] with permission.)

As was demonstrated in the above examples, both reflection and transmission cannot be complete in two-dimensional space, but the switching was shown to be made quite effective. The constriction model used here is rather special, but this enables us to formulate the problem analytically and is believed to present some essentail features of the multi-dimensional systems. Actually, the "completeness" is destroyed in multi-dimensional systems because of the nonseparability and the contribution from different transmission energy

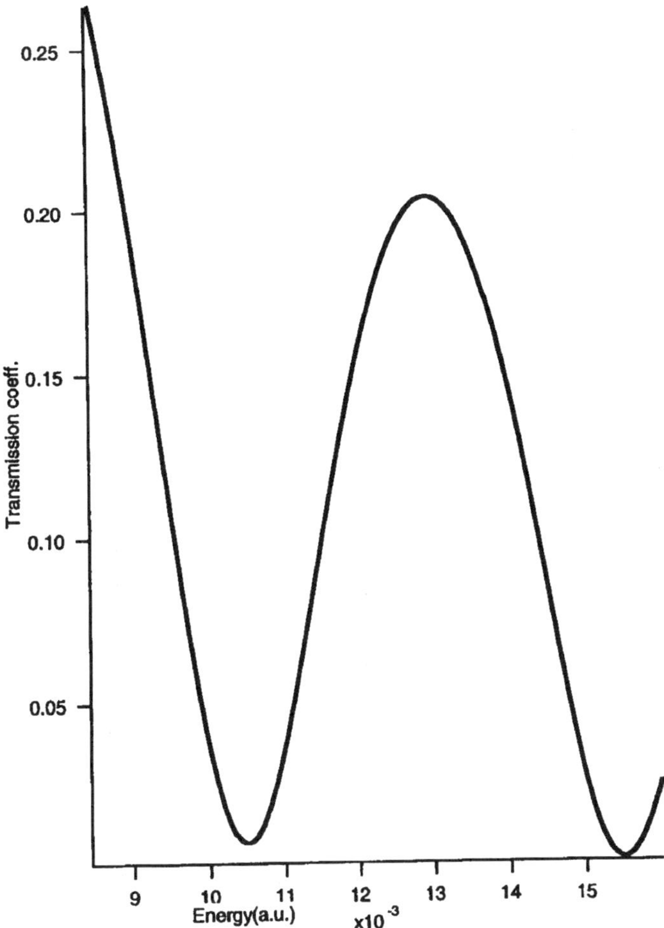

Fig. 12.16. Transmission coefficient in the case of one unit of impurity, the parameters of which are given in the text. (Taken from Ref. [59] with permission.)

components corresponding to the eigenstates in the direction perpendicular to the transmission. It is obvious to better keep the separability of the system as much as possible. It is also better to reduce the number of contributing eigen states in the perpendicular direction. In this sense it is recommended to use the lowest complete reflection, since the transmission of the other energy components is generally small there. Furthermore, in order to use the complete

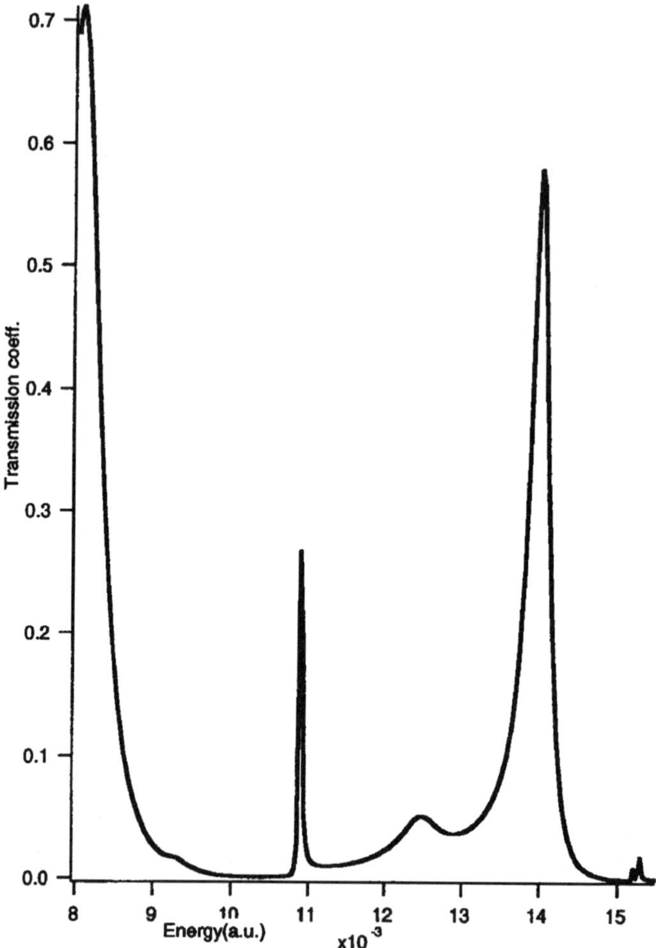

Fig. 12.17. The same as Fig. 12.16 for the two-unit system. (Taken from Ref. [59] with permission.)

reflection phenomenon effectively, it is desirable to design the system so that the level separation of the eigenstates in the perpendicular direction becomes equal to or close to the difference of the complete reflection positions.

As was mentioned before, the potential functions and the various parameters used here are just a mere example, and the theory presented here is generally applicable to other various cases, whenever the complete reflection

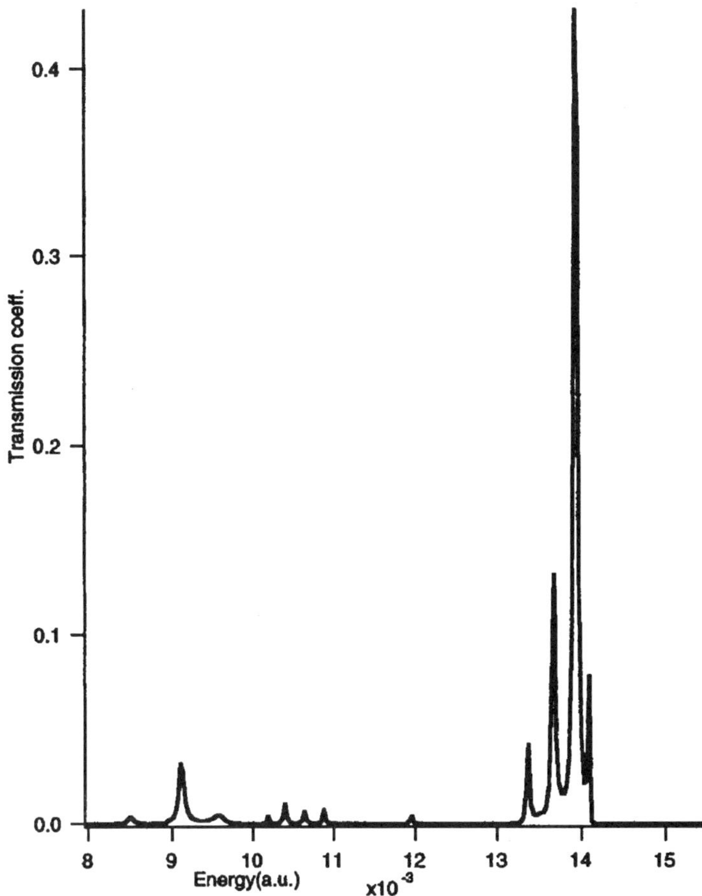

Fig. 12.18. Seven-unit system with two impurities. The two impurities are at the right-end of the system. This corresponds to the seven-unit system in Fig. 12.15, two out of which are replaced by the impurities here. The broad peak around $E \sim 1 \times 10^{-2}$ is almost switched off. (Taken from Ref. [59] with permission.)

phenomenon of the NT-typepotential can be effectively utilized. Unfortunately, it is not straightforward to scale all these parameters to be dimensionless; but the following arguments may be useful for designing such an effective switching in a different range of various parameters as that considered here. Since it is better to use the lowest complete reflection position U_*, i.e., $E_{\mathrm{tr}}/U_* \gtrsim 1$. Qualitatively speaking, the electronic excitation energy of molecule provides a rough measure of U_*. The representative translational energy E_{tr} may be given by $E - E_{n=1}$, where E is the initial total energy and $E_{n=1}$ represents the lowest excitation energy in the x-direction. On the other hand, since it is not appropriate to have many eigenstates in the x-direction contributing to the process, $E_{n_{\max}}/U_* \sim 1$ may be required, where $E_{n_{\max}}$ is again given by Eq. (12.76) and $n_{\max}$ is the maximum quantum number of open channels $(n_{\max} \sim 5$ or so$)$.

Chapter 13

Control of Nonadiabatic Processes by an External Field

Control of molecular processes or chemical dynamics, in general, is one of the most active fields in chemical physics recently, and attracts much attention theoretically as well as experimentally [25, 26]. This is basically because the recent remarkable progress in laser technology has opened new possibilities of realizing it. Several approaches have been proposed so far. One of these is the coherent control method originated by Brumer and Shapiro [205]. The essence of this method is to utilize the quantum mechanical interference effect to control the branching of a prepared state into possible decay channels with use of the phase and intensity of two laser pulses. Second example is the pump-dump scheme proposed by Tannor and Rice [206]. Their original idea was to use a laser pulse to create a localized wavepacket on a bound excited state. When the wavepacket comes to an appropriate position of the excited state, then the second laser pulse is used to make a transition to a final desirable state. The use of chirped (or frequency swept) laser pulse, originally proposed by Chelkowski and Bandrauk [207] is also one of the effective methods and has been studied by many authors [208–211]. Most of them, however, utilize the so called adiabatic rapid passage (ARP) [208, 212]. That is to say, by sweeping the field slowly, the system is kept to stay on the same adiabatic state. Methods to find the best pulse shape of lasers to control wave packet dynamics (optimal control theory) have been formulated and discussed by Kosloff *et al.* [213], Rabitz and coworkers [214], Kohler *et al.* [215], and Fujimura and coworkers [216]. The pulse shapes are optimized under various conditions. It is usually the case, however, that the optimized pulses have complicated structures in frequency as well as time domains. The so-called π-pulse method has also been proposed to utilize the phase effectively [217, 218]. In the above mentioned approaches, nonadiabatic transitions, in general, play important roles, because molecular processes induced by an external field, not only transitions among electronic states but also rovibrational transitions, may be considered as nonadiabatic transitions with the external field regarded as the adiabatic parameter. For instance, a rovibrational transition induced by a chirped laser field can be regarded as a nonadiabatic transition among adiabatic Floquet states [219, 220] by taking the laser frequency as the adiabatic parameter.

In this chapter, two new methods of controlling molecular processes are discussed. Both are based on the idea of controlling nonadiabatic transitions with use of the Floquet (or dressed) state picture. The first is to sweep an external field periodically at avoided crossings and to control nonadiabatic transitions as we desire [221–223]. The second is to use the intriguing phenomenon of complete reflection in the NT-type of potential curve crossing [224]. The first one are discussed in the following Secs. 13.1–13.3, and the second one is explained in Sec. 13.4.

13.1.   Control of Nonadiabatic Transitions by Periodically Sweeping External Field

Here, the new idea of controlling nonadiabatic transition which was first introduced by Teranishi and Nakamura [221] is explained and generalized with minimal mathematical description. In order to explain the essence of the idea, we take a simple two-state problem in which the Hamiltonian matrix depends on the fieldparameter F which may be intensity of an external field or the laser frequency.

For convenience, let us consider the simple Landau–Zener–Stueckelberg (LZS) type nonadiabatic transition for the moment. In the diabatic state representation, the diagonal terms, i.e. the diabatic states, cross with each other as a function of the time-dependent parameter F and the off-diagonal elements represent the diabatic coupling between them. The adiabatic states, which are obtained by diagonalizing this diabatic Hamiltonian matrix, avoid crossing at the original diabatic crossing point (solid lines in Fig. 13.1). The nonadiabatic transition, i.e. a transition between these adiabatic curves, is induced by the time-dependence of the matrix elements, and is well known to occur locally in a small region of F near the avoided crossing point. If the diabatic states are linear functions of time t and the diabatic coupling is

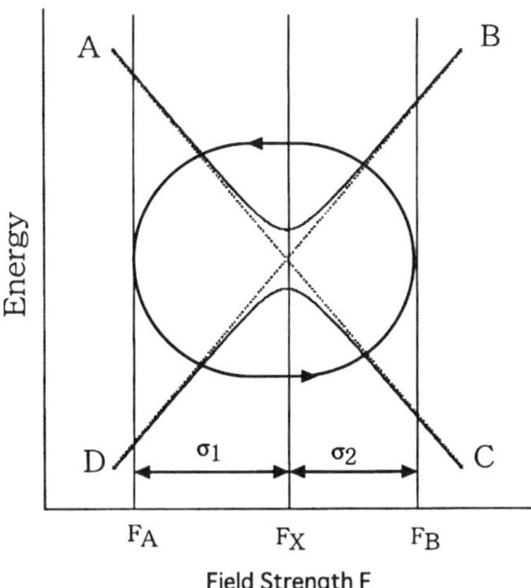

Fig. 13.1. Schematic two diabatic (dotted lines) and two adiabatic (solid lines) potentials. The external field oscillates between F_A and F_B, striding the avoided crossing point F_X. σ_1 and σ_2 represent the phases accumulated in the designated regions, and can be controlled by changing F_A and F_B. (Taken from Ref. [222] with permission.)

constant, then the nonadiabatic transition probability is given simply by the famous Landau–Zener formula, $p_{\mathrm{LZ}} = \exp[-2\pi V^2/\hbar\alpha]$, where V is the diabatic coupling strength proportional to the square root of the field intensity I, and α is the slope difference of the diabatic states proportional to the speed of field sweeping at F_X, i.e. $\alpha \propto \dot{F}_X \equiv (dF/dt)_{F_X}$.

As mentioned above, the adiabatic parameter F is either the field intensity I or the laser frequency ω. The probability p_{LZ} takes a value in between 0 and 1, and becomes lager (smaller) with increasing (decreasing) $\dot{F}_X$ or with decreasing (increasing) intensity I. Thus, the simplest idea of controlling the nonadiabatic transition would be to change p_{LZ} as we wish by changing the field intensity and/or the field sweeping speed. The adiabatic rapid passage (ARP) is such an example. It is not easy at all, however, to directly control p_{LZ} in a wide range. Unrealistically strong field, or unreasonably rapid or slow sweeping is required.

Instead, if we change the field parameter F periodically over a certain range (F_A, F_B), we can control transitions between any pair of two states of A and D in Fig. 13.1 as we like. This can be realized by controlling phases associated with the field oscillation. Take, for instance, the transition from A to D in

Fig. 13.1 by sweeping the field F for one period, i.e. F_A to F_B and back to F_A. Then the overall transition probability is given by

$$P_{AD} = 4p(1-p)\sin^2\psi, \qquad (13.1)$$

where p is the nonadiabatic transition probability to which p_{LZ} gives a rough approximation, and ψ is the phase difference between the two possible paths $A \to C \to D$ and $A \to B \to D$. The phase ψ can be controlled by changing the sweep range and thus the overall probability P_{AD} can be controlled to a desirable value between 0 and $4p(1-p)$. If p is equal to 0.5, we can control the transition as we wish. Even when p is not equal to 0.5, we can control P_{AD} to any desirable value as we wish by sweeping the field more than one period. That is to say, n-time periodical sweeping of the field creates 2^n multiple paths between the initial and final states and their phase interference can be controlled by changing the amplitude and the number n of the oscillation. If we want to control a transition from a state on the left side (A or D) to a state on the right (B or C) in Fig. 13.1, then we have to sweep the field $(n + 1/2)$ periods. The necessary conditions for the control parameters, i.e. $\dot{F}_X$, (F_A, F_B), and n, can be formulated analytically and evaluated accurately with use of the new theory of time-dependent curve crossing problems developed in Ref. [71] based on the Zhu–Nakamura theory for the time-independent LZS problems described in Chap. 5. This theory enables us to treat even the case $F_X \geq F_B$ in Fig. 13.1.

If p is very close to unity and we still want to realize an overall adiabatic passage, i.e. a transition from A to A (or D to D) or from A to B (or D to C) in Fig. 13.1 with unit probability, a large number of oscillations (n) are required as may be easily conjectured. The large number of oscillation is also required when p is very small and we want to realize an overall transition from A to C (or D to B) in Fig. 13.1 with unit probability. In this case, however, it would probably be possible to increase p substantially by simply reducing the field intensity to diminish the diabatic coupling strength. When F is the laser frequency ω, it is convenient to use the Floquet state representation in which the dressed states are drawn as a function of ω [24, 220]. The frequency sweeping in this case is similar to the idea of chirped laser [207]. In the present case, however, we sweep the field periodically and can find appropriate values of the control parameters explicitly with the help of analytical theory. Thus the present method is quite effective and actually can make the control perfect in principle.

The appropriate sweeping velocity as one of the control parameters varies in a wide range from problem to problem. In the case of the spin tunneling in a magnetic field discussed in Ref. [221] the time variation of the magnetic field may be carried out in a macroscopic time-scale. In the case of laser control of molecular processes, on the other hand, the control is realized in the time-scale of laser pulse, i.e. in nano- to femto-second time-scale.

The above mentioned basic idea can, of course, be applied to a general multi-level system. Transitions at each avoided crossing can be controlled perfectly, in principle, and we can even specify a path from an initial state to any desired final state. One difficulty of the present method, however, arises when the density of avoided crossings is high and the oscillation amplitudes at each crossing overlap to each other. In this case it is better to treat the bunch of avoided crossings as a whole, although its analytical treatment naturally becomes less straightforward. As mentioned above, the number of field oscillation becomes large, when the nonadiabatic transition probability p, for which we can use our new theory [71] which is much better than the Landau–Zener formula, is very close to unity and yet we want an overall adiabatic passage. A large number of oscillation could easily cause an experimental error and is better to be avoided. In this case we can think of using the non-crossing Rosen–Zener–Demkov (RZD) type nonadiabatic transition by employing the laser intensity as the adiabatic parameter F. The adiabatic potential curves become like Fig. 13.2. The nonadiabatic transition probability p_{RZ} in the original RZD model is given by $p_{\mathrm{RZ}} = [1 + \exp(\pi \Delta E/\hbar\beta)]^{-1}$, where ΔE is the constant energy separation of two diabatic states and β is the exponent of the square root of intensity, i.e. $\sqrt{F} \propto e^{\beta t}$. The analytical expressions of nonadiabatic transition probability and phases are naturally different from those of the LZS case, but the basic qualitative idea of sweeping field and the analytical treatment of the whole process are the same as before. It should be noted that F_X in Fig. 13.2 is the real part of the complex crossing point of the adiabatic potentials. The frequency ω is fixed at the value of F_X in Fig. 13.1, namely $\omega = \omega_X$. $p \simeq 1$ means that ΔE in Fig. 13.2 that is equal to the level separation at $\omega = \omega_X$ is small and thus $p_{\mathrm{RZ}} \simeq 0.5$. This is a big advantage of the RZD case, and what we have to do is to earn the necessary phase by sweeping the intensity. In the limit of $\Delta E = 0$, we recover the so called π-pulse [218]. It should be noted that the overall transition probability after one period of oscillation is again given by Eq. (13.1) with p replaced by p_{RZ}. As can be easily thought of from the above discussions, in the case of laser we may utilize both

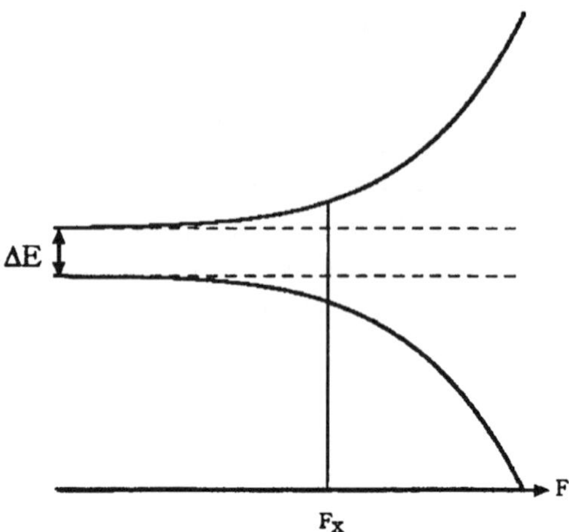

Fig. 13.2.   Schematic two diabatic (dotted lines) and two adiabatic (solid lines) potentials of the Rosen–Zener–Demkov model. Point F_X corresponds to the real part of the complex crossing point of two adiabatic potentials. (Taken from Ref. [222] with permission.)

intensity and frequency as the adiabatic parameters at the same time, and find an appropriate path of the adiabatic parameter F in the two-dimensional (I, ω) space. Nonadiabatic transitions are also not necessarily restricted only to the LZS and the RZD cases, but can be any other general type such as the exponential models [4, 67, 69], although the availability of analytical theory is limited.

13.2.   Basic Theory

In this section, we present the mathematical formulation of the idea presented in the previous section. The most basic quantity is the following transition matrix I which describes a transition from F_A to F_B in Fig. 13.1 (this I should not be confused with the intensity I) [221]:

$$I = \begin{bmatrix} \sqrt{1-p}\,e^{i(\phi_S+\sigma_1/2+\sigma_2/2)} & -\sqrt{p}\,e^{i(\sigma_0-\sigma_1/2+\sigma_2/2)} \\ \sqrt{p}\,e^{-i(\sigma_0-\sigma_1/2+\sigma_2/2)} & \sqrt{1-p}\,e^{-i(\phi_S+\sigma_1/2+\sigma_2/2)} \end{bmatrix}. \tag{13.2}$$

The (i, j) element of this matrix provides the transition amplitude for the transition $j \rightarrow i$ ($i, j = 1, 2$) when the field changes from F_A to F_B, where $i = 1$ (2) represents the lower (upper) adiabatic state. The backward transition from F_B to F_A can be described by the transpose of this matrix, I^t. Here p is the same as before, representing the nonadiabatic transition probability by one passage through the transition region, ϕ_S (Stokes phase) (see Eq. (5.97)) and σ_0 are the phase factors due to the nonadiabatic transition and their explicit expressions are given before. The phases σ_1 and σ_2 are the phase factors which describe, respectively, the adiabatic propagation in the region (F_A, F_X) and (F_X, F_B) in Fig. 13.1. They are given by

$$\sigma_1 = \int_{F_A}^{F_X} \Delta E(F) \frac{dt}{dF} dF = \int_{t_A}^{t_X} \Delta E(t) dt \,, \tag{13.3}$$

and

$$\sigma_2 = \int_{F_X}^{F_B} \Delta E(F) \frac{dt}{dF} dF = \int_{t_X}^{t_B} \Delta E(t) dt \,, \tag{13.4}$$

where $\Delta E(F)$ is the adiabatic energy difference as a function of the adiabatic parameter F. The time t_α ($\alpha = A, B, X$) is the time at which $F(t_\alpha) = F_\alpha$ is satisfied. The final overall transition matrix after n periods of oscillation between F_A and F_B is expressed as

$$T_n = T^n \,, \tag{13.5}$$

where T is the transition matrix for one period of oscillation and is given by

$$T \equiv I^t I = \begin{bmatrix} \{p + (1-p)e^{2i\psi}\}e^{-i\sigma} & -2i\sqrt{p(1-p)}\sin\psi \\ -2i\sqrt{p(1-p)}\sin\psi & \{p + (1-p)e^{-2i\psi}\}e^{i\sigma} \end{bmatrix} \tag{13.6}$$

with $\psi \equiv \phi_S + \sigma_0 + \sigma_2$ and $\sigma \equiv 2\sigma_0 + \sigma_2 - \sigma_1$.[a] From Eqs. (13.5) and (13.6), the final transition probability after n periods of oscillation can be written as

$$P_{12}^{(n)} \equiv |(T_n)_{12}|^2 = |(T^n)_{12}|^2 \,. \tag{13.7}$$

Using the Lagrange–Sylvester formula, we obtain

$$T_n = \frac{\lambda_+ \lambda_- (\lambda_-^{n-1} - \lambda_+^{n-1})}{\lambda_+ - \lambda_-} E + \frac{\lambda_+^n - \lambda_-^n}{\lambda_+ - \lambda_-} T \,, \tag{13.8}$$

---

[a]We would like to take this opportunity to point out that Eqs. (1), (3), and (4), and the definitions of ψ and σ in Ref. [221] are not correct and should be replaced by the corresponding expressions given here.

where E is the unit matrix and $\lambda_\pm$ are the eigenvalues of T, which are given by

$$\lambda_\pm = e^{\pm i\xi}, \tag{13.9}$$

where

$$\cos\xi = (1-p)\cos(2\psi - \sigma) + p\cos(\sigma). \tag{13.10}$$

The unitarity of the matrix T requires ξ to be real. Equation (13.10) implies that the nonadiabatic transition probability p should satisfy

$$\frac{1 - |\cos\xi|}{2} \le p \le \frac{1 + |\cos\xi|}{2}. \tag{13.11}$$

From Eqs. (13.7) and (13.8), we can write $P_{12}^{(n)}$ as

$$P_{12}^{(n)} = 4\frac{\sin^2(n\xi)}{\sin^2\xi}p(1-p)\sin^2\psi. \tag{13.12}$$

Thus the conditions for $P_{12}^{(1)} = 0$ are $p = 0$, $p = 1$, or $\sin\psi = 0$, and the condition for $P_{12}^{(n)} = 0 (n > 1)$ is $\sin(n\xi) = 0$.

The conditions for $P_{12}^{(n)} = 1$ may be derived as follows. When $P_{12}^{(n)} = 1$ is satisfied, the diagonal terms of the transition matrix $T^{(n)}$ are zero and the transition matrix $T^{(2n)} = T^{(n)}T^{(n)}$ becomes diagonal $(P_{12}^{(2n)} = 0)$. Thus the condition for $P_{12}^{(n)} = 1$ may be divided into the following two equations:

$$\sin^2(n\xi) = 1 \tag{13.13}$$

and

$$4p(1-p)\sin^2\psi = \sin^2\xi. \tag{13.14}$$

From Eqs. (13.11) and (13.12), we can estimate the number of oscillation n for a given p as the minimum integer which satisfies the following condition (see Fig. 13.3):

$$\sin^2\frac{\pi}{2n} \le 4p(1-p). \tag{13.15}$$

For a given n, Eq. (13.13) gives ξ, and ψ can be determined according to Eq. (13.14) for given ξ and p. Now, from the definitions of ψ and σ, we can find the proper values of the two phase factors σ_1 and σ_2 to achieve $P_{12}^{(n)} = 1$, which can be adjusted by changing F_A and F_B.

Similar analysis can be done for n and half periods of oscillation. In this case, the final transition probability $P_{12}^{(n+1/2)}$ is given by

$$P_{12}^{(n+1/2)} \equiv |I(I^t I)^n|^2 \,. \tag{13.16}$$

For both $P_{12}^{(n+1/2)} = 0$ and $P_{12}^{(n+1/2)} = 1$, the transition matrix after $2n+1$ period $T^{(2n+1)} = (I(I^t I)^n)^t I(I^t I)^n$ becomes diagonal. Thus we have

$$\sin((2n+1)\xi) = 0 \,. \tag{13.17}$$

Substituting Eq. (13.17) into the condition $P_{12}^{(n+1/2)} = 0$ with help of Eq. (13.16), we obtain

$$4(1-p)\sin^2(\psi - \sigma) = \frac{\sin^2 \xi}{\sin^2(n\xi)} \,. \tag{13.18}$$

On the other hand, for $P_{12}^{(n+1/2)} = 1$ we have

$$4p\sin^2(\psi - \sigma) = \frac{\sin^2 \xi}{\sin^2(n\xi)} \,. \tag{13.19}$$

In the case of both $P_{12}^{(n+1/2)} = 0$ and $P_{12}^{(n+1/2)} = 1$, n may be estimated as the minimum integer which satisfies the following condition (see Fig. 13.3):

$$\sin^2 \frac{\pi}{2n+1} \leq 4p(1-p) \,. \tag{13.20}$$

The phases σ_1 and σ_2 are found just in the same way as in the case $P_{12}^{(n)} = 1$.

Now we can achieve the unit (or zero) overall transition probability for any p, ϕ_S and σ_0 by adjusting the number of oscillation n (or n and half) and the width of the oscillation F_A and F_B. In a real experiment, however, it may be difficult to adjust the parameters exactly. Thus, it is necessary to pay some attention to the stability of the final transition probability against fluctuations in laser intensity and frequency. Let us first look into the sensitivity of the final transition probability against variations in p and ψ. In the case of one period of oscillation, the overall transition probability is given by (see Eq. (13.1))

$$P_{12}^{(1)} = 4p(1-p)\sin^2 \psi \,, \tag{13.21}$$

and $P_{12}^{(1)} = 1$ is achieved when $p = 0.5$ and $\psi = \pi/2 + m\pi$. If we assume the required accuracy of $P_{12}^{(1)}$ is 97%, p should be in the range $0.4134 < p < 5.866$ for the accurate ψ, while ψ may contain about 11% relative error against $\pi/2$

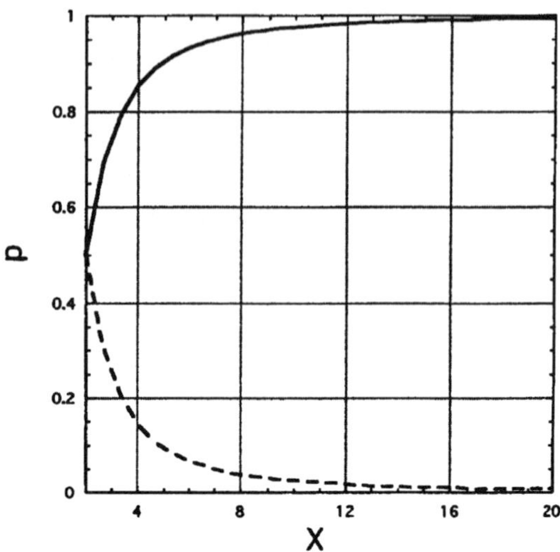

Fig. 13.3. The solid and dotted curves represent $(1 + |\cos(\pi/(X))|)/2$ and $(1 - |\cos(\pi/(X))|)/2$, respectively. The range between these two curves represents the range of p to fulfill Eq. (13.11), namely the range in which the complete control is achievable by n-times of field oscillation. In the case of n-period $X = 2n$; while $X = 2n + 1$ for the n and half periods of oscillation of the field. (Taken from Ref. [222] with permission.)

for the accurate p. When we increase the number of oscillation, the error in both ψ and p may be magnified mainly due to the term $\sin^2(n\xi)$ in Eq. (13.12). Sensitivities of p and ψ against the errors in the intensity and the frequency depend on the type of nonadiabatic transition. This will be discussed in the following subsections, in each of which we use a particular type of nonadiabatic transition.

If, for some reasons, one of the boundary, say F_B, cannot be varied in a range wide enough to satisfy the control conditions like Eqs. (13.13) and (13.14), we can utilize such a double-period of oscillation that F oscillates between F_{A1} and F_B for the first period and between F_{A2} and F_B for the second period as is depicted in Fig. 13.4. Even in this case, exactly the same analysis as that given above holds with the definitions of the matrix I and several parameters changed as described below. The mathematical expression of the I-matrix itself can be kept the same as that given by Eq. (13.2), because it is a general expression of unitary matrix. In the present double periodic case, however, this I-matrix describes the sequence of transitions $F_{A1} \to F_X \to$

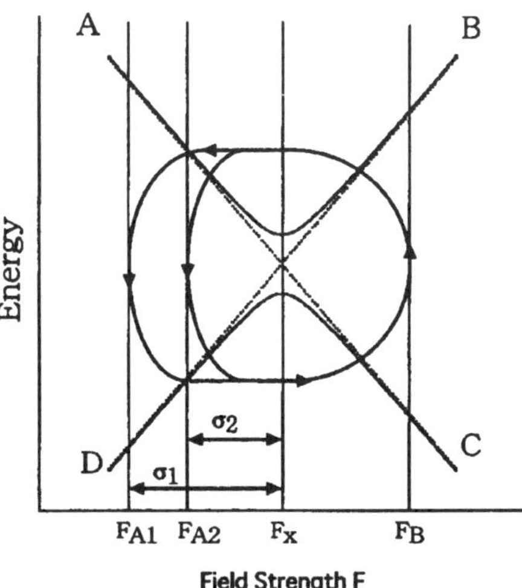

Fig. 13.4. Schematic diagram of double-period oscillation of external field. Two diabatic (dotted lines) and two adiabaic (solid lines) potentials are shown. The external field oscillates as: $F_{A1} \to F_X \to F_B \to F_X \to F_{A2} \to F_X \to F_B \to F_X \to F_{A1}$ striding the avoided crossing point F_X twice. Phase $\sigma_1(\sigma_2)$ can be controlled by changing $F_{A1}(F_{A2})$. (Taken from Ref. [222] with permission.)

$F_B \to F_X \to F_{A2}$, and its transpose I^t describes the inverse process, $F_{A2} \to F_X \to F_B \to F_X \to F_{A1}$. Thus the transition matrix for one double-period of oscillation is expressed by Eq. (13.6), where σ_1 and σ_2 are redefined by

$$\sigma_1 = \int_{F_{A1}}^{F_X} \Delta E(F) \frac{dt}{dF} dF = \int_{t_{A1}}^{t_X} \Delta E(t) dt \tag{13.22}$$

and

$$\sigma_2 = \int_{F_{A2}}^{F_X} \Delta E(F) \frac{dt}{dF} dF = \int_{t_{A2}}^{t_X} \Delta E(t) dt\,, \tag{13.23}$$

while p, ϕ_S and σ_0 should be redefined in such a way that the following matrix J describes the process $F_X \to F_B \to F_X$ with the two times of nonadiabatic transitions at F_X included:

$$J = \begin{bmatrix} \sqrt{1-p}\,e^{i\phi_S} & -\sqrt{p}\,e^{i\sigma_0} \\ \sqrt{p}\,e^{-i\sigma_0} & \sqrt{1-p}\,e^{-i\phi_S} \end{bmatrix}. \tag{13.24}$$

This matrix J corresponds to the reduced transition matrix T^R in the quadratic potential model in Ref. [71]. As mentioned above, the parameters p, ϕ_S, and σ_0 which originally represent the quantities arising from one passage of F_X, now correspond to the two passages of the transition point. We can now achieve unit (or zero) final transition probability by adjusting F_{A1}, F_{A2} and the number of oscillation for given values of all the other parameters including F_B.

So far, we have presented only the general framework of our idea of controlling molecular nonadiabatic processes. As mentioned before, however, in some cases the necessary number n of oscillation becomes very large, which could cause an experimental error and is not very convenient. If the nonadiabatic transition probability p is close to 0.5, then a small number of oscillation is enough to achieve our desire. This suggests that it would be very useful to try to control p by changing the functionality of the adiabatic parameter $F(t)$. In the case of laser, the Hamiltonian matrix is given by

$$H = \begin{bmatrix} n\hbar\omega(t) + E_1 & \epsilon E(t)/2 \\ \epsilon E(t)/2 & m\hbar\omega(t) + E_2 \end{bmatrix}, \tag{13.25}$$

where $\omega(t)$ and $E(t)(= \sqrt{I(t)})$ are the laser frequency and the intensity of the electric field, respectively, as a function of time, and ϵ is the dipole transition matrix element. This tells that we can use both intensity I and frequency ω as adiabatic parameters at the same time and utilize various schemes of nonadiabatic transitions, not just sticking to the LZS type. In the subsequent subsections we will discuss various possibilities.

13.2.1.   *Usage of the Landau–Zener–Stueckelberg type transition*

If the system has a clear avoided crossing as a function of the adiabatic parameter, the Landau–Zener–Stueckelberg type transition can be utilized. The Hamiltonian of the simplest linear potential model of Landau–Zener is given by

$$H_{\mathrm{LZ}} = \begin{bmatrix} \alpha_1 t & V \\ V & \alpha_2 t \end{bmatrix}, \tag{13.26}$$

where V is the constant diabatic coupling and t is the time. In the case of laser, this model corresponds to constant intensity and linear sweeping of the frequency. This model is good enough to explain qualitative features of the control scheme. The nonadiabatic transition occurs at $t = 0$ with the transition probability p given by

$$p_{\mathrm{LZ}} = \exp\left(-2\pi\frac{V^2}{\hbar\alpha}\right) = \exp\left(-2\pi\frac{I\epsilon^2}{\hbar^2|l-m|\dot{\omega}}\right), \qquad (13.27)$$

where $\alpha \equiv |\alpha_1 - \alpha_2|$, I and $\dot{\omega}$ are the laser intensity and the sweep velocity of the frequency at the avoided crossing, respectively, l and m (can be negative) are the photon numbers, and ϵ is the dipole matrix element. The conventional adiabatic passage requires large laser intensity and small sweep velocity to make p_{LZ} very small. For instance, the adiabatic passage with $p_{\mathrm{LZ}} \leq 0.001$ requires

$$1.0994\hbar\dot{\omega}/\epsilon \leq I. \qquad (13.28)$$

For the values $\epsilon = 1.0$ (Åe) and $\dot{\omega} = 0.5$ (cm^{-1}/ps), I must be larger than 1.0 (TW/cm^2).

In order to accomplish the passage in a reasonably short time scale with relatively large $\dot{\omega}$, a large intensity I or a large number of oscillation is required. If $p_{\mathrm{LZ}} = 0.5$ can be attained without difficulty, then one period of oscillation enables us to achieve exactly zero or unit final transition probability. The required intensity in this case is given by

$$I = 0.1103\hbar\dot{\omega}/\epsilon. \qquad (13.29)$$

Namely, one period of oscillation requires the intensity by one order smaller compared to the case of one passage for the same ϵ and $\dot{\omega}$. Furthermore, ten periods of oscillation require the following condition (see Eq. (13.12)):

$$0.982 \times 10^{-3}\hbar\dot{\omega}/\epsilon \leq I \leq 0.810\hbar\dot{\omega}/\epsilon. \qquad (13.30)$$

This requires only one-thousandth of the intensity required in the case of one passage. Many periods of oscillation, however, requires high accuracy of p and phases as discussed in the previous section. When p is large, it is sensitive to the error in the exponent which is proportional to the intensity. In the case of adiabatic passage, on the other hand, p is relatively stable against the error in the exponent, since the exponent is large. When $p \simeq 0.5$, about 15% of error in the exponent yields the fluctuation in the range $0.45 < p_{\mathrm{LZ}} < 0.65$.

So far, our discussion is based on the simple model Eq. (13.26). For finding the actual parameters, however, it is much better to use the sophisticated theory developed in Ref. [71], because the theory is applicable to general functonality of ω, even if the two diabatic potentials touch each other or avoid crossing.

13.2.2.   *Usage of the Rosen–Zener–Demkov type transition*

The Hamiltonian of the Rosen–Zener–Demkov model in the diabatic representation is given by

$$
H_{\mathrm{RZ}} = \begin{bmatrix} -\Delta/2 & Ae^{\beta t} \\ Ae^{\beta t} & \Delta/2 \end{bmatrix} .
\tag{13.31}
$$

This model describes the process with constant frequency and exponentially rising intensity in the case of laser. The nonadiabatic transition occurs at $t = \log(\Delta/A)/\beta$, and the transition probability p is given by

$$
p_{\mathrm{RZ}} = \frac{1}{1 + e^{2\delta}} ,
\tag{13.32}
$$

where

$$
\delta \equiv \frac{\pi\Delta}{2\hbar\beta} .
\tag{13.33}
$$

The analytical expression of I-matrix is given by Eq. (3.36). As is clearly seen from Eqs. (13.32) and (13.33), the range of p is $0 \leq p \leq 0.5$, and $p \to 0$ when $\Delta/\beta \to \infty$; while $p \to 0.5$ when $\Delta/\beta \to 0$. It should be noted that p_{RZ} does depend only on Δ/β and not on the laser intensity A, and that $p \simeq 0.5$ can be achieved even with very small intensity (small A) in short time with large β. Thus, by adjusting the phases, we can achieve unit final transition probability with use of one laser pulse of the shape $E(t) = 2A\,\mathrm{sech}(\beta t)/\epsilon$, namely by one period of oscillation. This process of one pulse with $\Delta = 0$ corresponds to the π-pulse. The condition $\Delta = 0$ leads to $p_{\mathrm{RZ}} = 0.5$ and thus $P_{12}^{(n)}$ given by Eq. (13.12) with $n = 1$ reduces to $\sin^2\psi = 1$, which coincides with the condition of the area of π-pulse. It should be noted that our theory is quite general, and that the condition for unit final transition probability can be attained for any frequency, if we use more than one pulses.

The sensitivity of p_{RZ} against an error in δ is largest when $p_{\mathrm{RZ}} = 0.5$, and decreases as p_{RZ} decreases. In the case of π-pulse ($\Delta = 0$), a constant shift of the frequency Δ from zero yields relatively large effect on p_{RZ} when β or pulse width is small. If the pulse width is, say, several pico seconds, about several cm^{-1} of shift in Δ yields $p_{\mathrm{RZ}} \simeq 0.45$. It should be noted, however, that a small fluctuation of Δ in time (not a constant shift) causes a large error in the nonadiabatic transition probability p. Numerical examples are shown in the next section.

We can choose either the LZS type (oscillation of the frequency in the case of laser) or the RZD type (oscillation of the intensity) depending on the experimental conditions. One thing we should keep in mind is that the LZS type requires large intensity to achieve $p = 0.5$, while the RZD (including the π-pulse) requires large intensity to satisfy the phase condition. In other words, the LZS does not require large intensity to satisfy the phase condition and the RZD does not require that to satisfy $p = 0.5$. Thus we may think of a hybrid of LZS and RZD which enables us to achieve $p = 0.5$ by changing the intensity and to accumulate enough phase by changing the frequency. We will discuss this in the next subsection.

13.2.3. *General case*

The well known model which contains time-variations in both diabatic energy and diabatic coupling is the exponential model [4, 67, 69]. The Hamiltonian of this model is given by

$$H_{\exp} = \begin{bmatrix} U_1 + V_1 e^{-\beta t} & V e^{-\beta t} \\ V e^{-\beta t} & U_2 + V_2 e^{-\beta t} \end{bmatrix}.$$ (13.34)

In the case of laser, this model describes the process of exponentially changing intensity and frequency with the same exponent. The nonadiabatic transition probability in this model is given by [69] (see Sec. 5.4 also)

$$p_{\exp} = \frac{\exp(-\pi\delta_2)\sinh(\pi\delta_1)}{\sinh[\pi(\delta_1 + \delta_2)]},$$ (13.35)

where

$$\delta_1 = \frac{U_1 - U_2}{2\hbar\beta}\left\{1 + \frac{(V_1 - V_2)/2V}{\sqrt{1 + [(V_1 - V_2)/2V]^2}}\right\}$$ (13.36)

and

$$\delta_2 = \frac{U_1 - U_2}{2\hbar\beta}\left\{1 - \frac{(V_1 - V_2)/2V}{\sqrt{1 + [(V_1 - V_2)/2V]^2}}\right\}. \qquad (13.37)$$

In the same way as in the RZD case, we can achieve $p = 0.5$ within a relatively short time by small intensity. The necessary phase, on the other hand, may be accumulated by changing not only the intensity but also the frequency. This means that large intensity is not necessary to accumulate the necessary phase, and we can achieve unit transition probability with small intensity by one period of oscillation (one pulse). The control scheme with use of this exponential model may provide one of the most effective ones (rapid transition with small intensity). A model with different exponents for the diabatic energy and the diabatic coupling would be more versatile and useful, although there is no analytical theory available yet. It should, however, be noted that the transition probability p_{exp} is rather sensitive to the functionalities of both intensity and frequency as a function of time, and a small experimental error might affect the final overall transition probability. Comparative studies on the effectiveness and the stability of the various control schemes described here are made in the next section together with the presentation of numerical examples.

13.3.   Numerical Examples

13.3.1.   *Spin tunneling by a magnetic field*

The idea mentioned above is not restricted only to laser field but can be applied to any time dependent external field. Here, as an example of multilevel crossing, the quantum tunneling of the magnetization of $Mn_{12}Ac$ in a magnetic field [225–227] is considered. Figure 13.5 shows the corresponding three-level system, the Hamiltonian of which is taken from Eq. (1) of Ref. [227].

In this figure, the energy is scaled by the anisotropy energy D. Figure 13.6 shows the time evolution of the state probability from the point "a" on state 1 to "b" on state 3 via the two avoided crossings A and B. At the avoided crossing $A(B)$ four-(five-) period oscillation of the filed are applied. This is shown in Fig. 13.6(a). The probability of the state 1 (P_1) becomes zero after the four periods, as is seen in Fig. 13.6(b). The probability P_2 reaches unity when P_1 becomes zero, and after five periods at B it becomes zero (Fig. 13.6(c)) at

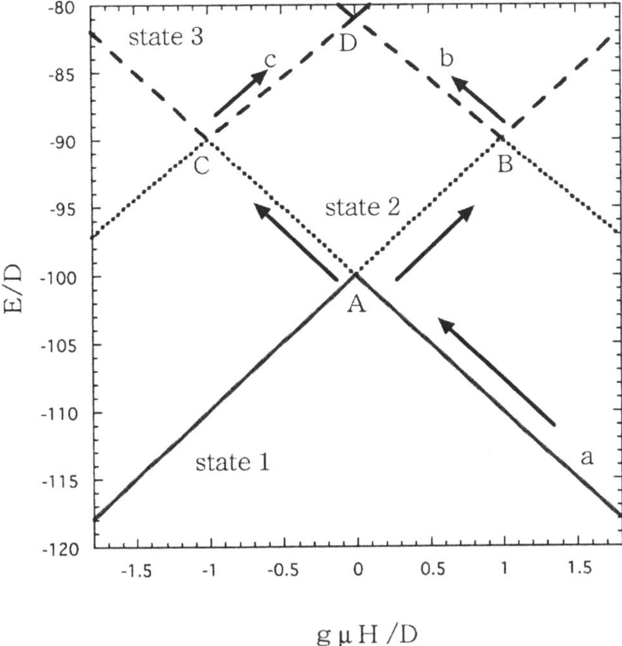

Fig. 13.5. Adiabatic spin states as a function of an external magnetic field, $g\mu_\beta H$: Three lowest levels (scaled by the anisotropy energy D) are shown, where g is the Landé g-factor, μ_β is the Bohr magneton, and H is the magnetic field. The nonadiabatic probabilities p at avoided crossings $A - C$ are 0.039, 0.977, and 0.977, respectively. (Taken from Ref. [221] with permission.)

which time P_3 reaches unity (Fig. 13.6(d)). Figure 13.7 demonstrates another path from "a" on state 1 to "c" on state 3 via two avoided crossings A and C. In this case we have applied $4 + 1/2$ (five) periods of oscillation of the field at the avoided crossing $A(C)$. This example clearly demonstrates that we can choose any path to reach any specified final state with unity probability. Figures 13.6 and 13.7 are the results of the numerical calculations of the coupled Schrödinger equations, but it is confirmed that the semiclassical theory descrived in Chap. 6 [71] gives results almost indistinguishable from Figs. 13.6 and 13.7 except for humps and dips which appear when the probability jumps abruptly. This guarantees that we do not have to solve multichannel coupled equations numerically, and that we can formulate all necessary conditions of control analytically.

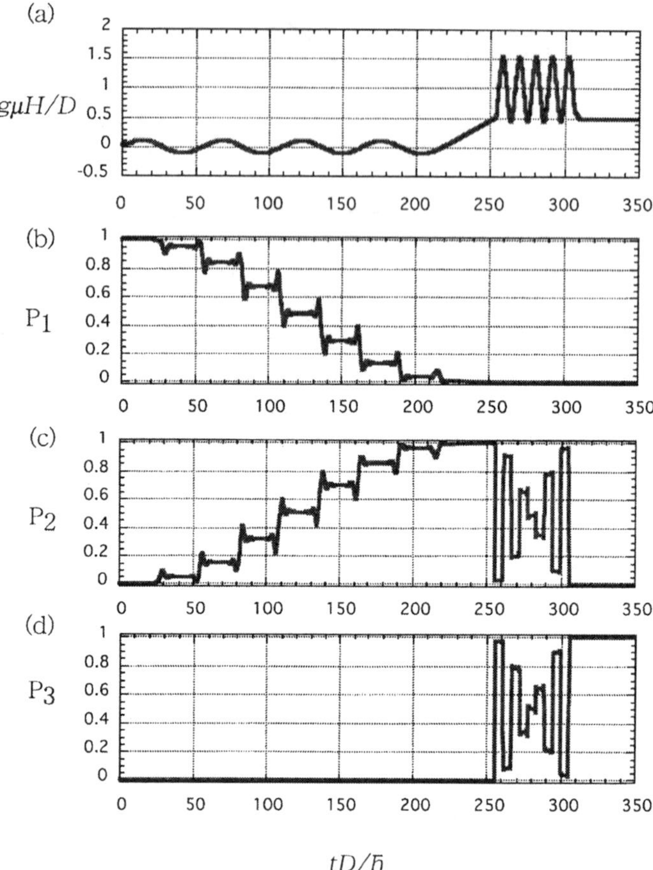

Fig. 13.6.   Controlled nonadiabatic processes, starting from "a" on state 1 and ending at "b" on state 3 via avoided crossings A and B (see Fig. 13.5). (a) Variation of the external magnetic field as a function on time, $tD/\hbar$. The first four-period oscillation around $g\mu_\beta H/D = 0$ corresponds to the control at the avoided crossing A. The second five-period oscillation around $g\mu_\beta H/D = 1.0$ corresponds to the control at B. (b) Time evolution of the probability P_1 for the system to be staying on the state 1. (c) Time evolution of P_2. (d) Time evolution of P_3. (Taken from Ref. [221] with permission.)

13.3.2.   *Vibrational and tunneling transitions by laser*

Let us take, as an example, a laser-induced ring-puckering isomerization of trimethylenimine which was discussed by Sugawara and Fujimura in Ref. [216]. This problem may be reduced to a double well problem in which the left (right)

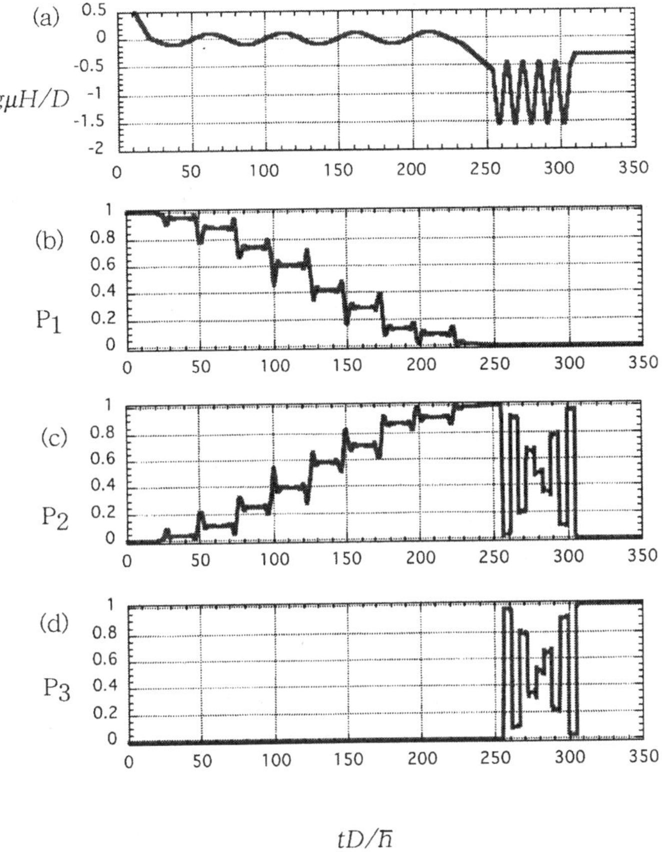

Fig. 13.7. The same as Fig. 13.6 for a process from a to c. (Taken from Ref. [221] with permission.)

well corresponds to the isomer A (B), and the isomerization from A to B occurs through tunneling from the left well to the right well (see Fig. 13.8(a)). We try to control this isomerization with use of the various types of laser pulses. All the parameters to determine the potential system are taken from Ref. [216].

The Floquet state diagram as a function of laser frequency ω [cm^{-1}] with constant intensity ($I = 0.1$ [TW/cm^2]) is shown in Fig. 13.8(b). There appear a lot of avoided crossings, where the energy gap is proportional to the

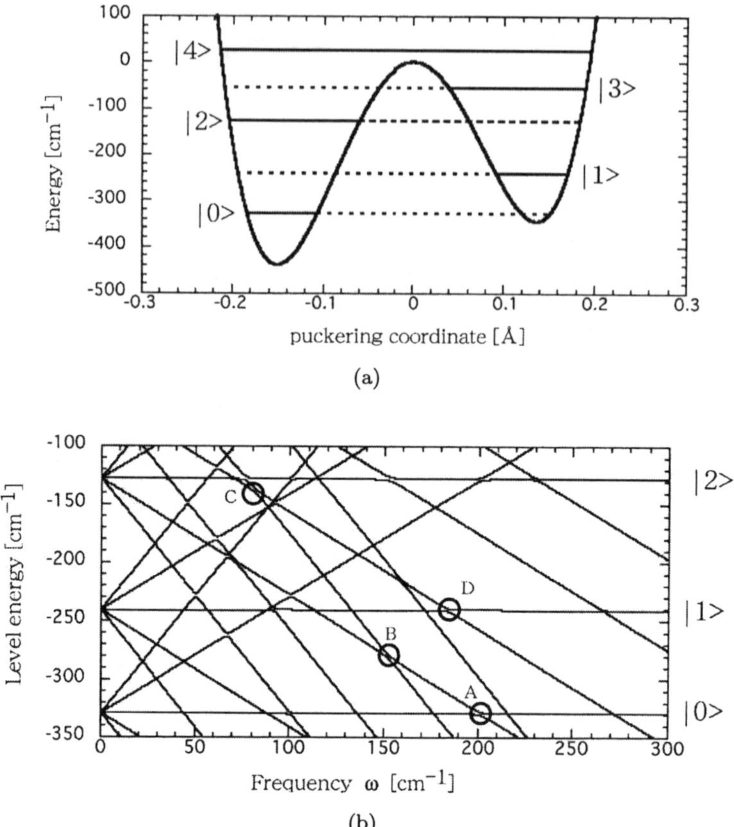

Fig. 13.8.   (a) A double well potential model of ring-puckering isomerization of trimethylen-imine (see Ref. [216]). The origin of potential energy is taken at the barrier top. The horizontal solid line represents the main portion of the eigenfunction. (b) Floquet state diagam, i.e., vibrational levels of (a) as a function of laser frequent ω (cm^{-1}) with fixed intensity ($I = 0.1$ [TW/cm^2]). The gap at each avoided crossing is proportional to the transition dipole moment between the corresponding two states. (Taken from Ref. [222] with permission.)

laser intensity I and the square of the transition dipole moment between the corresponding two states. We can treat each avoided crossing separately unless the laser intensity is extremely strong and avoided crossings overlap with each other.

Nonadiabatic transitions among the Floquet states induced by the variation of intensity and/or frequency can be described by the nonadiabatic

Floquet theory [219, 220] and we can employ the various analytical theories of nonadiabatic transition to analyze them. In the following subsections we demonstrate control of vibrational transitions and isomerization numerically with use of the various theoretical schemes presented in the previous section. The present section is designed so that the reader does not necessarily have to refer to the mathematics. In the following numerical calculation, the optimum values of the control parameters are found by analytical theories as much as possible, and the actual time variations of transition probabilities are solved numerically.

13.3.2.1. *Landau–Zener–Stueckelberg type transition*

Let us first consider the vibrational transition $|0\rangle \to |2\rangle$ via the avoided crossing A in Fig. 13.8(b) with use of the LZS type curve crossing model. In this case the complete control can be achieved by one period of oscillation with reasonable values of the laser intensity and the sweep velocity. Figure 13.9(b) and 13.9(c) show the frequency and intensity as a function of time. The frequency is taken to be a quadratic function of time, i.e. $\omega(t) = at^2 + b$, and the analytical theory developed in Sec. 6.2 (Ref. [71]) has been used. The resonance frequency ω_X corresponding to the avoided crossing is $\omega_X = 202.6$ [cm^{-1}]. Two functional forms, constant (solid line) and $4A^2 \, \mathrm{sech}^2(\beta t)/\epsilon^2$ (dash line), are assumed for the laser intensity (Fig. 13.9(c)). The frequencies shown in Fig. 13.9(b) are the solutions of our control theory corresponding to the intensities given in Fig. 13.9(c). Figure 13.9(a) shows the time-variation of the transition probability for the process $|0\rangle \to |2\rangle$ with use of the avoided crossing A in Fig. 13.8(b). Nonadiabatic transitions occur twice, at each of which $p = 0.5$ is achieved. Unit transition probability is realized finally in the two cases. In the case of the intensity of pulse shape (dash line in Fig. 13.9(c)), not only the frequency but also the area of the intensity pulse contribute to the phase. Thus the corresponding frequency in Fig. 13.9(b) is slightly smaller than the solid line. In this LZS type of nonadiabatic transition, functionality of the intensity is not important, but the intensity at the avoided crossing is critical. This is the reason why the difference in frequency is so small in the two cases. The dotted line is to show the effects of intensity variation on the final result. Although the nonadiabatic transition probability p is reduced to 0.45 by small shift in intensity, the final transition probability is not so bad, since the phase is accurate.

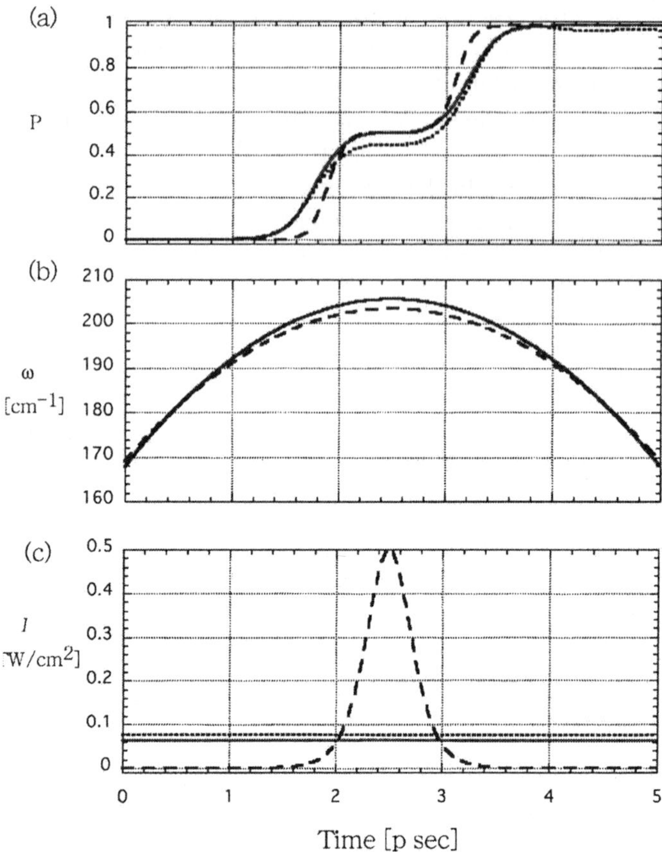

Fig. 13.9.   Controlled nonadiabatic process from $|0\rangle$ to $|2\rangle$ with use of the LZS type nonadiabatic transition at the avoided crossing "A" at 202.6 cm^{-1} in Fig. 13.8(b). (a) Time-evolution of the transition probability. (b) Variation of laser frequency as a function of time. (c) Variation of laser intensity as a function of time. The solid line is the case of constant intensity and quadraric variation of frequency, and the dash line corresponds to the pulse shape intensity and quadratic variation of frequency. In both cases complete control is attained. The dotted line shows the sensitivity to the constant shift of frequency. The final probability is about 0.975 in this case. (Taken from Ref. [222] with permission.)

If we sweep the frequency more than once at the avoided crossing, we can naturally achieve the final unit transition probability with smaller intensity I, but it may take longer time. It should be noted that the required intensity for the case of constant intensity is much smaller (less than one tenth) than that

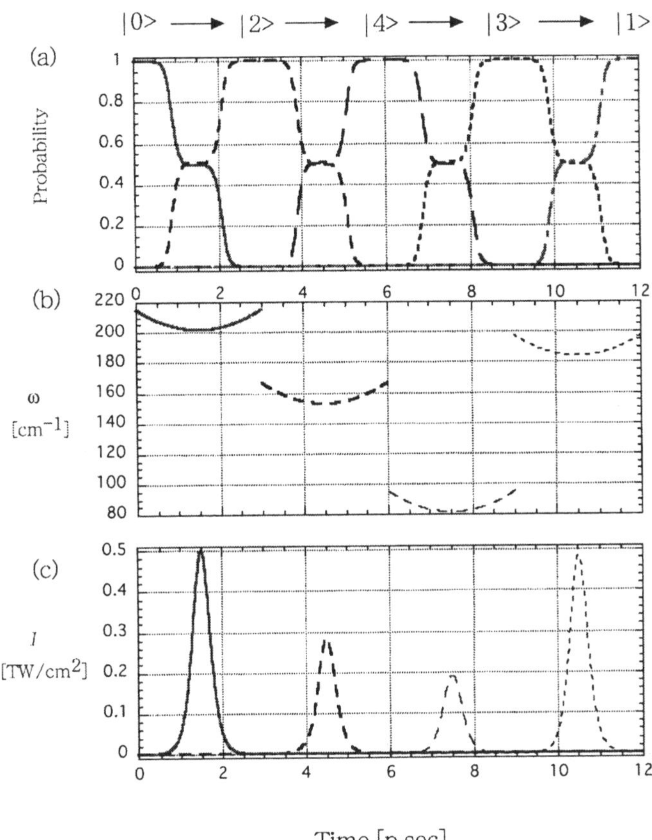

Fig. 13.10. Controlled isomerization process induced by the sequence of the LZS type transitions, $|0\rangle \to |2\rangle \to |4\rangle \to |3\rangle \to |1\rangle$. The corresponding avoided crossings are designated as A–D in Fig. 13.8(b). The corresponding resonance frequencies are 202.6, 153.6, 82.6, and 185.2 cm⁻¹, respectively. (a) Time-evolution of the probability. At each stage complete transition is attained. Twelve Floquet states are taken into account for the calculation. The time variations of frequency and intensity are shown in (b) and (c), respectively. (Taken from Ref. [222] with permission.)

required by π-pulse (discussion will be made later in relation to Fig. 13.11). It should also be noted that the constant intensity can of course be cut off outside the transition region.

For a transition between two states with a small transition moment ϵ, a larger intensity or a smaller sweep velocity is required to satisfy $p = 0.5$. This

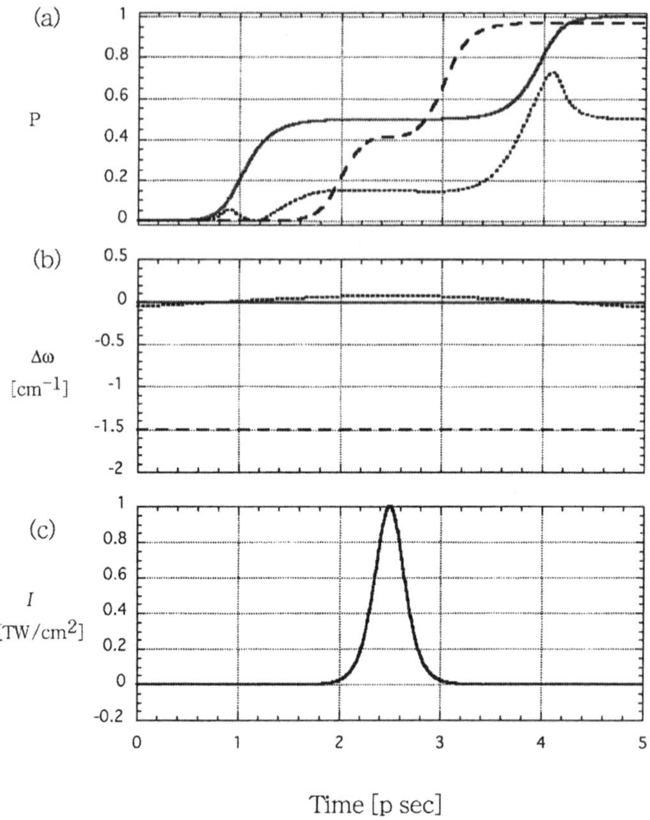

Fig. 13.11.   The same as Fig. 13.9 for the case of π-pulse. (a) Time-evolution of the transition probability. Time variations of frequency and intensity ar shown in (b) and (c), respectively. $\Delta\omega$ represents the shift from the resonance frequency $\omega_X = 202.6$ cm^{-1}. The dash line shows the case of constant shift in frequency. The solid line shows the case of exact one. The dotted line shows the large effect of small time variation of the frequency around resonance. It should be noted that the small fluctuation of frequency in (b) gives a large effect on the transition probability [dotted line in (a)]. (Taken from Ref. [222] with permission.)

means that the direct isomerization from $|0\rangle$ to $|1\rangle$ requires very large intensity or a very long transition time (very slow sweeping). For the isomerization, it is thus better to use an indirect process which is composed of the transitions of relatively large transition moments [216]. It should be noted that the square of the transition moment for the direct process $|0\rangle \to |1\rangle$ is about four orders of magnitude smaller than that for $|0\rangle \to |2\rangle$. Figure 13.10 shows an example of such indirect isomerization: $|0\rangle \to |2\rangle \to |4\rangle \to |3\rangle \to |1\rangle$, where four pulses are applied corresponding to these four transitions. That is to say, the first pulse achieves the complete transition $|0\rangle \to |2\rangle$, and the second one does $|2\rangle \to |4\rangle$, and so on. The corresponding avoided crossings are designated as A–D in Fig. 13.8(b). To obtain Fig. 13.8(a), 12 Floquet states are taken into account. Here, we have used exactly the same shape of $w(t)$ at four avoided crossings. This can be done by simply multiplying a certain constant to the intensity I. Since the essential qualitative features such as the characteristics of various types of nonadiabatic transitions we want to address here do not depend on the transitions, we consider the transition $|0\rangle \to |2\rangle$ for a while.

13.3.2.2.   *Rosen–Zener–Demkov type transition*

Let us next consider the control of the transition $|0\rangle \to |2\rangle$ by the RZD type of transition. Figure 13.11 is an example of the so called π-pulse (solid lines) with parameters chosen so that the overall transition time between the corresponding Floquet states becomes the same order as that in Fig. 13.9. Again, nonadiabatic transitions occur twice with the transition probability $p = 0.5$, and the final overall transition probability is controlled to be unity with use of the phase accumulated between the two transitions. In the case of π-pulse, the condition of $p = 0.5$ is attained by fixing the frequency at the resonance frequency ω_X, or the frequency at the avoided crossing. The phase condition is satisfied by adjusting the area of the intensity pulse. These conditions are relatively simple and seem to be easily realized compared to the LZS case. The control by π-pulse, however, is less effective compared to the case of quadratic chirp mentioned previously. As is seen from Figs. 13.9 and 13.11, the control by π-pulse requires stronger intensity. The necessary constant intensity in the LZS case (see Fig. 13.9(c)) is ten times lower than the peak height in Fig. 13.11(c). It should be noted that a small variation of the frequency could cause unexpected transitions and errors, if the frequency is close to the resonance. Dash and dotted lines in Fig. 13.11 demonstrate the sensitivity of the π-pulse method to frequency variations. The dash line shows the effects of constant shift of the frequency from the resonance. As discussed in the previous section, frequency can contain a constant shift error up to several cm^{-1}, if the phase due to the intensity variation is accurate, and the intensity can have 10% of error, if the frequency is exact at the resonance. Much more shocking one is the large effects of very small time-dependent fluctuation in frequency, as is demonstrated by the dotted line. This small fluctuation induces curve crossing type nonadiabatic transitions between the closely lying states effectively, and causes a big effect in the final result, as seen in Fig. 13.11(a). In order to avoid this instability, it might be worthwhile to use the off-resonant case explicitly. In the case of off-resonance, the nonadiabatic transition probability p is smaller than 0.5, and more than one pulses, i.e. more than one periods of oscillation, are required. Figure 13.12 shows an example of two pulses. The frequency is kept constant at 5 cm^{-1} (solid line) and the various parameters are chosen so as for the transition time to be the same order as that in Figs. 13.9 and 13.11. As is seen from this figure in comparison with Fig. 13.11, this scheme is very effective. The nonadiabatic transition probability for one passage is about 0.2, and after four nonadiabatic transitions the final transition probability reaches unity. This can be achieved with much smaller intensity compared to π-pulse. This is because the necessary phase can be accumulated not only by the intensity but also by the frequency because of the off-resonance. In the case of off-resonance, however, we should adjust not only the height and shape of the pulse $I(t)$ but also the interval of two pulses, because the nonadiabatic transition probability $p_{\rm RZ}$ depends on the exponent of $I(t)$, i.e. β in Eq. (13.31) and the interval of two pulses determine the phase σ_2 (see Fig. 13.1). It might be difficult to adjust the shape of $I(t)$ accurately, but the control by off-resonant pulses is very attractive. It requires very small intensity. Stability against frequency fluctuation is also satisfactory, as is demonstrated by the dash and dotted lines in Fig. 13.12.

13.3.2.3.   *General case*

Finally, let us consider the control with use of the variation of both intensity and frequency by taking the exponential model of Eq. (13.34) as an example. Figure 13.13 (solid line) shows the control of the transition $|0\rangle \rightarrow |2\rangle$ by one pulse (period) with two nonadiabatic transitions of $p = 0.5$. It should be noted that the required intensity for the same order of transition time as before is quite small and the frequency is not necessary to be kept at resonance; while, as demonstrated before, the LZS type requires the larger intensity and RZD requires the frequency to be close to resonance to achieve $p = 0.5$. Due to the off-resonance, small fluctuation of ω does not cause any appreciable errors as in the case of π-pulse. The necessary phase can be accumulated by means of

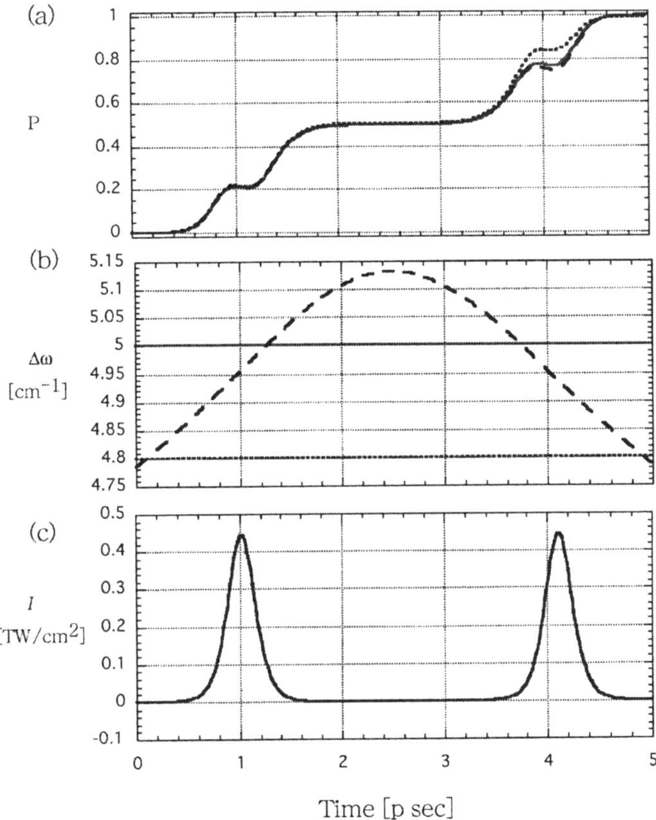

Fig. 13.12. The same as Fig. 13.9 for the case of off-resonant two pulses. (a) Time-evolution of the probability. Time variations of frequency and intensity are shown in (b) and (c), respectively. $\Delta\omega$ represents the shift from the resonance frequency $\omega_X = 202.6$ cm^{-1}. The solid line is the case of complete control. The dash and dotted lines demonstrate the stability of the method against time variation of frequency (dash line) and the constant shift in frequency (dotted line). (Taken from Ref. [222] with permission.)

both frequency and intensity, which is the reason why the required intensity can be so small. Dash and dotted lines in Fig. 13.13 show the stability of the method against the variations of intensity and frequency. Thus, the exponential model may provide quite an efficient control method compared to LZS and RZD. By taking the vibrational transition $|0\rangle \rightarrow |2\rangle$, we have explained the characteristics of various types of nonadiabatic transitions and proved that

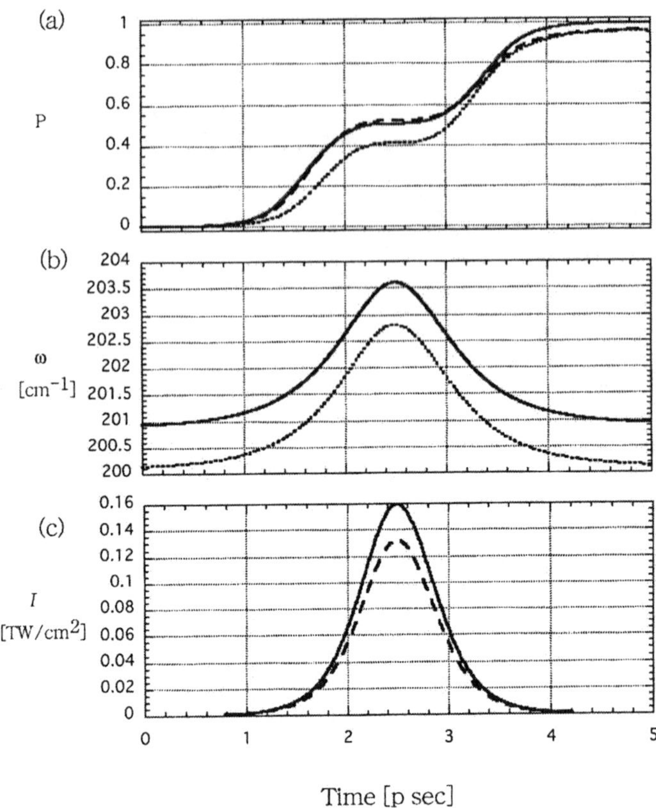

Fig. 13.13.   The same as Fig. 13.12 for the case of exponential model. The solid line is the case of complete control. Frequency is swept around the avoided crossing "A" at $\omega_X = 202.6$ cm^{-1}. The dash and dotted lines demonstrate the stability of the method against the error in the exponent of intensity (dash line) and the constant shift in frequency from the solid line (dotted line). (Taken from Ref. [222] with permission.)

the exponential model presents the most effective way of control. As was mentioned before, this does not depend on the transitions, and thus is also true for the tunneling transition $|0\rangle \rightarrow |1\rangle$. Figure 13.14 demonstrates this. As is well known, it is far better to use the detour $|0\rangle \rightarrow |2\rangle \rightarrow |4\rangle \rightarrow |3\rangle \rightarrow |1\rangle \rightarrow$ in the same way as in Fig. 13.10 than to take the direct path $|0\rangle \rightarrow |1\rangle$, since the dipole moment for the latter is about two orders of magnitude smaller than that of $|0\rangle \rightarrow |2\rangle$. As seen in Fig. 13.14, the required intensity is much smaller than that in Fig. 13.10. If the laser intensity could be kept constant, however,

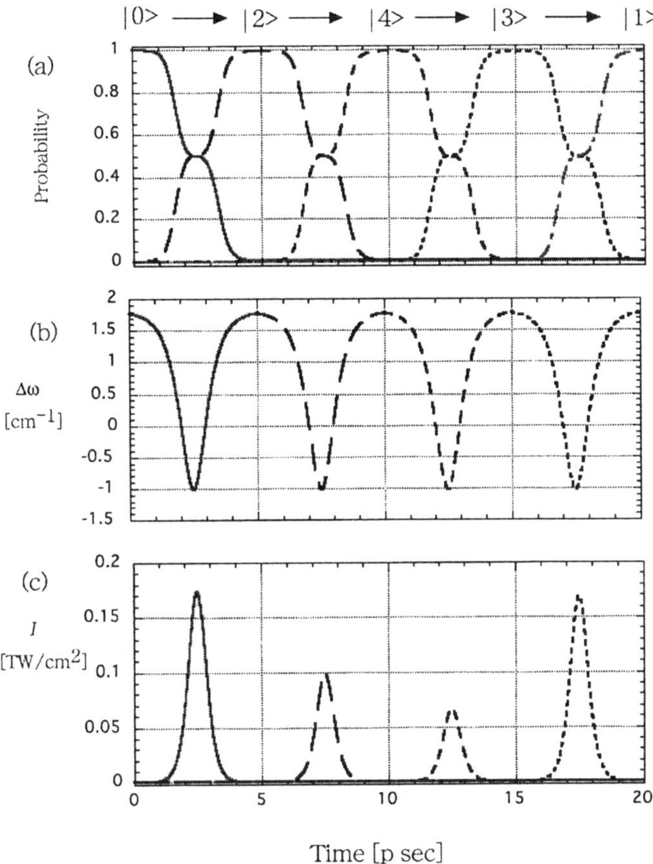

Fig. 13.14. The same as Fig. 13.10, i.e., control of the isomerization process $|0\rangle \to |2\rangle \to |4\rangle \to |3\rangle \to |1\rangle$. The exponential model is employed. Twelve Floquet states are taken into account to obtain (a). It should be noted that the intensity is reduced by about a factor of 3 compared to Fig. 13.10. In (b), the frequency shift $\Delta\omega$ from the resonance is shown instead of ω itself in order to clearly show its exponential form. The shape is taken to be the same for the four transitions. (Taken from Ref. [222] with permission.)

we could further reduce the necessary maximum intensity, as was demonstrated in Fig. 13.9. Figure 13.15 shows this type of control of the isomerization.

A more general model, like an exponential model with different exponents for frequency and intensity, for which no analytical theory is unfortunately available yet, is expected to provide a more efficient scheme. It should be

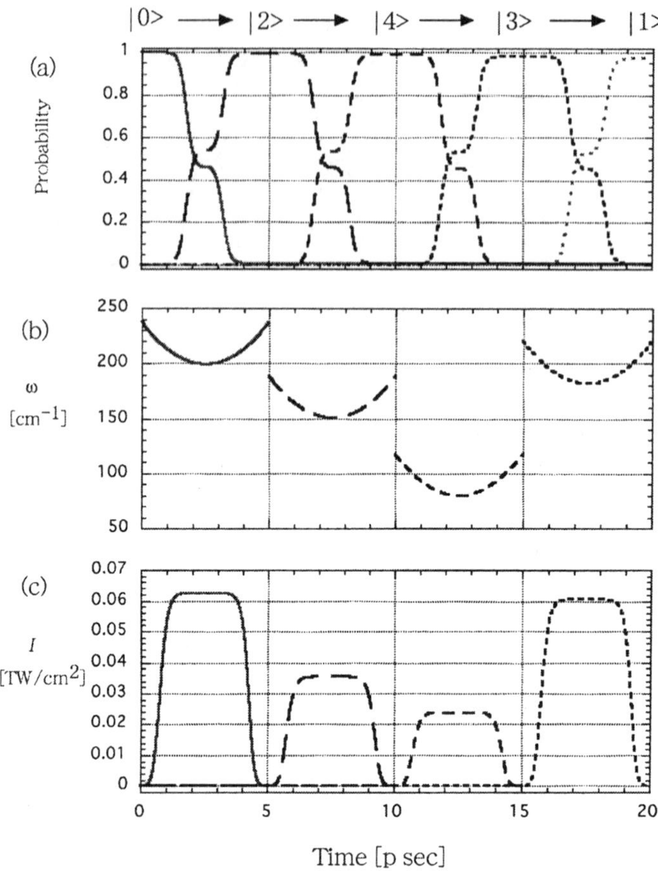

Fig. 13.15.   The same as Fig. 13.10, i.e., control of the isomerization process $|0\rangle \to |2\rangle \to |4\rangle \to |3\rangle \to |1\rangle$. The constant intensity and the quadratic variation of frequency are utilized. (Taken from Ref. [222] with permission.)

noted, however, that the exponential model in general requires accurate shaping of both intensity and frequency pulses, since the nonadiabatic transition probability p_{exp} depends explicitly on the exponents of both intensity and frequency.

So far we have discussed several control schemes with the help of analytical theories. We can choose one of them depending on the molecular process and availability of lasers. If accurate pulse shaping of both intensity and frequency is possible, the scheme with the exponential type of nonadiabatic transition is the best. If pulse shaping of intensity is available, but not for frequency, the RZD type works relatively well. In this case, the off-resonant RZD type is recommended. If accurate shaping of intensity is not attainable, the resonant RZD (π-pulse) or the LZS methods are recommended.

Generally speaking, molecular processes in a laser field can be considered as a sequence of nonadiabatic transitions and adiabatic propagations, and we can treat the whole control problem analytically, if the nonadiabatic transitions are separated from each other in time. This indicates that we may construct a control scheme for a general shape of pulse, even if no analytical theory for each nonadiabatic transition is available. Once we know the phase change and the probability change by one pulse (one period of oscillation), we can design an efficient way of control based on the analytical scheme developed in Sec. 13.2. The optimum conditions for one pulse, namely the conditions to satisfy $\sin^2 \psi = 1$ and $p = 0.5$, may be found numerically or even experimentally.

13.4. Laser Control of Photodissociation with Use of the Complete Reflection Phenomenon

With use of the complete reflection phenomenon, one and two-dimensional models of photodissociation are discussed here for a triatomic molecule ABC consisting of a bound ground electronic state and a dissociative excited electronic state with two dissociation channels, A+BC and AB+C [224]. In the one-dimensional model, complete and selective dissociation into any desired channel can be realized by adjusting the initial vibrational excited state and the laser frequency. In the two-dimensional case, the control cannot be complete, but quite selective dissociation into any channel is possible. Even such a dissociation can be realized that a potential barrier hinders the ordinary photodissociation along the electronically excited state potential surface. This control scheme cannot be very robust, because the method is based on the phase interference; but this can present a new intriguing selective control of molecular photodissociation when the conditions with respect to potential energy surface topography and initial vibrational state are appropriately satisifed.

In order to demonstrate the present idea clearly, numerical calculations of wave packet propagation are carried out for a one-dimensional potential system shown in Fig. 13.16.

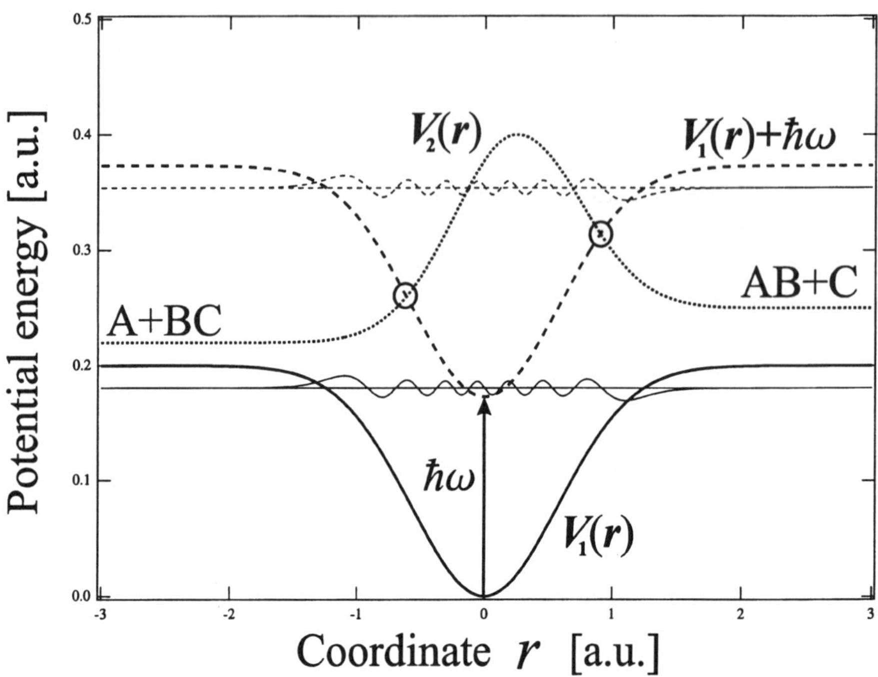

Fig. 13.16. One-dimensional model of the control scheme. Solid line: ground electronic state $V_1(r)$, dotted line: the excited electronic state $V_2(r)$. Two circles represent the NT-type crossings created between the excited electronic state and the dressed ground electronic state (dashed line). The 14th vibrational eigenstate (thin line) and its dressed state (thin dashed line) of $V_1(r)$ are also depicted. (Taken from Ref. [224] with permission.)

The potential functions employed are

$$V_1(r) = 0.2(1 - e^{-1.5r^2}) \tag{13.38}$$

and

$$V_2(r) = \begin{cases} 0.15e^{-2.0(r-0.25)^2} + 0.25 & (r > 0.25) \\ 0.18e^{-2.0(r-0.25)^2} + 0.22 & (r \leq 0.25) \end{cases} \tag{13.39}$$

in atomic units. The molecule-laser interaction is taken to be

$$V_{\text{int}}(t) = -\mu E(t) = \mu\sqrt{I}\cos(\omega t + \delta), \tag{13.40}$$

where μ is the transition dipole moment between the two electronic states and $E(t)$ is the stationary laser field with the frequency ω and the intensity I. The one-dimensional time-dependent Schrödinger equation to be solved is

$$i\hbar\frac{\partial}{\partial t}\begin{bmatrix}\Psi_1(r,t)\\ \Psi_2(r,t)\end{bmatrix}=\begin{bmatrix}-\dfrac{\hbar^2}{2m}\dfrac{d^2}{dr^2}+V_1(r) & -\mu E(t)\\ -\mu E(t) & -\dfrac{\hbar^2}{2m}\dfrac{d^2}{dr^2}+V_2(r)\end{bmatrix}\begin{bmatrix}\Psi_1(r,t)\\ \Psi_2(r,t)\end{bmatrix}, \quad (13.41)$$

where $\Psi_1(r,t)(\Psi_2(r,t))$ is the nuclear wavefunction on the ground (excited) electronic state. Equation (13.41) is solved by using the split operator method [228] with the fast Fourier transform:

$$\begin{bmatrix}\Psi_1(r,t+\Delta t)\\ \Psi_2(r,t+\Delta t)\end{bmatrix}=\exp\left(-\frac{i}{\hbar}\frac{\hat{V}}{2}\Delta t\right)\exp\left(-\frac{i}{\hbar}\hat{K}\Delta t\right)$$

$$\times\exp\left(-\frac{i}{\hbar}\frac{\hat{V}}{2}\Delta t\right)\begin{bmatrix}\Psi_1(r,t)\\ \Psi_2(r,t)\end{bmatrix}+o(\Delta t^3), \quad (13.42)$$

where $\hat{V}$ and $\hat{K}$ are the 2×2 matrices,

$$\hat{V}=\begin{bmatrix}V_1(r) & -\mu E(t)\\ -\mu E(t) & V_2(r)\end{bmatrix} \quad (13.43)$$

$$\hat{K}=\begin{bmatrix}-\dfrac{\hbar^2}{2m}\dfrac{d^2}{dr^2} & 0\\ 0 & -\dfrac{\hbar^2}{2m}\dfrac{d^2}{dr^2}\end{bmatrix}. \quad (13.44)$$

The dissociation flux is integrated over time at a certain asymptotic position to obtain the corresponding dissociation probability. In order to prevent the unphysical reflection of the wave packet at the edges, the negative imaginary potential (absorption potential) is put at both ends [229]. We have prepared the 14th vibrational eigenstate of the ground electronic state (the quantum number $v = 13$) as an initial state and used the following parameters for the wave packet calculation: m (reduced mass) $= 1.0$ a.m.u., $\mu = 1.0$ a.u., $I = 1.0$ TW/cm^2, Δt (time step) $= 1.0$ a.u., and Δr (spatial grid size) $= 0.016$ a.u. The total number of the spatial grid points is 512. The complete reflection positions can be analytically predicted as is shown in Fig. 13.17(a). Once the laser intensity I is selected, the shapes of the adiabatic potential curves, i.e.

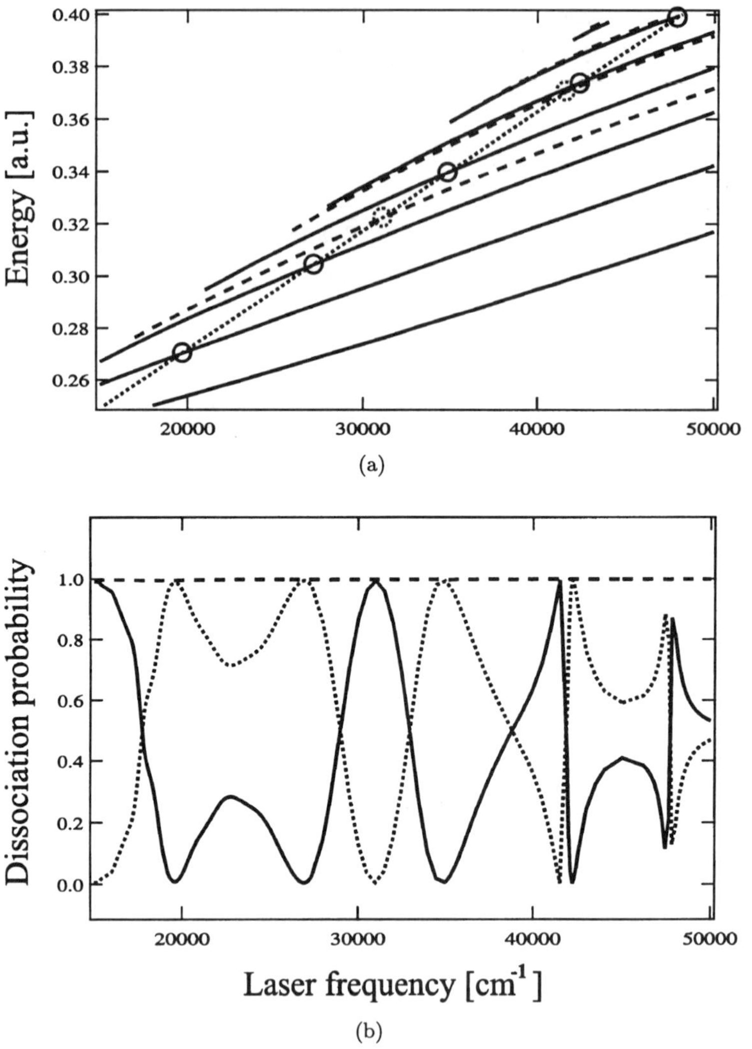

Fig. 13.17. (a) Analytical prediction of the complete reflection positions. Solid (dashed) lines: complete reflection positions at the left-side (right-side) channel. Dotted line: the dressed 14th vibrational eigen state of $V_1(r)$. Solid (dashed) circles represent the complete reflection of the vibrational state at the left-side (right-side) crossing. (b) Dissociation probability as a function of laser frequency. Solid line: dissociation into the right-side (AB+C) channel, dotted line: dissociation into the reft-side (AB+C) channel, dashed line: sum of the two dissociation probabilities to guarantee the unitarity. (Taken from Ref. [224] with permission.)

the ground electronic state shifted up by one photon energy $V_1(r) + \hbar\omega$ and the excited electronic state $V_2(r)$, are determined for a fixed laser frequency ω and then the positions of complete reflection can be easily estimated by solving Eq. (5.159) (see also Eq. (11.7)). The solid (dashed) line in Fig. 13.17(a) represents the complete reflection position at the left-side (right-side) channel. The dotted line represents the relevant vibrational level, which is shifted up by one photon energy $\hbar\omega$, namely a linear function of the frequency ω. Thus the crossing points marked by solid (dotted) circles give the positions of the complete reflection for this vibrational state on the left (right) side. The calculated dissociation probabilities against laser frequency are depicted in Fig. 13.17(b). In order to concentrate on the important dissociation dynamics, here, we have assumed that the initial vibrational state was prepared. The solid (dotted) line represents the dissociation into the right (left) channel; the zero probability positions of the solid (dotted) line coincide with the solid (dotted) circles in Fig. 13.17(a). The highest dip at $\omega \sim 48000$ cm^{-1} is not complete because of tunneling. As is demonstrated above, within the one-dimensional model our control scheme can be perfect and a molecule can be made to dissociate into any channel as we desire by adjusting the laser frequency for a given vibrational state. The initial vibrational state should, however, be appropriately an excited one, since the phase ψ should satisfy Eq. (5.159).

Let us next consider a two-dimensional model. As an example, we take the HOD molecule with two dissociation channels: H+OD and HO+D. We consider the ground electronic state $\tilde{X}$ and the excited electronic state $\tilde{A}$. The bending and rotational motions are neglected for simplicity with the bending angle fixed at the equilibrium structure of the ground electronic state, i.e. at $104.52°$. The ground electronic state potential $V_1(r_H, r_D)$ (shown by dotted lines in Fig. 13.18(a)) is taken to be two coupled Morse oscillators [230],

$$V_1(r_H, r_D) = D(1 - e^{-\gamma(r_H - r_0)^2}) + D(1 - e^{-\gamma(r_D - r_0)^2})$$

$$- B\frac{(r_H - r_0)(r_D - r_0)}{1 + e^{A((r_H - r_0) + (r_D - r_0))}}, \tag{13.45}$$

where r_H (r_D) represents the H–O (D–O) bond length, and $D = 0.2092$ hartree, $\gamma = 1.1327$ a.u.$^{-1}$, $r_0 = 1.81$ a.u., $A = 3.0$ a.u.$^{-1}$ and $B = 0.25$ hartree/a.u.^{2}. The last term represents the mode coupling, and the parameters A and B are assumed to be larger than the values used in Ref. [230] ($A = 1.0$ a.u.$^{-1}$ and $B = 0.00676$ hartree/a.u.^{2}) in order to demonstrate the present control scheme

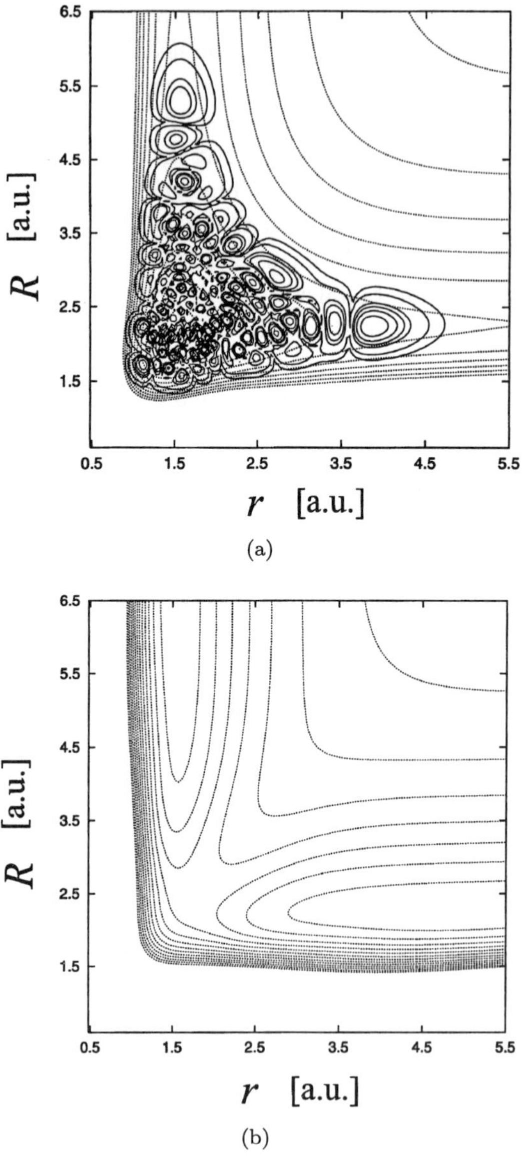

Fig. 13.18.   (a) Control plots of the ground electronic state of HOD (dotted line). The contour spacing is 8500 cm^{-1}. The density of the 145th vibrational eigenstate is superimposed (solid line). (b) Contour plots of the excited electronic state of HOD. The contour spacing is 5000 cm^{-1}. (Taken from Ref. [224] with permission.)

more clearly. The excited electronic state $V_2(r_{\mathrm{H}}, r_{\mathrm{D}})$ shown in Fig. 13.18(b) is taken to be the analytical function of Engel *et al.* [231], which was fitted to the *ab initio* calculations by Staemmler *et al.* [232]. The mass-scaled Jacobi coordinates r for O–H distance and R for OH–D distance are introduced,

$$r = \left(\frac{m_{\mathrm{OH}}}{m_{\mathrm{D,OH}}} \right)^{1/4} |\mathbf{r}_{\mathrm{H}}|, \tag{13.46}$$

$$R = \left(\frac{m_{\mathrm{D,OH}}}{m_{\mathrm{OH}}} \right)^{1/4} \left| \mathbf{r}_{\mathrm{D}} - \frac{m_{\mathrm{H}}}{m_{\mathrm{H}} + m_{\mathrm{O}}} \mathbf{r}_{\mathrm{H}} \right|, \tag{13.47}$$

where $\mathbf{r}_{\mathrm{H}}$ ($\mathbf{r}_{\mathrm{D}}$) is the vector from O atom to H (D) atom, $m_{\mathrm{OH}} = m_{\mathrm{O}} m_{\mathrm{H}}/(m_{\mathrm{O}} + m_{\mathrm{H}})$, $m_{\mathrm{D,OH}} = m_{\mathrm{D}}(m_{\mathrm{O}} + m_{\mathrm{H}})/(m_{\mathrm{D}} + m_{\mathrm{O}} + m_{\mathrm{H}})$, and m_{H}, m_{O} and m_{D} are the mass of H, O and D, respectively. With use of these coordinates (r, R) the two-dimensional time-dependent Schrödinger equation is written as

$$i\hbar \frac{\partial}{\partial t} \begin{bmatrix} \Psi_1(r, R, t) \\ \Psi_2(r, R, t) \end{bmatrix}$$

$$= \begin{bmatrix} -\dfrac{\hbar^2}{2m} \left(\dfrac{\partial^2}{\partial r^2} + \dfrac{\partial^2}{\partial R^2} \right) + V_1(r, R) & -\mu E(t) \\ -\mu E(t) & -\dfrac{\hbar^2}{2m} \left(\dfrac{\partial^2}{\partial r^2} + \dfrac{\partial^2}{\partial R^2} \right) + V_2(r, R) \end{bmatrix}$$

$$\times \begin{bmatrix} \Psi_1(r, R, t) \\ \Psi_2(r, R, t) \end{bmatrix}. \tag{13.48}$$

Here m is the reduced mass of the system,

$$m = \sqrt{\frac{m_{\mathrm{H}} m_{\mathrm{O}} m_{\mathrm{D}}}{m_{\mathrm{H}} + m_{\mathrm{O}} + m_{\mathrm{D}}}}. \tag{13.49}$$

$\Psi_1(r, R, t)(\Psi_2(r, R, t))$ represents the wavefunction of the ground (excited) electronic state. Equation (13.48) is solved by using the split operator method with the two-dimensional fast Fourier transform in the same way as before. The dissociation flux is integrated over time in a certain asymptotic region before the negative imaginary potentials which are put at both ends. The transition dipole moment μ is assumed to be 1.0 a.u. and the stationary laser field $E(t)$ is taken to be $\sqrt{I} \cos(\omega t + \delta)$. An initial state is prepared at the 145th vibrational eigenstate of $V_1(r, R)$ by solving the two-dimensional eigenvalue problem by

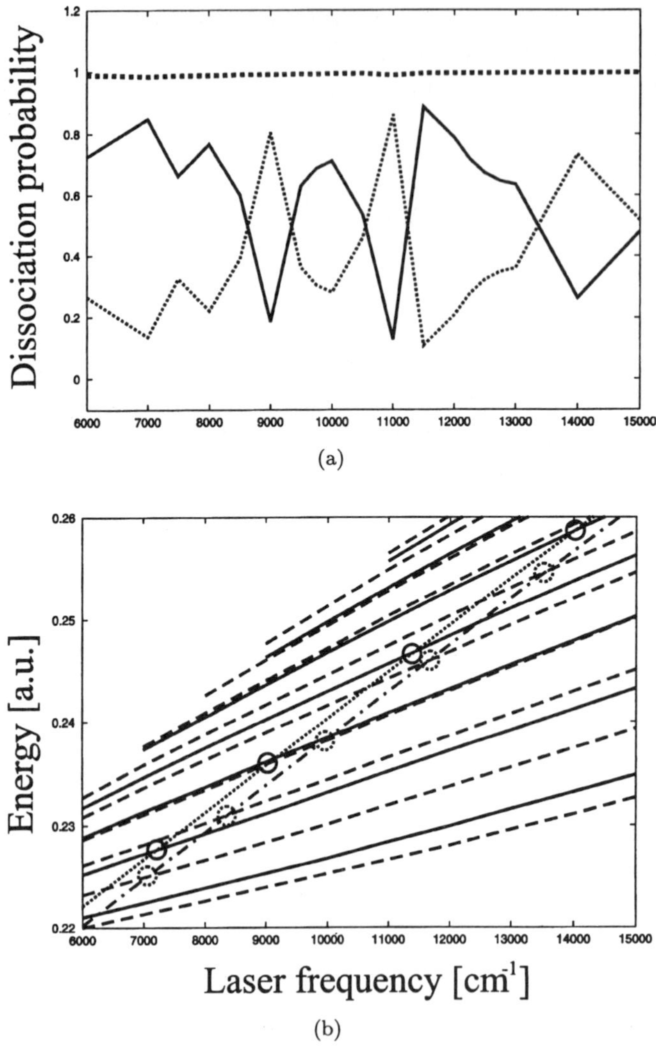

Fig. 13.19. (a) Dissociation probability against laser frequency in the case of the 145th vibrational state of HOD. Solid (dotted) line: dissociation into the H+OD (HO+D) channel. Dashed line: sum of the two dissociation probabilities. (b) Analytical prediction of the complete reflection positions. Solid (dashed) lines: the complete reflection positions in the H+OD (HO+D) channel. Dotted (dash-dotted) line: the vibrational state $v = 17$ of the O–H bond ($v = 23$ of the O–D bond). The solid (dotted) circles represent the complete reflection positions when the 145th vibrational eigenstate is prepared as an initial state on the H+OD (HO+D) side. (Taken from Ref. [224] with permission.)

the DVR (discrete variable representation) method [233]. This initial state is mainly composed of the 17th vibrational state of the O–H bond and the 23rd vibrational state of the O–D bond and spreads into both bonds due to the coupling, as is depicted by solid lines in Fig. 13.18(a). The laser intensity I and the various grid sizes are taken to be 1.0 TW/cm^2, $\Delta t = 1.0$ a.u., $\Delta r = 0.029$ a.u. and $\Delta R = 0.034$ a.u., respectively. The total number of the spatial grid points is 256×256. Figure 13.19(a) depicts the calculated dissociation probabilities against laser frequency. The solid (dotted) line represents the dissociation probability into the H+OD (HO+D) channel. The dissociation probabilities change alternatively as a function of the laser frequency ω, and the dissociation into the H+OD (HO+D) channel is preferential when the laser frequency ω is ~ 7000 cm^{-1}, 8000 cm^{-1}, 10000 cm^{-1}, and 11500 cm^{-1} (9000 cm^{-1}, 11000 cm^{-1} and 14000 cm^{-1}).

The control is not perfect this time because of multidimensionality, but is still quite selective. The molecule can be made to dissociate preferentially into any channel as we desire by choosing the laser frequency and the vibrational state appropriately. The complete reflection positions in the H+OD (HO+D) channel can be roughly estimated analytically by taking a one-dimensional cut of the potential energy surface along the minimum energy path of the O–H (O–D) bond into account and using the one-dimensional formula Eq. (5.159). The vibrational state is taken to be $v = 17(23)$ for the O–H (O–D) channel. Figure 13.19(b) depicts these estimates. The solid (dashed) lines represent the complete reflection positions in the H+OD (HO+D) channel. The dotted (dash-dotted) line shows the vibrational state $v = 17$ of the O–H bond ($v = 23$ of the O–D bond) shifted up by one photon energy. Thus the solid (dotted) circles indicate the positions of complete reflection in the H+OD (HO+D) channel. As is seen in Fig. 13.19(a), the dips of calculated dissociation probabilities correspond well to the complete reflection positions predicted analytically. Exceptions are the dip at $\omega \sim 7200$ cm^{-1} and the peak at $\omega \sim 13500$ cm^{-1} in the H+OD dissociation channel. The former is shallow and shifted to higher frequency (~ 7500 cm^{-1}) in Fig. 13.19(a), and the latter has almost disappeared. This is due to the topography of the potential energy surface, representing the difficulty of multi-dimensionality.

The wave packet dynamics on the excited state are shown in Figs. 13.20. The wave packet is depicted by solid lines at various times. The dashed lines in these figures represent the crossing seam lines between the dressed ground state and the excited state. Figure 13.20(a) corresponds to the laser frequency

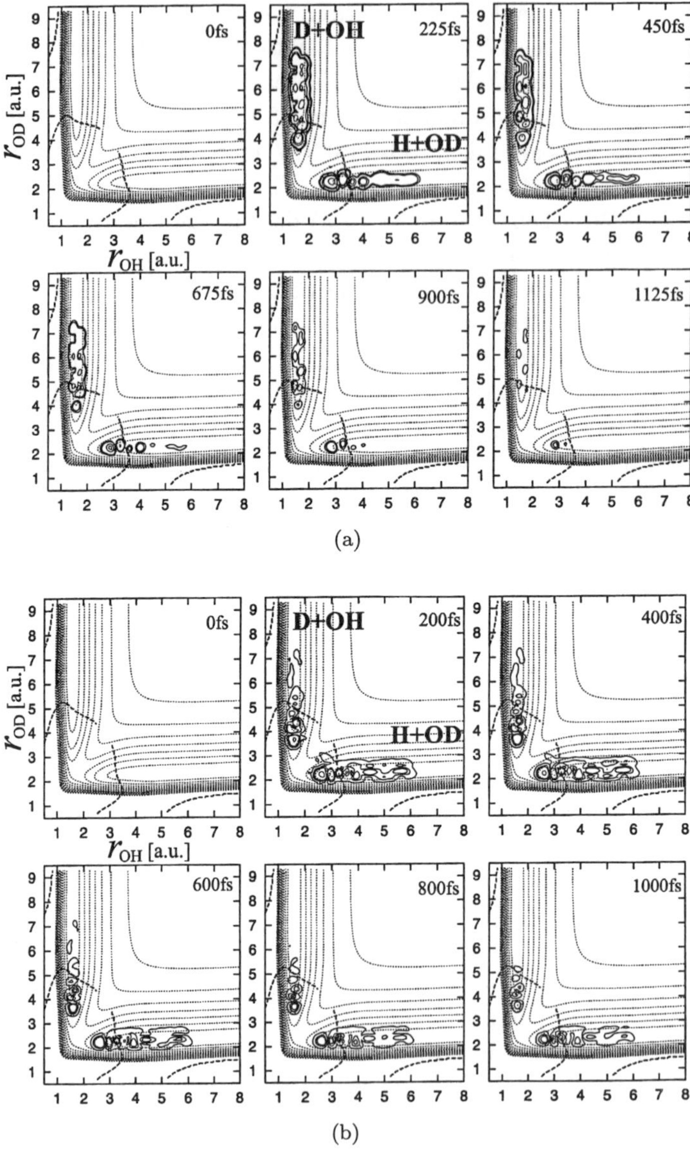

Fig. 13.20.　Time-dependent behavior of the excited state wave packet for the laser frequency (a) 9000 cm^{-1} and (b) 11500 cm^{-1} (the contours of the density by solid line). The contours of the excited electronic state are superimposed (dotted line). The dashed lines represent the crossing seams induced by the laser field. (Taken from Ref. [224] with permission.)

9000 cm^{-1}, and thus the wave packet moves out into the HO+D channel. Figure 13.20(b) corresponds to the laser frequency 11500 cm^{-1}, and the wave packet almost dissociates into the H+OD channel (see Fig. 13.19(a)).

As the above results demonstrate, the selective dissociation based on the complete reflection phenomenon can be realized even in two-dimensional systems. The control naturally cannot be perfect like in the one-dimensional case, but can still be quite effective. The dissociation into a certain channel is stopped by the complete reflection phenomenon and the reflected wave packet is transfered into the other channel due to the mode-coupling via the ground electronic state and is finally dissociated into the latter channel. No analytical theory exists for two-dimensional problems, but the one-dimensional theory can be used to some extent in the above-mentioned way to roughly estimate the appropriate conditions. The favorable conditions of the selective control in the two-dimensional system may be summarized as follows: The mode-coupling in the ground electronic state should be localized around the equilibrium position and should be negligible in the region of crossing seam created by the laser field. Otherwise the mode-coupling potential destroys the complete reflection condition. In other words, the initial vibrational state should have the one mode character around the region of crossing seam, and the L-shape wavefunction like that in Fig. 13.18(a) is favorable.

Our control scheme can also be applied to such a two-dimensional system that the excited electronic state has a potential barrier which is shifted very much to one of the channels and prohibits the dissociation into that channel. As such a model system, we have employed the same potentials as those used in the previous subsection and slightly modified the excited state potential so that the saddle point is located in the HO+D channel (see Fig. 13.21). The only one term of the excited electronic state, $0.2443589 \times 10^2 \times (S_1 S_3^2 + S_1^2 S_3)$ in Ref. [231], is changed to $0.1443589 \times 10^2 \times (S_1 S_3^2 + 0.8 S_2^2 S_3)$. With this modification the dissociation into the HO+D channel is not possible anymore along the excited state potential surface, if the initial vibrational state is localized in the H+OD channel. The same initial state and method of wave packet propagation are used as those in the previous subsection. The calculated dissociation probabilities against the laser frequency are shown in Fig. 13.22(a). The solid (dotted) line stands for the dissociation probability into the H+OD (HO+D) channel. When the laser frequency is either one of ~ 10000 cm^{-1}, ~ 12500 cm^{-1}, ~ 17000 cm^{-1} and ~ 19000 cm^{-1}, the dissociation into the H+OD channel is preferential, while the dissociation into the HO+D channel

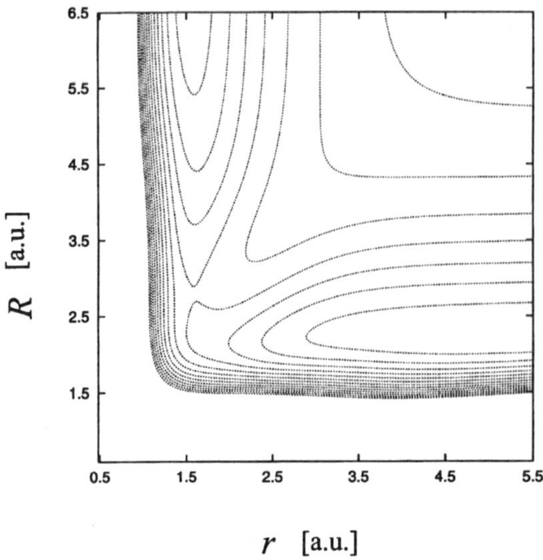

Fig. 13.21.   Contour plots of the modified excited electronic state of HOD. The saddle point is located in the HO+D channel. The contour spacing is 5000 cm^{-1}. (Taken from Ref. [224] with permission.)

is preferential when the laser frequency is set at ~ 9000 cm^{-1}, ~ 11000 cm^{-1} or ~ 13500 cm^{-1}. As is clearly seen, the control is quite selective and the dissociation even into the non-dissociative HO+D channel is possible by adjusting the laser frequency appropriately. The analytical prediction of the complete reflection positions is shown in Fig. 13.22(b) in the same way as in Fig. 13.19(b). The solid (dashed) lines depict the complete reflection positions in the H+OD (HO+D) channel, and the dotted (dash-dotted) line represents the vibrational state $v = 17$ of the O–H bond ($v = 23$ of the O–D bond) shifted up by one photon energy. The solid (dotted) circles show the position of the complete reflection in the H+OD (HO+D) channel. The dissociation probability dips in Fig. 13.22(a) coincide quite well with these analytical predictions. Some of the dips ($\omega \sim 16000$ cm^{-1} and 18000 cm^{-1}) and the peak at $\omega \sim 15000$ cm^{-1} are, however, not complete because of the multi-dimensional topography of potential energy surface.

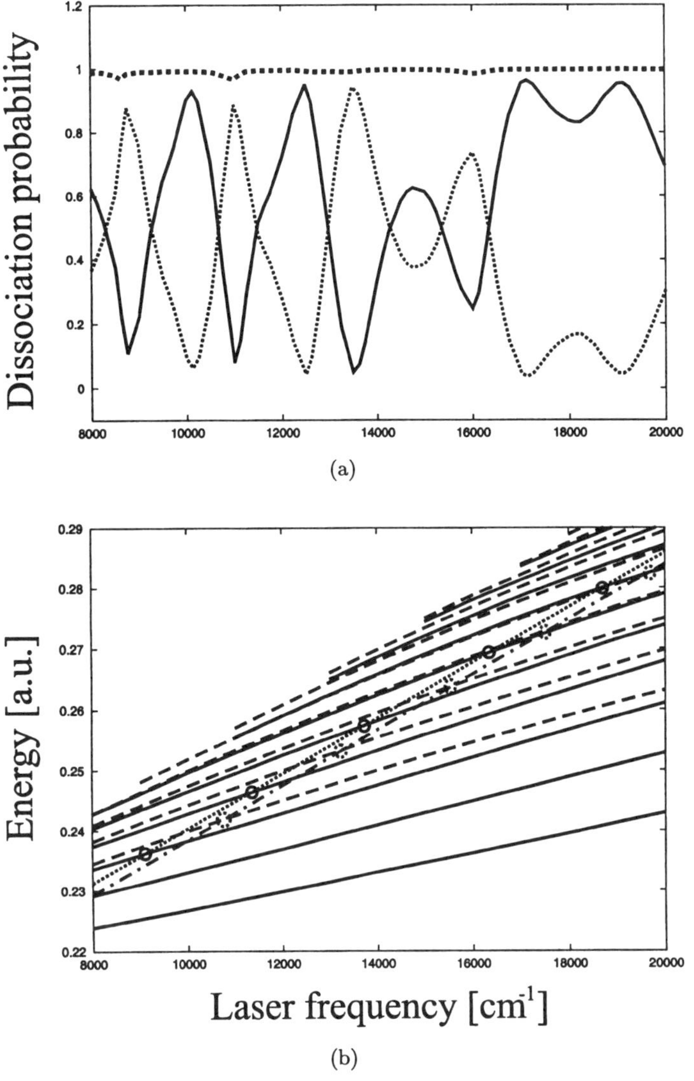

Fig. 13.22. (a) Dissociation probability against laser frequency in the case of the 145th vibrational state of HOD. Solid (dotted) line: dissociation into the H+OD (HO+D) channel. Dashed line: sum of the two dissociation probabilities. (b) Analytical prediction of the complete reflection positions. Solid (dashed) lines: the complete reflection positions in the H+OD (HO+D) channel. Dotted (dashed-doted) line: the vibrational state $v = 17$ of the O–H bond ($v = 23$ of the O–D bond). The solid (dotted) circles represent the complete reflection positions for the 145th vibrational eigenstate. (Taken from Ref. [224] with permission.)

Chapter 14

Conclusions: Future Perspectives

As was pointed out frequently, nonadiabatic transitions play very important roles in various fields of natural sciences and, in a sense, they are controlling this world. Thus, it is very significant that the LZS (Landau–Zener–Stueckelberg) type problems, which constitute the most fundamental parts of nonadiabatic transitions, have been solved completely.

Naturally, it is highly desirable that practically important big physical, chemical, and biological systems can be treated nicely. Since it is obvious that fully quantum mechanically accurate treatments of such systems are impossible, it is strongly required that some sort of useful semiclassical methodologies are to be developed. In order to do that we have to rely on classical trajectories somehow and to figure out some clever ways of using them to incorporate various quantum mechanical effects properly. Such semiclassical theories should be tested in small systems in which accurate quantum mechanical studies are possible. In order to take into account the effects of nonadiabatic transitions within such type of semiclassical frameworks, the basic theory such as the Zhu–Nakamura theory for the LZS problems would be very useful. Not only the classically forbidden transitions but also all the effects of phases can be incorporated into the properly chosen classical trajectories. The simplest version of that is to use the original framework of the TSH (trajectory surface hopping) method. In this case, only nonadiabatic transition probabilities are considered. By properly treating the classically forbidden transitions, namely nonadiabatic transitions at energies lower than the crossing points, it is expected that averaged physical quantities such as the cumulative reaction probabilities can be quite accurately estimated [170]. More sophisticated way of treatment would be semiclassical propagation method based on IVR (initial value representation) [163, 164] and the cellular Gaussian wave packet propagation method [165, 166]. The nonadiabatic transition matrix which contains the dynamical phases due to nonadiabatic transition can be incorporated into these frameworks without difficulty. The various phases can be taken into account by this treatment. Furthermore, the effects of phases, which have never been clarified in multidimensional dynamics, can be analyzed in detail and the validity of TSH can also be fully investigated. The recently developed *ab initio* type treatment of dynamics such as the spawning method [234] could also be made more efficient by using the analytical theories of nonadiabatic transitions. Conical intersections, which are well known to play important roles in various chemical dynamics, can also be dealt with by the semiclassical propagation methods with nonadaiabtic transtions along trajectories treated by the analytical theories and with the geometrical phases added into the propagation.

Another very important application of the theories of nonadiabatic transitions is control of molecular processes by external fields. Curve crossings or nonadiabatic transitions can be induced by applying an external field;thus by manipulating the field, the nonadiabatic transitions and the associated molecular dynamic processes could be controlled. In this field of science, the theories of nonadiabatic transitions, both time-independent and time-dependent, and the concept of nonadiabatic transition would play significant roles. Control of chemical dynamics by lasers would be supposed to be a very active and splendid field, but various new viewpoints and dimensions could be open up with use of the other type of fields also. For instance, by manipulating magnetic as well as radio frequency fields, new effective ways of controlling spin polarization in NMR [235, 236] could be established.

Finally, future possible developments of the basic theories of nonadiabatic transitions should also be discussed. In comparison with the LZS problems, the theory for the non-crossing RZD problems is not complete yet and should be further developed. For instance, the case that the two parallel potentials are slanted and the energetical threshold exists cannot be treated accurately. A more challenging subject is to develop a unified theory which can cover both LZS and RZD types of transitions uniformly. This would be very useful for applications, since we would not have to pay any attention to classify the type of transitions. Although much efforts have been paid in this direction with use of the exponential potential models [4, 67–70], it is still a long way to go.

The most difficult thing is to develop an intrinsically multidimensional theory without relying on trajectories. Real multidimensionality effects could be clarified only by such kind of theory. Since no useful mathemaics is available yet as far as the author knows, however, this would probably be almost hopeless at this moment unfortunately. Another important and interesting subject is to develop a basic theory in the two-dimensional (t, x) space, where t is time and x is a spatial coordinate. It is pitty that there is no basic analytical theory at all in spite of the fact that interactions between a time-dependent strong external field and molecular quantum dynamics in coordinate space x are attracting much interest recently, as explained in Chap. 13. The only possible way to treat these problems at this moment is to solve the time-dependent Schrödinger equation numerically with use of wave packets.

Appendix A

Final Recommended Formulas for General Time-Independent Two-Channel Problem

The final recommended formulas of the two-channel Zhu–Nakamura theory are summarized below. As can be understood from the summary below, the theory has the following nice features and can be easily utilized for various applications:

(1) All the probabilities are expressed in simple analytical forms.
(2) All the necessary phases are provided in compact analytical forms.
(3) All the basic parameters can be directly estimated only from adiabatic potentials on the real axis.

This means that

(1) no non-unique diabatization is needed,
(2) no complex calculus is necessary,
(3) no nonadiabatic coupling information is required, and furthermore,
(4) the theory works for whole range of energy and coupling strength.

A.1. Landau–Zener Type

The transition matrix I_X at the avoided crossing point R_0 is defined by

$$\begin{pmatrix} C \\ D \end{pmatrix} = I_X \begin{pmatrix} A \\ B \end{pmatrix}, \tag{A.1}$$

where A and B (C and D) are the coefficients of the wave functions at $R = R_0 + 0$ ($R = R_0 - 0$). The nonadiabatic transition matrix I_X is given by

$$I_X = \begin{pmatrix} \sqrt{1 - p_{\mathrm{ZN}}}\, e^{i(\psi_{\mathrm{ZN}} - \sigma_{\mathrm{ZN}})} & -\sqrt{p_{\mathrm{ZN}}}\, e^{i\sigma_0^{\mathrm{ZN}}} \\ \sqrt{p_{\mathrm{ZN}}}\, e^{-i\sigma_0^{\mathrm{ZN}}} & \sqrt{1 - p_{\mathrm{ZN}}}\, e^{-i(\psi_{\mathrm{ZN}} - \sigma_{\mathrm{ZN}})} \end{pmatrix}, \qquad (A.2)$$

where p_{ZN} represents the nonadiabatic transition probability for one passage of avoided crossing, and $\psi_{\mathrm{ZN}}, \sigma_{\mathrm{ZN}}$, and σ_0^{ZN} are certain phases. The explicit expressions of these quantities are given below. In the final formulas given below some empirical corrections are introduced in order to make them cover better the whole range of coupling strength. Thus these formulas can be readily utilized.

The basic parameters a^2 and b^2 are originally defined in terms of the diabatic potentials as

$$a^2 = \frac{\hbar^2}{2\mu} \frac{F(F_1 - F_2)}{8V_X^3} \qquad (A.3)$$

$$b^2 = (E - E_X)\frac{F_1 - F_2}{2FV_X}, \qquad (A.4)$$

where μ is the mass, F_j $(j = 1, 2))$ is the slope of the jth diabatic potential with $F = \sqrt{|F_1 F_2|}$, V_X is the diabatic coupling and E_X is the energy at the crossing point. Without loss of generality $F_1 - F_2$ is assumed to be positive. When adiabatic potentials, $E_j(R)(j = 1, 2)$, are directly available, these parameters a^2 and b^2 can be estimated directly from them as (Fig. A.1)

$$a^2 = \sqrt{d^2 - 1}\,\frac{\hbar^2}{\mu(T_2^{(0)} - T_1^{(0)})^2(E_2(R_0) - E_1(R_0))} \qquad (A.5)$$

$$b^2 = \sqrt{d^2 - 1}\,\frac{E - (E_2(R_0) + E_1(R_0))/2}{(E_2(R_0) - E_1(R_0))/2}, \qquad (A.6)$$

where

$$d^2 = \frac{[E_2(T_1^{(0)}) - E_1(T_1^{(0)})][E_2(T_2^{(0)}) - E_1(T_2^{(0)})]}{[E_2(R_0) - E_1(R_0)]^2}. \qquad (A.7)$$

Even when the diabatic potentials are available, it is better to use above expressions defined in terms of adiabatic potentials, since it is confirmed that they are more accurate. It should be noted that the Landau–Zener formula is

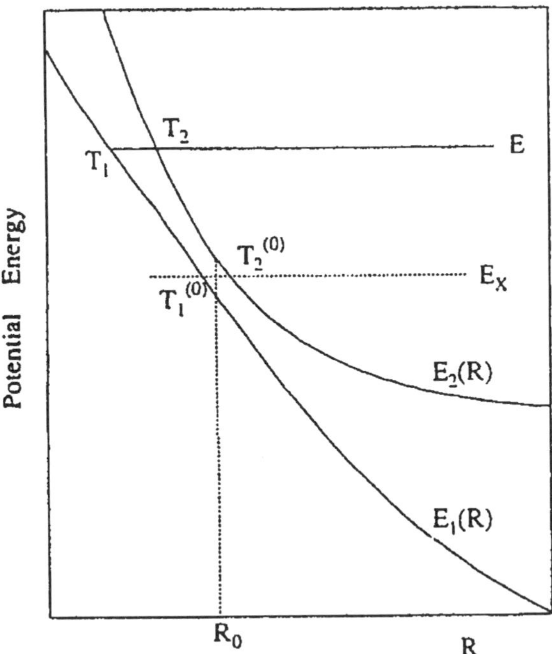

Fig. A.1. (The same as Fig. 2.1(a).) Landau–Zener type curve crossing. (Taken from Ref. [183] with permission.)

given by

$$p_{\mathrm{LZ}} = \exp\left[-\frac{\pi}{4a|b|}\right].\tag{A.8}$$

A.1.1. $E \geq E_X$ *(crossing energy)* $(b^2 \geq 0)$

The various quantities in Eq. (A.2) are given by

$$p_{\mathrm{ZN}} = \exp\left[-\frac{\pi}{4a|b|}\left(\frac{2}{1+\sqrt{1+b^{-4}(0.4a^2+0.7)}}\right)^{1/2}\right],\tag{A.9}$$

$$\sigma_0^{\mathrm{ZN}} = \frac{\sqrt{2}\pi}{4\sqrt{a^2}}\frac{F_-^C}{F_+^2+F_-^2},\tag{A.10}$$

$$\psi_{\mathrm{ZN}} = \sigma_{\mathrm{ZN}} + \phi_{\mathrm{S}}, \tag{A.11}$$

$$\sigma_{\mathrm{ZN}} = \int_{T_1}^{R_0} K_1(R)\,dR - \int_{T_2}^{R_0} K_2(R)\,dR + \sigma_0^{\mathrm{ZN}}, \tag{A.12}$$

$$K_j(R) = \sqrt{\frac{2\mu}{\hbar^2}(E - E_j(R))}, \tag{A.13}$$

$$\phi_{\mathrm{S}} = -\frac{\delta_{\mathrm{ZN}}}{\pi} + \frac{\delta_{\mathrm{ZN}}}{\pi}\ln\left(\frac{\delta_{\mathrm{ZN}}}{\pi}\right) - \arg\Gamma\left(i\frac{\delta_{\mathrm{ZN}}}{\pi}\right) - \frac{\pi}{4}, \tag{A.14}$$

$$\delta_{\mathrm{ZN}} = \frac{\sqrt{2\pi}}{4\sqrt{a^2}}\frac{F_+^C}{F_+^2 + F_-^2} \equiv \delta_0^{\mathrm{ZN}}, \tag{A.15}$$

$$F_\pm = \sqrt{\sqrt{(b^2 + \gamma_1)^2 + \gamma_2} \pm (b^2 + \gamma_1)}$$
$$+ \sqrt{\sqrt{(b^2 - \gamma_1)^2 + \gamma_2} \pm (b^2 - \gamma_1)}, \tag{A.16}$$

$$F_+^C = F_+\left(b^2 \longrightarrow \left[b^2 - \frac{0.16 b_x}{\sqrt{b^4 + 1}}\right]\right) \tag{A.17}$$

$$F_-^C = F_-\left(\gamma_2 \longrightarrow \frac{0.45\sqrt{d^2}}{1 + 1.5 e^{2.2 b_x |b_x|^{0.57}}}\right), \tag{A.18}$$

$$b_x = b^2 - 0.9553, \qquad \gamma_1 = 0.9\sqrt{d^2 - 1}, \qquad \gamma_2 = 7\sqrt{d^2}/16. \tag{A.19}$$

A.1.2. $E \le E_X$ $(b^2 \le 0)$

The nonadiabatic transition probability for one passage is given by

$$p_{\mathrm{ZN}} = [1 + B(\sigma_{\mathrm{ZN}}/\pi)\exp(2\delta_{\mathrm{ZN}}) - g\sin^2(\sigma_{\mathrm{ZN}})]^{-1}, \tag{A.20}$$

where

$$B(x) = \frac{2\pi x^{2x} e^{-2x}}{x\Gamma^2(x)}, \tag{A.21}$$

$$\delta_{\mathrm{ZN}} = \int_{T_1}^{R_0} |K_1(R)| dR - \int_{T_2}^{R_0} |K_2(R)| dR + \delta_0^{\mathrm{ZN}} , \qquad (A.22)$$

$$\sigma_{\mathrm{ZN}} = \sigma_0^{\mathrm{ZN}} \qquad (A.23)$$

and

$$g = \frac{3\sigma_{\mathrm{ZN}}}{\pi \delta_{\mathrm{ZN}}} \ln(1.2 + a^2) - \frac{1}{a^2} . \qquad (A.24)$$

The local wave number $K_j(R)$ is defined by Eq. (A.13). The quantities σ_0^{ZN} and δ_0^{ZN} are given by Eqs. (A.10) and (A.15). The phase ψ_{ZN} in Eq. (A.2) is given by

$$\psi_{\mathrm{ZN}} = \arg(U) \qquad (A.25)$$

where

$$\mathrm{Re}\, U = \cos(\sigma_{\mathrm{ZN}}) \left\{ \sqrt{B(\sigma_{\mathrm{ZN}}/\pi)} e^{\delta_{\mathrm{ZN}}} - h \sin^2(\sigma_{\mathrm{ZN}}) \frac{e^{-\delta_{\mathrm{ZN}}}}{\sqrt{B(\sigma_{\mathrm{ZN}}/\pi)}} \right\}, \qquad (A.26)$$

$$\mathrm{Im}\, U = \sin(\sigma_{\mathrm{ZN}}) \left\{ B(\sigma_{\mathrm{ZN}}/\pi) e^{2\delta_{\mathrm{ZN}}} - h^2 \sin^2(\sigma_{\mathrm{ZN}}) \cos^2(\sigma_{\mathrm{ZN}}) \frac{e^{-2\delta_{\mathrm{ZN}}}}{B(\sigma_{\mathrm{ZN}}/\pi)} \right.$$

$$\left. + 2h \cos^2(\sigma_{\mathrm{ZN}}) - g \right\}^{1/2} \qquad (A.27)$$

with

$$h = 1.8(a^2)^{0.23} e^{-\delta_{\mathrm{ZN}}} \qquad (A.28)$$

A.1.3. *Total scattering matrix*

The scattering matrix in the adiabatic representation is given by

$$S_{nm} = S_{nm}^{\mathrm{R}} \exp[i\eta_n + i\eta_m] , \qquad (A.29)$$

even if the energy E is lower than E_X. Here η_n is the elastic scattering phase by the nth adiabatic potential and S^{R} is the reduced scattering matrix. The total S-matrix is given by

$$S = P_{\infty X} I_X^t P_{XTX} I_X P_{X\infty} = Q I^t I Q = Q S^{\mathrm{R}} Q , \qquad (A.30)$$

where $I_X^t(I^t)$ is the transpose of $I_X(I)$,

$$(P_{\infty X})_{nm} = (P_{X\infty})_{nm}$$

$$= \delta_{nm} \exp\left[i \int_X^\infty (K_{1,2}(R) - K_{1,2}(\infty))dR - iK_{1,2}(\infty)X\right], \quad (A.31)$$

$$(P_{XTX}) = \delta_{nm} \exp\left[2i \int_{T_{1,2}}^X K_{1,2}(R)dR + i\frac{\pi}{2}\right], \quad (A.32)$$

$$Q_{nm} = \delta_{nm} \exp[i\eta_n] \quad (A.33)$$

$$\eta_{1,2} = \lim_{R\to\infty} \left[\int_{T_{1,2}}^R K_{1,2}(R)dR - K_{1,2}(R)R + \frac{\pi}{4}\right], \quad (A.34)$$

$$I = \begin{pmatrix} \sqrt{1 - p_{\mathrm{ZN}}}\, e^{i(\psi_{\mathrm{ZN}} - \sigma_{\mathrm{ZN}})} & -\sqrt{p_{\mathrm{ZN}}}\, e^{i\sigma_{\mathrm{ZN}}} \\ \sqrt{p_{\mathrm{ZN}}}\, e^{-i\sigma_{\mathrm{ZN}}} & \sqrt{1 - p_{\mathrm{ZN}}}\, e^{-i(\psi_{\mathrm{ZN}} - \sigma_{\mathrm{ZN}})} \end{pmatrix}, \quad (A.35)$$

and

$$S^{\mathrm{R}} = I^t I. \quad (A.36)$$

A.2.   Nonadiabatic Tunneling Type (see Fig. A.2)

The I_X-matrix for $E \geq E_b$ ($b^2 \geq 1$) is given by

$$I_X = \begin{pmatrix} \sqrt{1 - p_{\mathrm{ZN}}}\, e^{i\phi_{\mathrm{S}}} & \sqrt{p_{\mathrm{ZN}}}\, e^{i\sigma_0^{\mathrm{ZN}}} \\ -\sqrt{p_{\mathrm{ZN}}}\, e^{-i\sigma_0^{\mathrm{ZN}}} & \sqrt{1 - p_{\mathrm{ZN}}}\, e^{-i\phi_{\mathrm{S}}} \end{pmatrix}, \quad (A.37)$$

where

$$p_{\mathrm{ZN}} = \exp\left[-\frac{\pi}{4ab}\left(\frac{2}{1 + \sqrt{1 - b^{-4}(0.72 - 0.62a^{1.43})}}\right)^{1/2}\right], \quad (A.38)$$

$$\sigma_0^{\mathrm{ZN}} = \frac{R_b - R_t}{|R_b - R_t|} \frac{2\sqrt{1 - \gamma^2}}{3a} \frac{[2b^2 + \sqrt{b^4 - 1}]}{\sqrt{b^2 + 1} + \sqrt{b^2 - 1}}, \quad (A.39)$$

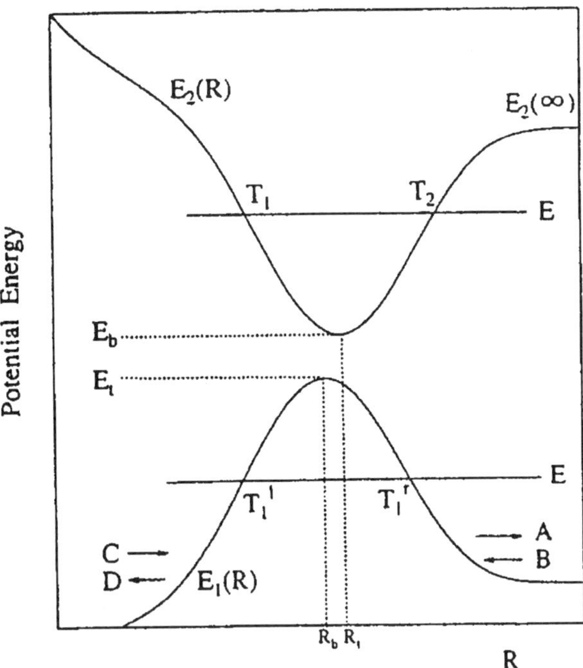

Fig. A.2. (The same as Fig. 2.1(b).) Nonadiabatic tunneling type curve crossing. (Taken from Ref. [183] with permission.)

and ϕ_S is the same as Eq. (A.14). At $E > E_2(\infty)$, this I_X can be used in the same way as I_X in the LZ-case.

The total $(S-)$ and the reduced (S^R-) scattering matrices are defined as follows:

$$S_{mn} = S^R_{mn} \exp[i(\eta_m + \eta_n)], \tag{A.40}$$

where

$$\eta_1 = \lim_{R \to \infty} \left[\int_{T_1^r}^R K_1(R)dR - K_1(R)R + \frac{\pi}{4} \right], \tag{A.41}$$

$$\eta_2 = \lim_{R \to -\infty} \left[-\int_{T_1^l}^R K_1(R)dR + K_1(R)R + \frac{\pi}{4} \right], \tag{A.42}$$

$$S^{\mathrm{R}} = \frac{1}{1 + U_1 U_2} \begin{pmatrix} e^{i\Delta_{11}} & U_2 e^{i\Delta_{12}} \\ U_2 e^{i\Delta_{12}} & e^{i\Delta_{22}} \end{pmatrix} \tag{A.43}$$

and

$$U_2 = 2i\,\mathrm{Im}(U_1)/(|U_1|^2 - 1)\,. \tag{A.44}$$

Note that T_1^l and T_1^r should be replaced by R_t for $E \geq E_b$ and the state 1(2), i.e. the indices of the S-matrix, corresponds to the right (left) side of the barrier. The parameters b^2 and a^2 originally defined by Eqs. (A.3) and (A.4) can also be estimated directly from adiabatic potentials by

$$b^2 = \frac{E - (E_b + E_t)/2}{(E_b - E_t)/2} \tag{A.45}$$

and

$$a^2 = \frac{(1 - \gamma^2)\hbar^2}{\mu(R_b - R_t)^2(E_b - E_t)}\,, \tag{A.46}$$

where

$$\gamma = \frac{E_b - E_t}{E_2(R_b + R_t/2) - E_1(R_b + R_t/2)}\,. \tag{A.47}$$

When $R_b = R_t$, $\gamma = 1$ and

$$a^2 = \frac{\hbar^2}{4\mu(E_b - E_t)}\left[\left.\frac{\partial^2 E_2(R)}{\partial R^2}\right|_{R=R_b} - \left.\frac{\partial^2 E_1(R)}{\partial R^2}\right|_{R=R_t} \right]\,. \tag{A.48}$$

The transfer matrix N for $E < E_2(\infty)$ is defined by

$$\begin{pmatrix} C \\ D \end{pmatrix} = N \begin{pmatrix} A \\ B \end{pmatrix}\,, \tag{A.49}$$

where $A(B)$ is the coefficient of the wave running to the right (left) along $E_1(R)$ on the right side of the barrier, and C (D) is the coefficient of the wave running to the right (left) along $E_1(R)$ on the left side of the barrier. The N-matrix is related to the S^{R}-matrix as

$$N_{11} = 1/S_{12}^{\mathrm{R}}, \qquad N_{12} = S_{22}^{\mathrm{R}}/S_{12}^{\mathrm{R}}, \qquad N_{21} = N_{12}^*, \qquad N_{22} = N_{11}^*\,. \tag{A.50}$$

A.2.1. $E \geq E_b$

The various quantities in Eq. (A.43) are given as follows:

$$\Delta_{12} = \sigma_{ZN} , \tag{A.51}$$

$$\Delta_{11} = 2 \int_{T_1}^{R_b} K_2(R)dR - 2\bar{\sigma}_0 , \tag{A.52}$$

$$\Delta_{22} = 2 \int_{R_b}^{T_2} K_2(R)dR + 2\bar{\sigma}_0 , \tag{A.53}$$

$$\bar{\sigma}_0 = \left(\frac{R_b - R_t}{2}\right) \left\{ K_1(R_t) + K_2(R_b) + \frac{1}{3}\frac{[K_1(R_t) - K_2(R_b)]^2}{K_1(R_t) + K_2(R_b)} \right\}, \tag{A.54}$$

$$U_1 = i\sqrt{1 - p_{ZN}} \exp[i(\sigma_{ZN} - \phi_S)] , \tag{A.55}$$

$$\sigma_{ZN} = \int_{T_1}^{T_2} K_2(R)dR , \tag{A.56}$$

and

$$\delta_{ZN} = \frac{\pi}{8ab}\frac{1}{2}\frac{\sqrt{6 + 10\sqrt{1 - 1/b^4}}}{1 + \sqrt{1 - 1/b^4}} . \tag{A.57}$$

The nonadiabatic transition probability p_{ZN} is given by Eq. (A.38). The overall transmission probability is given by

$$P_{12} = \frac{4\cos^2(\sigma_{ZN} - \phi_S)}{4\cos^2(\sigma_{ZN} - \phi_S) + (p_{ZN})^2/(1 - p_{ZN})} . \tag{A.58}$$

It should be noted that when $\psi_{ZN} = \sigma_{ZN} - \phi_S = (n + 1/2)\pi (n = 0, 1, 2, \ldots)$, P_{12} becomes zero. Namely, the intriguing phenomenon of complete reflection occurs.

A.2.2. $E_b \geq E \geq E_t$

The necessary quantities to define S^R in Eq. (A.43) are given as follows:

$$\Delta_{12} = \sigma_{ZN}, \qquad \Delta_{11} = \sigma_{ZN} - 2\bar{\sigma}_0 \quad \text{and} \quad \Delta_{22} = \sigma_{ZN} + 2\bar{\sigma}_0 , \tag{A.59}$$

$$\bar{\sigma}_0 = -\frac{1}{3}(R_t - R_b)K_1(R_t)(1 + b^2)\,, \tag{A.60}$$

$$U_1 = i[\sqrt{1 + W^2}e^{i\phi} - 1]/W\,, \tag{A.61}$$

$$\phi = \sigma_{\mathrm{ZN}} + \arg\Gamma\left(\frac{1}{2} + i\frac{\delta_{\mathrm{ZN}}}{\pi}\right) - \frac{\delta_{\mathrm{ZN}}}{\pi}\ell n\left(\frac{\delta_{\mathrm{ZN}}}{\pi}\right) + \frac{\delta_{\mathrm{ZN}}}{\pi} - g_4\,, \tag{A.62}$$

$$g_4 = 0.34\frac{a^{0.7}(a^{0.7} + 0.35)}{a^{2.1} + 0.73}(0.42 + b^2)\left(2 + \frac{100b^2}{100 + a^2}\right)^{0.25}\,, \tag{A.63}$$

$$\sigma_{\mathrm{ZN}} = -\frac{1}{\sqrt{a^2}}\left[0.057(1 + b^2)^{0.25} + \frac{1}{3}\right](1 - b^2)\sqrt{5 + 3b^2}\,, \tag{A.64}$$

$$\delta_{\mathrm{ZN}} = \frac{1}{\sqrt{a^2}}\left[0.057(1 - b^2)^{0.25} + \frac{1}{3}\right](1 + b^2)\sqrt{5 - 3b^2}\,, \tag{A.65}$$

$$W = \frac{g_3}{a^{2/3}}\int_0^\infty \cos\left[\frac{t^3}{3} - \frac{b^2}{a^{2/3}}t - \frac{g_1}{2a^{2/3}}\frac{t}{g_2 + a^{1/3}t}\right]dt, \tag{A.66}$$

$$g_3 = 1 + \frac{0.38}{a^2}(1 + b^2)^{1.2 - 0.4b^2}\,, \tag{A.67}$$

$$g_2 = 0.61\sqrt{2 + b^2}\,, \tag{A.68}$$

$$g_1 = \frac{\sqrt{a^2} - 3b^2}{\sqrt{a^2} + 3}\sqrt{1.23 + b^2}\,. \tag{A.69}$$

The overall transmission probability is given by

$$P_{12} = \frac{W^2}{1 + W^2}\,. \tag{A.70}$$

A.2.3. $E \leq E_t$

The required quantities in Eq. (A.43) are given by

$$\Delta_{12} = \Delta_{11} = \Delta_{22} = -2\sigma_{\mathrm{ZN}}\,, \tag{A.71}$$

$$\mathrm{Im}\, U_1 = \sin(2\sigma_c)\left\{\frac{0.5\sqrt{a^2}}{1+\sqrt{a^2}}\sqrt{B\left(\frac{\sigma_c}{\pi}\right)}e^{-\delta_{\mathrm{ZN}}} + \frac{e^{\delta_{\mathrm{ZN}}}}{\sqrt{B(\sigma_c/\pi)}}\right\}, \tag{A.72}$$

$$\mathrm{Im}\, U_1 = \cos(2\sigma_c)\sqrt{\frac{(\mathrm{Re}\, U_1)^2}{\sin^2(2\sigma_c)} + \frac{1}{\cos^2(2\sigma_c)}} - \frac{1}{2\sin(\sigma_c)}\left|\frac{\mathrm{Re}\, U_1}{\cos(\sigma_c)}\right|, \tag{A.73}$$

$$\sigma_c = \sigma_{\mathrm{ZN}}(1 - 0.32 \times 10^{-2/a^2}e^{-\delta_{\mathrm{ZN}}}), \tag{A.74}$$

$$\delta_{\mathrm{ZN}} = \int_{T_1^l}^{T_1^r} |K_1(R)|dR\,, \tag{A.75}$$

and

$$\sigma_{\mathrm{ZN}} = \frac{\pi}{8a|b|}\frac{1}{2}\frac{\sqrt{6 + 10\sqrt{1 - 1/b^4}}}{1 + \sqrt{1 - 1/b^4}}\,, \tag{A.76}$$

where the function $B(x)$ is given by Eq. (A.21). The overall transmission probability is equal to

$$P_{12} = \frac{B(\sigma_c/\pi)e^{-2\delta_{\mathrm{ZN}}}}{[1 + (0.5\sqrt{a^2}/\sqrt{a^2}+1)B(\sigma_c/\pi)e^{-2\delta_{\mathrm{ZN}}}]^2 + B(\sigma_c/\pi)e^{-2\delta_{\mathrm{ZN}}}}\,. \tag{A.77}$$

It should be noted that when $a^2 \to 0$, we have

$$P_{12} = \frac{e^{-2\delta_{\mathrm{ZN}}}}{1 + e^{-2\delta_{\mathrm{ZN}}}}\,, \tag{A.78}$$

which agrees with the ordinary single potential barrier penetration probability.

The I-, I_X-, and S^{R}-matrices in this Appendix can be directly applied to various practical problems. To know how to use them in each practical problem, it would be helpful to refer to various chapters in the text.

Appendix B

Time-Dependent Version of the Zhu–Nakamura Theory

The above solutions derived in the time-independent framework can be easily transferred to the solutions in the corresponding time-dependent scheme. The exact solutions in the time-independent linear potential model, for instance, provide the exact solutions for the time-dependent quadratic potential model, covering not only the two-crossing case, but also tangentially touching and diabatically avoided crossing cases. The following replacements of the parameters are good enough for this transfer.

$$a^2 \Leftrightarrow \alpha = \frac{\sqrt{d^2 - 1}\,\hbar^2}{2V_0^2(t_t^2 - t_b^2)}\,, \tag{B.1}$$

$$b^2 \Leftrightarrow \beta = -\sqrt{d^2 - 1}\,\frac{t_b^2 + t_t^2}{t_t^2 - t_b^2}\,, \tag{B.2}$$

and

$$\sigma + i\delta = \frac{1}{\hbar}\left[\int_0^{t_t} E_-(t)dt - \int_0^{t_b} E_+(t)dt + \sqrt{\frac{\beta}{\alpha}} + \Delta_1\right] \tag{B.3}$$

with

$$\Delta_1 = \frac{T_X - (t_b + t_t)/2}{\sqrt{\alpha(\beta^2 + i)}(t_b - t_t)}\sqrt{\frac{d^2}{d^2 - 1}} + \frac{1}{2\sqrt{\alpha}}\int_0^i \left(\frac{1 + t^2}{t + \beta}\right)^{1/2} dt\,, \tag{B.4}$$

$$V_0 = \frac{1}{2}(E_+(t_x) - E_-(t_x))\,, \tag{B.5}$$

and

$$d^2 = \frac{[E_+(t_b) - E_-(t_b)][E_+(t_t) - E_-(t_t)]}{[E_+(t_x) - E_-(t_x)]^2} \, , \tag{B.6}$$

where T_x is the position at which $E_+(t) - E_-(t)$ becomes minimum, and $t_b(t_t)$ is the bottom (top) of the adiabatic potential $E_+(t)(E_-(t))$ with $E_+(t) > E_-(t)$. It should be noted that the NT-case in the time-independent scheme does not show up in the time-dependent framework. This is simply because the time is unidirectional. Thus the above replacement should be applied only to the LZ-case.

Bibliography

1. L. D. Landau, *Phys. Zts. Sov.* **2**, 46 (1932).

2. C. Zener, *Proc. Roy. Soc.* **A137**, 696 (1932).

3. E. C. G. Stueckelberg, *Hel. Phys. Acta.* **5**, 369 (1932).

4. E. E. Nikitin and S. Ya. Umanskii, *Theory of Slow Atomic Collisions* (Springer, Berlin, 1984).

5. M. S. Child, *Semiclassical Mechanics with Molecular Applications* (Clarendon, Oxford, 1991).

6. E. S. Medvedev and V. I. Osherov, *Radiationless Transitions in Polyatomic Molecules*, Springer Series in Chemical Physics, Vol. 57 (Springer, Berlin, 1994).

7. H. Nakamura, *Int. Rev. Phys. Chem.* **10**, 123 (1991).

8. H. Nakamura, *Adv. Chem. Phys.* **LXXXII**, 243 (1992).

9. H. Nakamura, *Dynamics of Molecules and Chemical Reactions*, eds. R. E. Wyatt and J. Z. H. Zhang (Marcel Dekker, New York, 1996), Chap. 12.

10. H. Nakamura, *Ann. Rev. Phys. Chem.* **48**, 299 (1997).

11. D. S. F. Crothers, *Adv. Phys.* **20**, 405 (1971).

12. M. S. Child, *Molecular Collision Theory* (Academic Press, London, 1974).

13. B. C. Eu, *Semiclassical Theories of Molecular Scattering* (Springer-Verlag, Berlin, 1984).

14. Yu. N. Demkov and V. N. Ostrovsky, *Zero Range Potentials and Their Applications in Atomic Physics* (Plenum, New York, 1988).

15. J. C. Tully, *Dynamics of Molecular Collisions, Part B*, ed. W. H. Miller (Plenum, New York, 1976), p. 217.

16. N. Rosen and C. Zener, *Phys. Rev.* **40**, 502 (1932).

17. Yu. N. Demkov, *J. Exp. Theo. Phys.* (USSR) **18**, 138 (1964).

18. M. Baer, *Theory of Chemical Reaction Dynamics Vol. II*, ed. M. Baer (CRC Press, Boca Raton, Florida, 1985), Chap. 4.

19. A. Thiel, *J. Phys.* **G16**, 867 (1990).

20. B. Imanishi and W. von Oertzen, *Phys. Rep.* **155**, 29 (1987).

21. A. Yoshimori and M. Tsukada (ed.), *Dynamical Processes and Ordering on Solid Surfaces* (Springer, Berlin, 1985).

22. R. Engleman, *Non-Radiative Decay of Ions and Molecules in Solids* (North-Holland, Amsterdam, 1979).

23. D. Kleppner, M. G. Littman and M. L. Zimmerma, *Rydberg Atoms in Strong Fields, Rydberg States of Atoms and Molecules*, eds. R. F. Stebbings and F. B. Dunning (Cambridge University Press, Cambridge, 1983), p. 73.

24. S. I. Chu, *Adv. Chem. Phys.* **73**, 739 (1989).

25. R. J. Gordon and S. A. Rice, *Ann. Rev. Phys. Chem.* **48**, 601 (1997).

26. A. D. Bandrauk ed., *Molecules in Laser Fields* (Marcel Dekker Inc., New York, 1994).

27. For example, G. Blatter and D. A. Browne, *Phys. Rev.* **B37**, 3856 (1988).

28. For example, E. Ben-Jacob and Y. Gefen, *Phys. Lett.* **108A**, 289 (1985).

29. Y. Gefen, E. Ben-Jacob and A. O. Caldeira, *Phys. Rev.* **B36**, 2770 (1987).

30. R. Ramaswamy and R. A. Marcus, *J. Chem. Phys.* **74**, 1385 (1981).

31. P. Gaspard, S. A. Rice and K. Nakamura, *Phys. Rev. Lett.* **63**, 930 (1989).

32. J. Zakrzewski and M. Kus, *Phys. Rev. Lett.* **67**, 2749 (1991).

33. B. Schwarzschild, *Phys.* Today (June 17, 1986).

34. J. N. Bahcall and M. H. Pinsonneault, *Rev. Mod. Phys.* **64**, 885 (1992).

35. J. Michl and V. Bonacic-Koutecky, *Electronic Aspects of Organic Photochemistry* (John Wiley and Sons, New York, 1990).

36. S. S. Shaik and P. C. Hiberty, *Theoretical Models of Chemical Bonding, Part 4*, ed. Z. B. Maksic (Springer, Berlin, 1991), p. 269.

37. V. Sundström, ed., *Femtochemistry and Femtobiology: Ultrafast Reaction Dynamics at Atomic-Scale Resolution, Nobel Symposium 101* (Imperial College Press, 1996).

38. L. J. Butler, *Ann. Rev. Phys. Chem.* **49**, 125 (1998).

39. D. DeVault, *Quantum-Mechanical Tunneling in Biological Systems* (Cambridge University Press, Cambridge, 1984).

40. J. R. Bolton, N. Mataga and G. McLendon, eds., *Electron Transfer in Inorganic, Organic, and Biological Systems, Advances in Chemistry Series 228* (American Chemical Society, Washington DC, 1991).

41. R. A. Marcus and N. Sutin, *Biochimica et Biophysica Acta* **811**, 265 (1985).
42. H. Nakamura, *Chemistry Today* **8**, 36 (1996) [in Japanese].
43. The comic story "The God of Death", in *The Complete Collection of Enshoh Vol. 4* (Sei-A-Boh Publish. Tokyo, 1967), p. 177 [in Japanese].
44. L. Pichl, H. Nakamura and H. Deguchi, Joint Conference on Information Sciences 2000 Proceedings (AIM, Atlantic City, 2000), Vol. 2, p. 1002 (2000).
45. L. D. Landau and E. M. Lifshitz, *Quantum Mechanics — Nonrelativistic Theory* (Pergamon Press, New York, 1965).
46. G. Herzberg, *Molecular Spectra and Molecular Structure I: Spectra of Diatomic Molecules* (D. Van Nostrand, Princeton, 1950).
47. J. Heading, *An Introduction to Phase Integral Methods* (Methuen, London, 1962).
48. M. A. Evgrafov and M. V. Fedoryuk, *Asymptotic Behavior as $\lambda \to \infty$ of the Solution of the Equation $w''(z) - p(z, \lambda)w(z) = 0$ in the Complex z-Plane* (Russ. Math. Surv. 21), p. 1 (1996).
49. Y. Sibuya, *Global Theory of a Second-Order Linear Ordinary Differential Equation with a Polynomial Coefficient* (North-Holland, Amsterdam, 1975).
50. C. Zhu and H. Nakamura, *J. Math. Phys.* **33**, 2697 (1992).
51. C. Zhu, H. Nakamura, N. Re and V. Aquilanti, *J. Chem. Phys.* **97**, 1892 (1992).
52. C. Zhu and H. Nakamura, *J. Chem. Phys.* **97**, 8497 (1992).
53. C. Zhu and H. Nakamura, *J. Chem. Phys.* **98**, 6208 (1993).
54. C. Zhu and H. Nakamura, *J. Chem. Phys.* **101**, 4855 (1994).
55. F. L. Hinton, *J. Math. Phys.* **20**, 2036 (1979).
56. C. Zhu and H. Nakamura, *Comp. Phys. Comm.* **74**, 9 (1993).
57. H. Nakamura, *J. Chem. Phys.* **97**, 256 (1992).
58. S. Nanbu, H. Nakamura and F. O. Goodman, *J. Chem. Phys.* **107**, 5445 (1997).
59. H. Nakamura, *J. Chem. Phys.* **110**, 10253 (1999).
60. C. Zhu and H. Nakamura, *J. Chem. Phys.* **108**, 7501 (1998).
61. C. Zhu and H. Nakamura, *J. Chem. Phys.* **101**, 10630 (1994).
62. C. Zhu and H. Nakamura, *J. Chem. Phys.* **102**, 7448 (1995).
63. C. Zhu and H. Nakamura, *J. Chem. Phys.* **109**, 4689 (1998).
64. V. I. Osherov and A. I. Voronin, *Phys. Rev.* **A49**, 265 (1994).

65. Y. Luke, *Mathematical Functions and Their Applications* (Academic, New York, 1975).
66. L. Pichl, H. Nakamura and J. Horacek, *J. Chem. Phys.* **113**, 906 (2000).
67. V. I. Osherov and H. Nakamura, *J. Chem. Phys.* **105**, 2770 (1996).
68. V. I. Osherov and H. Nakamura, *Phys. Rev.* **A59**, 2486 (1999).
69. V. I. Osherov, V. G. Ushakov and H. Nakamura, *Phys. Rev.* **A57**, 2672 (1998).
70. L. Pichl, V. I. Osherov and H. Nakamura, *J. Phys.* **A**; *Math. Gen* **33**, 3361 (2000).
71. Y. Teranishi and H. Nakamura, *J. Chem. Phys.* **107**, 1904 (1997).
72. F. O. Goodman and H. Nakamura, *Prog. Surf. Sci.* **50**, 389 (1995).
73. S. P. Mikheyev and A. Yu. Smirnov, *Sov. J. Nucl. Phys.* **42**, 913 (1985).
74. L. Wolfenstein, *Phys. Rev.* **D17**, 2369 (1978).
75. M. Bruggen, W. C. Haxton and Y. Z. Qian, *Phys. Rev.* **D51**, 4018 (1995).
76. A. B. Balantekin and J. F. Beacom, *Phys. Rev.* **54**, 6323 (1996).
77. S. Toshev, *Phys. Lett.* **196**, 170 (1987).
78. K. A. Suominen and B. M. Garraway, Preprint Series in Theoretical Physics, Reseach Institute for Theoretical Phys., University of Helsinki, HU-TFT-91-23 (1991).
79. Yu. N. Demkov and M. Kunike, Vestnik Leningradskovo Univ. Fisicheskai Chemia **16**, 39 (1969).
80. H. Bateman, *Higher Transcendental Functions* (R. E. Krieger Publish, Florida, 1981).
81. F. T. Hioe and C. E. Carroll, *J. Opt. Soc. Am.* **B3**, 497 (1985).
82. F. T. Hioe and C. E. Carroll, *Phys. Rev.* **A32**, 1541 (1985).
83. A. Bambini and P. R. Berman, *Phys. Rev.* **A23**, 2496 (1981).
84. F. T. Hioe, *Phys. Rev.* **A30**, 2100 (1984).
85. M. S. Child, *J. Mol. Spec.* **53**, 280 (1974).
86. K. Mullen, E. Ben-Jacob, Y. Gefen and Z. Schuss, *Phys. Rev. Lett.* **62** 2543 (1989).
87. R. Landauer and M. Buttiker, *Phys. Rev. Lett.* **54**, 2049 (1986).
88. Y. Gefen and D. J. Thouless, *Phys. Rev. Lett.* **59**, 1572 (1987).
89. Y. Kayanuma, *J. Phys. Soc. Jpn.* **51**, 3526 (1982).
90. Y. Kayanuma, *J. Phys. Soc. Jpn.* **53**, 108 (1984).
91. Y. Kayanuma, *J. Phys. Soc. Jpn.* **53**, 118 (1984).
92. Y. Kayanuma, *J. Phys. Soc. Jpn.* **54**, 2037 (1985).
93. Y. Kayanuma and H. Nakayama, *Phys. Rev.* **B57**, 13099 (1998).

94. Y. Toyozawa and M. Inoue, *J. Phys. Soc. Jpn.* **21**, 1663 (1966).
95. Yu. N. Demkov and V. I. Osherov, *Soviet Phys. JETP* **26**, 916 (1968).
96. Yu. N. Demkov, P. B. Kurasov and V. N. Ostrovsky, *J. Phys.* **A28**, 4361 (1995).
97. Yu. N. Demkov and V. N. Ostrovsky, *J. Phys.* **B28**, 403 (1995).
98. C. E. Carroll and F. T. Hioe, *J. Opt. Soc. Am.* **B32**, 1355 (1985).
99. C. E. Carroll and F. T. Hioe, *J. Phys.* **A19**, 1151 (1986).
100. C. E. Carroll and F. T. Hioe, *J. Phys.* **A19**, 2061 (1986).
101. S. Brundobler and V. Elser, *J. Phys.* **A26**, 1211 (1993).
102. V. N. Ostrovsky and H. Nakamura, *J. Phys.* **A30**, 6939 (1997).
103. Yu. N. Demkov and V. N. Ostrovsky, *Phys. Rev.* **A61**, 32705 (2000).
104. C. Zhu and H. Nakamura, *Chem. Phys. Lett.* **258**, 342 (1996).
105. C. Zhu and H. Nakamura, *J. Chem. Phys.* **106**, 2599 (1997).
106. C. Zhu and H. Nakamura, *Chem. Phys. Lett.* **274**, 205 (1997).
107. C. Zhu and H. Nakamura, *J. Chem. Phys.* **107**, 7839 (1997).
108. H. Nakamura, *J. Chem. Phys.* **87**, 4031 (1987).
109. S. J. Seaton, *Rep. Prog. Phys.* **46**, 167 (1983).
110. F. H. Mies, *J. Chem. Phys.* **80**, 2514 (1984).
111. F. H. Mies and P. S. Julienne, *J. Chem. Phys.* **80**, 2526 (1984).
112. B. K. Homer and M. S. Child, *Farad. Disc. Chem. Soc.* **75**, 831 (1983).
113. P. S. Julienne and M. J. Krauss, *J. Mol. Spectr.* **56**, 270 (1975).
114. J. C. Tully, *Nonadiabatic Dynamics*, in *Modern Methods for Multidimensional Dynamics Computations in Chemistry*, ed. D. L. Thompson (World Scientific, Singapore, 1998), p. 34.
115. W. H. Miller, *Adv. Chem. Phys.* **25**, 69 (1974).
116. W. H. Miller, *Adv. Chem. Phys.* **30**, 77 (1975).
117. M. V. Berry and K. E. Mount, *Rep. Prog. Phys.* **35**, 315 (1972).
118. V. P. Maslov and M. V. Fedoriuk, *Semiclassical Approximation in Quantum Mechanics* (D. Reidel, Boston, 1981).
119. J. C. Tully, in *Modern Methods for Multi-Dimensional Dynamics Computations in Chemistry*, ed. D. L. Thompson (World Scientific, Singapore, 1998), and references therein.
120. M. Desouter-Lecomte, D. Dehareng, B. Leyh-Nihant, M. Th. Praet, A. J. Lorquet and J. C. Lorquet, *J. Phys. Chem.* **89**, 214 (1985).
121. W. Domcke and G. Stock, *Adv. Chem. Phys.* **100**, 1 (1997).
122. H. A. Jahn and E. Teller, *Proc. Roy. Soc.* **A161**, 220 (1937).

123. R. Englman, *The Jahn-Teller Effect in Molecules and Crystals* (Wiley-Interscience, New York, 1972).

124. I. B. Bersuker, *The Jahn-Teller Effect and Vibronic Interaction in Modern Chemistry* (Plenum, New York, 1984).

125. D. R. Yarkony, *Rev. Mod. Phys.* **68**, 985 (1996).

126. H. C. Longuet-Higgins, *Adv. Spectr.* **2**, 429 (1961).

127. G. Herzberg and H. C. Longuet-Higgins, *Disc. Faraday Soc.* **35**, 77 (1963).

128. C. A. Mead and D. G. Truhlar, *J. Chem. Phys.* **70**, 2284 (1979).

129. C. A. Mead, *J. Chem. Phys.* **78**, 807 (1983).

130. T. C. Thompson and C. A. Mead, *J. Chem. Phys.* **82**, 2408 (1985).

131. A. Kuppermann, *Dynamics of Molecules and Chemical Reactions*, eds. R. E. Wyatt and J. Z. Zhang (Marcel Dekker, 1996), p. 411.

132. M. Baer, *J. Chem. Phys.* **107**, 10662 (1997).

133. Z. Xu, M. Baer and A. C. Varandas, *J. Chem. Phys.* **112**, 2746 (2000).

134. J. Weiss, R. Schinke and V. A. Mandelshtam, *J. Chem. Phys.* **113**, 4588 (2000).

135. C. Petrongolo, *J. Chem. Phys.* **89**, 1297 (1988).

136. A. Shapere and F. Wilczek, *Geometric Phases in Physics, Advanced Series in Mathematical Physics, Vol. 5* (World Scientific, Singapore, 1989).

137. M. V. Berry, *Proc. Roy. Soc.* **A392**, 45 (1984).

138. J. C. Slonczewski and V. L. Moruzzi, *Physics* **3**, 237 (1967).

139. J. C. Slonczewskii, *Phys. Rev.* **131**, 1569 (1968).

140. E. E. Nikitin, *Dokl. Akad. Nauk* **183**, 319 (1968); *Theo. Chim. Acta* **12**, 308 (1968).

141. A. I. Voronin and V. I. Osherov, *Zhurn. Eksp. Teor. Fiz.* **66**, 135 (1974).

142. Y. Karni and E. E. Nikitin, *J. Chem. Phys.* **100**, 2027 (1994).

143. V. I. Voronin, S. P. Karkach, V. I. Osherov and V. G. Ushakov, *Zhur. Eks. Teor. Fiz.* **71**, 884 (1976).

144. O. Atabek and R. Lefebre, *J. Chem. Phys.* **97**, 3973 (1992).

145. C. Zhu, E. E. Nikitin and H. Nakamura, *J. Chem. Phys.* **104**, 7059 (1996).

146. J. C. Light and R. B. Walker, *J. Chem. Phys.* **65**, 4272 (1976); E. B. Stechel, R. B. Walker and J. C. Light, *J. Chem. Phys.* **69**, 3518 (1976).

147. G. V. Mil'nikov, C. Zhu, H. Nakamura and V. I. Osherov, *Chem. Phys. Lett.* **293**, 448 (1998).

148. F. T. Smith, *Phys. Rev.* **118**, 349 (1960).

149. A. Kupperman and P. G. Hipes, *J. Chem. Phys.* **84**, 5962 (1986); J. M. Launay and B. Lepetit, *Chem. Phys. Lett.* **144**, 346 (1998); G. C. Schatz, D. Sokolovski and J. N. L. Connor, in *Adv. Molec. Vibr. and Coll. Dynamics*, ed. J. M. Bowman (JAI Press, London, 1994), Vol. 2B, p. 1; D. E. Manolopoulos and D. C. Clary, in *Ann. Reports on Progress of Chemistry*, Sec/C, Vol. 86 (1989), p. 95.

150. R. T. Pack and G. A. Parker, *J. Chem. Phys.* **87**, 3888 (1987); **90**, 3511 (1989); G. A. Parker and R. T. Pack, *J. Chem. Phys.* **98**, 6883 (1992).

151. V. Aquilanti, S. Cavalli and D. D. Fazio, *J. Chem. Phys.* **109**, 3792 (1998).

152. O. I. Tolstikhin and H. Nakamura, *J. Chem. Phys.* **108**, 8899 (1998).

153. K. Nobusada, O. I. Tolstikhin and H. Nakamura, *J. Chem. Phys.* **108**, 8992 (1998).

154. C. Zhu, H. Nakamura and K. Nobusada, *Phys. Chem. Chem. Phys.* **2**, 557 (2000).

155. A. Persky and M. Broida, *J. Chem. Phys.* **81**, 4352 (1984).

156. A. Ohsaki and H. Nakamura, *Phys. Rep.* **187**, 1 (1990).

157. K. Nobusada, O. I. Tolstikhin and H. Nakamura, *J. Phys. Chem.* **102**, 9445 (1998).

158. G. V. Milnikov, O. I. Tolstikhin, K. Nobusada and H. Nakamura, *Phys. Chem. Chem. Phys.* **1**, 1159 (1999).

159. M. Broida and A. Persky, *Chem. Phys.* **130**, 129 (1989).

160. J. B. Delos, *Adv. Chem. Phys.* **65**, 161 (1986).

161. J. C. Tully and R. K. Preston, *J. Chem. Phys.* **55**, 562 (1971).

162. S. Chapmah, *Adv. Chem. Phys.* **82**, 423 (1992).

163. W. H. Miller, *J. Chem. Phys.* **53**, 3578 (1970).

164. X. Sun and W. H. Miller, *J. Chem. Phys.* **106**, 6346 (1997).

165. E. J. Heller, *J. Chem. Phys.* **94**, 2723 (1991).

166. A. R. Walton and D. E. Manolopoulos, *Mol. Phys.* **87**, 961 (1996).

167. M. L. Brewer, J. S. Hulme and D. E. Manolopoulos, *J. Chem. Phys.* **106**, 4832 (1997).

168. J. C. Tully, *J. Chem. Phys.* **93**, 1061 (1990).

169. M. D. Hack, A. W. Jasper, Y. L. Volobuev, D. W. Schwenke and D. G. Truhlar, *J. Phys. Chem.* **103**, 6309 (1999).

170. C. Zhu, K. Nobusada and H. Nakamura, *J. Chem. Phys.* **115**, 3031 (2001); C. Zhu, H. Kamisaka and H. Nakamura, *J. Chem. Phys.* **115**, 11036 (2001).

171. W. H. Miller and T. F. George, *J. Chem. Phys.* **56**, 5637 (1972).

172. A. Kormonicki, T. F. George and K. Morokuma, *J. Chem. Phys.* **65**, 48 (1976).

173. K. G. Kay, *J. Chem. Phys.* **100**, 4377 (1994).

174. E. J. Heller, *J. Chem. Phys.* **75**, 2923 (1981).

175. M. F. Herman and E. Kluk, *Chem. Phys.* **91**, 27 (1984).

176. E. Kluk, M. F. Herman and H. L. Davis, *J. Chem. Phys.* **84**, 326 (1986).

177. M. S. Child, *Mol. Phys.* **32**, 1495 (1976).

178. A. D. Bandrauk, E. A. Aubanel and J.-M. Gauthier, in *Molecules in Laser Fields*, ed. A. D. Bandrauk (Marcel-Dekker, New York, 1994), p. 109.

179. U. Fano, *Phys. Rev.* **124**, 1866 (1961).

180. H. Nakamura and C. Zhu, *Commun. Atom. and Molec. Phys.* **32**, 249 (1996).

181. C. Zhu, Y. Teranishi and H. Nakamura, *Adv. Chem. Phys.* **117**, 127 (2001).

182. A. Vardi and M. Shapiro, *Phys. Rev.* **A58**, 1352 (1998).

183. H. Nakamura, XXI-ICPEAC Invited Talks (Amer. Inst. Physics, Woodburry, New York, 2000), p. 495.

184. R. de L. Kronig and W. G. Penney, *Proc. Roy. London* **A130**, 499 (1931).

185. H. A. Kramers, *Physica* **2**, 483 (1935).

186. H. M. James, *Phys. Rev.* **76**, 1602 (1949).

187. W. Kohn, *Phys. Rev.* **115**, 809 (1959).

188. J. F. Reading and J. L. Sigel, *Phys. Rev.* **B5**, 556 (1972).

189. D. J. Griffiths and N. F. Taussig, *Am. J. Phys.* **60**, 883 (1992).

190. W. L. Sprung, H. Wu and J. Martorell, *Am. J. Phys.* **61**, 1118 (1993).

191. G. J. Clerk and B. H. J. McKeller, *Phys. Rev.* **C41**, 1198 (1990).

192. R. A. English and S. G. Davidson, *Phys. Rev.* **B49**, 8718 (1994).

193. F. L. Carter, *Molecular Electronic Devices* (Marcel-Dekker, New York, 1982), p. 51 and p. 121.

194. J. M. Kowalsky and J. L. Fry, *J. Math. Phys.* **28**, 2407 (1987).

195. D. Bohm, *Quantum Theory* (Prentice-Hall, Englewood Cliffs, N.J., 1951), p. 290.

196. E. H. Hauge and J. A. Stouneng, *Rev. Mod. Phys.* **61**, 917 (1989).

197. G. Bergman, *Phys. Rep.* **107**, 1 (1984).

198. A. Vibok and G. G. Balint-Kurti, *J. Phys. Chem.* **96**, 8712 (1989).

199. T. C. Allison, S. L. Mielke, D. W. Schwenke and D. G. Truhlar, *J. Chem. Soc., Faraday Trans.* **93**, 825 (1997).

200. E. Tekman and S. Ciraci, *Phys. Rev.* **B39**, 8772 (1989).

201. A. Szafer and A. D. Stone, *Phys. Rev. Lett* **62**, 300 (1989).

202. P. M. Morse and H. Feshbach, *Methods of Theoretical Physics, Part II* (McGraw-Hill, New York, 1953).

203. M. Abramowitz and A. Stegan, *Handbook of Mathematical Functions with Formulas, Graphs, and Mathematical Tables* (Dover, New York, 1965).

204. R. B. Shirts, *Trans. Math. Software* **19**, 389 (1993). The actual code is available on the web.

205. P. Brumer and M. Shapiro, *Annu. Rev. Phys. Chem.* **48**, 601 (1997).

206. D. J. Tannor and A. Rice, *J. Chem. Phys.* **83**, 5013 (1985).

207. S. Chelkowski and A. D. Bandrauk, *J. Chem. Phys.* **99**, 4279 (1993).

208. M. M. T. Loy, *Phys. Rev. Lett.* **32**, 814 (1974).

209. J. S. Melinger, S. R. Gabdhi and W. S. Warren, *J. Chem. Phys.* **101**, 6439 (1994).

210. S. Guerin, *Phys. Rev.* **A56**, 1458 (1997).

211. K. Mishima and K. Yamashita, *J. Chem. Phys.* **109**, 1801 (1998).

212. B. W. Shore, K. Bergman, A. K. Schiemann, J. Oreg and J. H. Eberly, *Phys. Rev.* **A45**, 5297 (1992).

213. R. Kosloff, S. A. Rice, P. Gaspard, S. Tersigni and D. J. Tannor, *Chem. Phys.* **139**, 201 (1989).

214. D. Neuhauser and H. Rabitz, *Acc. Chem. Res.* **26**, 496 (1993).

215. B. Kohler, J. L. Krause, F. Raksi, K. R. Wilson, V. V. Yakovlev and R. M. Whitnell, *Acc. Chem. Res.* **28**, 133 (1995).

216. M. Sugawara and Y. Fujimura, *J. Chem. Phys.* **100**, 5646 (1994).

217. L. Allen and J. H. Eberly, *Optical Resonances and Two-Level Atom* (Dover, NewYork, 1987).

218. M. Holthaus and B. Just, *Phys. Rev.* **A49**, 1950 (1994).

219. T. S. Ho and S. I. Chu, *Chem. Phys. Lett.* **141**, 315 (1987).

220. S. I. Chu, in *Advances in Multiphoton Processes and Spectroscopy*, Vol. 2 (World Scientific, Singapore, 1986).

221. Y. Teranishi and H. Nakamura, *Phys. Rev. Lett.* **81**, 2032 (1998).

222. Y. Teranishi and H. Nakamura, *J. Chem. Phys.* **111**, 1415 (1999).

223. Y. Teranishi, K. Nagaya and H. Nakamura, in *Quantum Control of Molecular Reaction Dynamics*, eds. R. J. Gordon and Y. Fujimura (World Scientific, Singapore, 2001) pp. 215–227.

224. K. Nagaya, Y. Teranishi and H. Nakamura, *J. Chem. Phys.* **113**, 6197 (2000).

225. L. Thomas *et al.*, *Nature* **383**, 145 (1996).

226. I. Ya. Korenblit and E. F. Shender, *Sov. Phys. JETP.* **48**, 937 (1987).

227. J. R. Friedman, M. P. Sarachik, J. Tejada and R. Ziolo, *Phys. Rev. Lett.* **76**, 3830 (1996).

228. C. Leforestier, R. H. Besseling, C. Cerjan, M. D. Feit, R. Friesner, A. Guldberg, A. Hammerich, G. Jolicard, W. Karrlein, H. D. Meyer, N. Lipkin, O. Roncero and R.Kosloff, *J. Comp. Phys.* **94**, 59 (1991).

229. A. Vibok and G. G. Balint-Kurti, *J. Chem. Phys.* **96**, 7615 (1992).

230. J. R. Reimers and R. O. Watts, *Mol. Phys.* **52**, 357 (1984).

231. V. Engel, R. Shinke and V. Staemmler, *J. Chem. Phys.* **88**, 129 (1998).

232. V. Staemmler and A. Palma, *Chem. Phys.* **93**, 63 (1985).

233. J. C. Light, I. P. Hamilton and J. V. Lill, *J. Chem. Phys.* **82**, 1400 (1985).

234. M. Ben-Nun, J. Quenneville and T. J. Martinez, *J. Phys. Chem.* **104**, 5161 (2000).

235. P. Hodgkinson and A. Pines, *J. Chem. Phys.* **107**, 8742 (1997).

236. M. H. Levitt, D. Suter and R. R. Ernst, *J. Chem. Phys.* **84**, 4243 (1986).

Index